ROADSIDE GEOLOGY
OF GEORGIA

Pamela J. W. Gore and
William Witherspoon

2013
Mountain Press Publishing Company
Missoula, Montana

Photos © 2013 by Pamela J. W. Gore and
William Witherspoon unless otherwise credited

Geologic road maps and many of the illustrations revised by Mountain Press
Publishing Company based on original drafts by the authors

Roadside Geology is a registered trademark
of Mountain Press Publishing Company

Library of Congress Cataloging-in-Publication Data

Gore, Pamela J. W.
Roadside geology of Georgia / Pamela J. W. Gore and William Witherspoon.
  p. cm.
Includes bibliographical references and index.
ISBN 978-0-87842-602-7 (pbk. : alk. paper)
1. Geology—Georgia—Guidebooks. 2. Georgia—Guidebooks.
I. Witherspoon, William D. II. Title.
QE101.G67 2013
557.58—dc23
                 2012051766

Printed in the United States by Versa Press

Mountain Press
PUBLISHING COMPANY
P.O. Box 2399 • Missoula, MT 59806 • 406-728-1900
800-234-5308 • info@mtnpress.com
www.mountain-press.com

This book is dedicated to Dr. Timothy M. Chowns. A native of England, Tim came to the University of West Georgia in 1973. His contributions to the study of geology of northwest Georgia, the Georgia coast, pre-Cretaceous rocks beneath the Coastal Plain, and sedimentary rocks have been timely and fascinating. For more than twenty years, Tim and his colleague Randy Kath have brilliantly and doggedly shouldered the task of keeping the Georgia Geological Society together and its annual field trip a thriving event in which geologists, students, and rock lovers climb on rocks, argue, and build friendships.

We usually think of people as ephemeral compared to rocks, but though the oddly folded layers of carbonate rock in this 2008 Rockmart quarry photo have been blasted away, Tim, as professor emeritus, has barely slowed down. We look forward to many more years of his geologic thinking, sparkling wit, and leadership.

CONTENTS

Acknowledgments   x

Introduction   1
  Geology Determines Landscape   2
  Plate Tectonics   3
  Georgia's Five Landscape Provinces   6
  Sedimentation   10
  Geologic Timescale   12
  A Word about Collecting   14

Sea Islands   15
  Longshore Drift and Tides   15
  Beaches   17
  Marshes   20
  Ancient and Modern Shorelines   21

  ROAD GUIDES OF THE SEA ISLANDS   21
  US 80: Savannah—Tybee Island   21
    North Beach Area   25
    South Beach Area   25
  Wassaw Island and Sapelo Island   27
    Gray's Reef National Marine Sanctuary   31
  St. Simons Island and Its Neighbors   31
  Jekyll Island   37
  Cumberland Island National Seashore   42

Coastal Plain   47
  Sedimentary Formations   47
  Physiographic Districts   52
    Fall Line Hills   52
    Fort Valley Plateau   53
    Vidalia Upland   54
    Tifton Upland and Valdosta Limesink Districts   55
    Dougherty Plain   55
    Bacon Terraces   56
    Barrier Island Sequence District   57
    Okefenokee Basin   57
  Landforms   60
    Pleistocene Sand Dune Fields   60
    The Mysterious Carolina Bays   60

Orangeburg Escarpment 62
Pelham Escarpment 62
It Came from Outer Space 63
Fossils 64
Sharks and Other Fish 66
Marine Reptiles 66
Dinosaurs and other Reptiles 67
Pleistocene-Age Vertebrates 68
Kaolin Deposits 69
Heavy Mineral Sand 73

ROAD GUIDES OF THE COASTAL PLAIN 74
US 27: Columbus—Florida State Line 74
Providence Canyon State Outdoor Recreation Area 76
Lumpkin to Florida State Line 78
I-75: Macon—Florida State Line 80
Georgia 49: Byron—Americus 84
Andersonville and Bauxite 86
Georgia 24: Milledgeville—Waynesboro 89
Kaolin Mines 89
Kaolin Processing Facility 91
Tennille Lime Sinks 91
Hidden Basins 92
Sandersville to Waynesboro 94
Shell Bluff 96
I-16: Macon—Savannah 96
US 341: Fall Line—Brunswick 100
Oaky Woods Wildlife Management Area 103
Perry to McRae 103
Little Ocmulgee State Park 105
McRae to Hazlehurst 105
Broxton Rocks Preserve 106
Hazlehurst to Jesup 107
Griffin Ridge Wildlife Management Area 108
Jesup to Brunswick 108
US 1: South Carolina State Line—Florida State Line 108
Ohoopee Dunes Natural Area 110
Swainsboro to Waycross 110
Okefenokee Swamp Park 111
Trail Ridge 112
US 82: Alabama State Line—Brunswick 113
Albany and the Flint River 113
Radium Springs 115
Albany to Brunswick 116
I-95: South Carolina State Line—Florida State Line 117

Valley and Ridge and Appalachian Plateau   121
    The Landscape in Relation to Anticlines and Synclines   122
    Sedimentary Rocks and Ancient Geography   126
    Fossils   128
      Fossils of Cambrian Age   128
      Fossils of Ordovician Age   128
      Fossils of Mississippian Age   130
      Fossils of Pennsylvanian Age   130

    ROAD GUIDES OF THE VALLEY AND RIDGE AND APPALACHIAN PLATEAU   **133**
      I-24 and I-59: Tennessee State Line—Alabama State Line   133
      I-75: Tennessee State Line—Cartersville (US 411)   137
        Ringgold Gap   137
        Tunnel Hill to Dalton   141
        Rocky Face Mountain at Dug Gap   142
        Resaca to Cartersville   143
        Cartersville Mining District   145
        Ladds Quarry and Etowah Mounds   148
      GA 136: Alabama State Line—Carters Dam   149
        Cloudland Canyon State Park   150
        Lookout Mountain to Carters Dam   153
      GA 157: Tennessee State Line—Cloudland   159
        Rock City Gardens   159
        Along Lookout Mountain   161
      US 27: Tennessee State Line—Cedartown   164
        Chickamauga Battlefield   166
        Chickamauga to Gore   168
        Chattanooga Shale   168
        Gore to Cedartown   170
        Cave Spring   171
        Slate Quarry near Cedartown   172

Blue Ridge–Piedmont   175
    Premetamorphic Character of the Blue Ridge–Piedmont Rocks   175
    Metamorphism   176
    Igneous Intrusions   179
    Weathering   179
    The Patchwork of Terranes Assembled in Paleozoic Time   181
    Earthquakes Due to Human Activity   185
    Pull-Apart Activity during the Mesozoic Era   188
    The Dahlonega Gold Belt: A Terrane Discovered by Prospectors   190

    ROAD GUIDES OF THE BLUE RIDGE–PIEDMONT   **191**
      I-20: Alabama State Line—Austell   191
        Pine Mountain Gold Museum   193

Villa Rica to Lithia Springs   193
I-75: Cartersville (US 411)—Kennesaw   195
  Cooper's Furnace Day Use Area and Allatoona Dam   197
  Lake Allatoona and Red Top Mountain State Park   199
  Allatoona Lake to Kennesaw   200
GA 5 (I-575 and GA 515 in part): Tennessee State Line—Kennesaw   201
  Blue Ridge to Jasper   203
  Marble Mining District   204
  Ball Ground to Kennesaw   208
GA 52: Chatsworth—Lula   209
  Fort Mountain State Park   212
  Fort Mountain to Amicalola Falls   213
  Amicalola Falls State Park   214
  Amicalola Falls to Lula   216
US 19: North Carolina State Line—Cumming   217
  Brasstown Bald   217
  Blairsville to Dahlonega   219
  Dahlonega   220
  Dahlonega to Cumming   221
GA 17 (GA 385 and Alt GA 17 in part): Hiawassee—Toccoa   223
  Smithgall Woods State Park   223
  Unicoi State Park and Anna Ruby Falls   223
  Helen to Toccoa   225
US 23/US 441 (becoming I-985/US 23): North Carolina
 State Line—Lake Lanier   227
  Black Rock Mountain State Park   229
  Clayton to Tallulah Falls   229
  Tallulah Gorge State Park   231
  Tallulah Falls to I-85   233
  Lake Lanier and Buford Dam   233
I-85: South Carolina State Line—I-985 (near Atlanta)   235
  Hurricane Shoals Park   235
  Nodoroc   235
  Earthquakes at Dacula   236
US 78: Snellville—Thomson   236
  Athens   236
  Athens to Washington   238
  Graves Mountain   238
  Lincolnton Metadacite   239
  Washington to Thomson   240
GA 72: Athens—South Carolina State Line   241
  Watson Mill Bridge State Park   241
  Elberton   243
  Georgia Guidestones   245
  Elberton to the South Carolina State Line   246
I-185: LaGrange—Columbus   247

GA 85: Atlanta—Columbus   249
  The Cove   249
  Warm Springs and Little White House Historic Site   253
  F. D. Roosevelt State Park   256
  Sprewell Bluff State Outdoor Recreation Area   257
  Manchester to Columbus   259
  Flat Rock Park   260
  The Fall Line and Rocky Shoals at Columbus   260
I-75: Atlanta—Macon   262
  Indian Springs State Park   264
  Barnesville and Thomaston Area   264
  High Falls State Park   264
  Forsyth to Macon   267
I-20: Atlanta—Augusta   267
  Appling Granite at Heggies Rock   272
  Augusta   274
US 441/GA 24: Madison—Milledgeville   276
  Rock Eagle   276
  Milledgeville   277
Around Atlanta   278
  Parks of Geologic Interest   281
    Stone Mountain Park   281
    Panola Mountain State Park   286
    Davidson-Arabia Nature Preserve   286
    Boat Rock Preserve   289
    Kennesaw Mountain National Battlefield Park   290
    Sweetwater Creek State Park   290
  Quarries and Road Cuts   291
    Gneiss and Schist   292
    Quartzite   295
    Amphibolite   295
    Soapstone Ridge   296
    Mylonite   297

Appendix: Museums and Exhibits   299

Glossary   305

References   319

Index   329

ACKNOWLEDGMENTS

The insights into Georgia geology in this book are the product of many geologists' investigations. We are especially fortunate that six researchers took time to divide up the task of reviewing the manuscript in their regions of expertise: Clark Alexander, Burt Carter, Tim Chowns, Randy Kath, Mike Roden, and Sam Swanson.

We also owe special thanks to Dr. Robert D. "Bob" Hatcher Jr. of the University of Tennessee, who has been raising the bar for understanding the geologic history of the Southeast for more than forty years. He provided digital map versions that are the basis for many of the Blue Ridge–Piedmont maps in the book and reviewed several drafts of the terrane discussion. We thank Philip Prince for his assistance in understanding landscapes and the Blue Ridge–Piedmont boundary. Other geologists who have provided comments on segments of the manuscript include Callan Bentley, Erv Garrison, David Schwimmer, and Hovey Smith. Many geoscientists have helpfully responded to our inquiries: John Anderson, Stan Bearden, Gale Bishop, Jon Bryan, Clint Barineau, Chris Capps, Jeff Chaumba, Habte Churnet, John Costello, Steve Fitzpatrick, Julian Gray, Tom Hanley, Scott Harris, Mike Higgins, Paul Huddlestun, Matt Huebner, Nan Huebner, Dick Ketelle, Tim Long, Tony Martin, Arthur Merschat, Jon Mies, Katayoun Mobasher, Billy Morris, Andy Newman, Paul Schroeder, Joe Summerour, Donald Thieme, and Nick Woodward. Thanks to the staff at Mountain Press: our editor, James Lainsbury; cartographer, Chelsea Feeney; and Jennifer Carey, who reviewed earlier drafts of this manuscript.

Additional readers have given pointers on removing stumbling blocks to the general reader: thanks to Miranda Gore, Don Lundy, Caitlyn Mayer, Marty Rosenberg, and Noah Witherspoon. Others who have helped in various ways include Robert Ables, Mike Clark, Chuck Cochran, Brad Gane, Paul Knowlton, Mary Larsen, Tony Madden, Cindy Reittinger, Jose Santamaria, and Cantey Smith.

Many people, including some of the above, have provided photos or other illustrations and are credited in individual figure captions. We would also like to thank Bill Hood and Kellyn Willis of the Elberton Granite Association for photos they provided.

We thank our employers, Georgia Perimeter College (GPC) and DeKalb County Schools' Fernbank Science Center, and are grateful for the initial support of a Writers Institute Faculty Fellowship to Pamela Gore from GPC.

Others have helped as well, and we apologize to anyone we may have omitted. We have done our best with the mountain of information about our state, and any errors are our own.

This book began as a project of Ed Albin of Fernbank Science Center. Ed recruited each of us but was unable to continue as a coauthor. We are grateful that he also recruited Steven Jaret, who got us started in locating digital map data and moving it into a publishable format.

Our spouses, Thomas Gore and Rina Rosenberg, have provided not only advice on wording and patience with a project that seemed to roll on year after year but have also been our drivers to places all over the state, ready to find a wide spot to pull off into with ten seconds notice. Clearly this book would not exist without them.

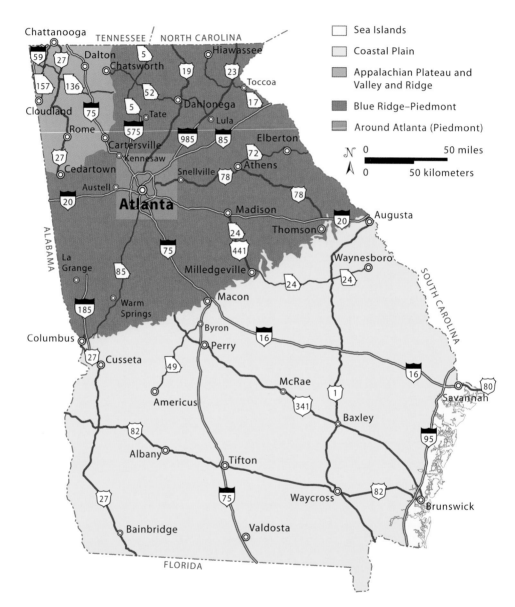

Roads and sections of this book.

INTRODUCTION

Ask a Georgian about geologic features in the state, and the answer you get will depend on where you are. If you are in Atlanta, you will probably hear about Stone Mountain, the white whaleback of a rock you can see from downtown office windows, 14 miles away on the eastern horizon. In the Blue Ridge Mountains, you will probably hear about Amicalola Falls, one of the highest cascading waterfalls (729 feet) east of the Mississippi River, or Tallulah Gorge, one the deepest gorges in the eastern United States. In south Georgia, you will probably hear about the Okefenokee Swamp, one of the largest wetlands in North America. In west Georgia, near Columbus, you will probably hear about Providence Canyon, often called Georgia's Little Grand Canyon, which formed due to erosion from human activity over the past 150 years. In northwestern Georgia, you will probably hear about Rock City Gardens atop Lookout Mountain, from which you are said to see seven states, or Cloudland Canyon, which is more than 800 feet deep with waterfalls. Wherever you are in Georgia, there is something spectacular to see.

In terms of museums (see the appendix), Georgia has some top honors. The Fernbank Museum of Natural History in Atlanta displays the world's largest dinosaur, *Argentinosaurus*. The Tellus Science Museum in Cartersville has one of the world's largest mineral collections and a fine fossil collection with many vertebrate skeletons. And the Georgia Aquarium in Atlanta is the largest aquarium in the world.

Georgia experienced the nation's first gold rush. In Dahlonega, where the rush began, you can visit an underground mine and see a building that opened as a branch of the U.S. Mint in 1837. Because of its vast resource of uniform, fine-grained granite, Elberton competes with a town in Vermont for the title Granite Capital of the World. Georgia marble is also prized; many of the monuments in Washington DC are made of Georgia marble, including the big statue of seated Abraham Lincoln in the Lincoln Memorial. Today, Georgia's richest mines are in the white clay kaolin, aptly dubbed "white gold." Georgia is the world's largest producer of kaolin, a billion-dollar enterprise and the largest mining industry in the state, and Sandersville is the Kaolin Capital of the World. Chances are good that a glossy magazine in your house, and the pages of this book, are coated with Georgia kaolin.

The state has a long history of mining. Soapstone Ridge, southeast of Atlanta, was the source of soapstone that Native Americans carved into bowls more than 3,000 years ago. Georgia had the first aluminum mine in the United States, northeast of Rome. And nearly all of Georgia is within 50 miles of one of the thousands of iron pits that dotted North America before the mass production of steel became economically feasible.

State seal and outline of Georgia carved into Elberton Granite in a floor tile at Confederate Hall, Stone Mountain Park.

GEOLOGY DETERMINES LANDSCAPE

The systematic search for mineral resources led to the production of the first geologic maps of Georgia during the nineteenth century. The geologic maps in this book are the result of more than a century of study by field geologists. Field geologists scan the countryside for outcrops, where the bedrock pokes through the soil at the surface. Elsewhere they rely on clues in soil formed by the weathering of bedrock to determine what rock lies below. Once the boundaries of a particular rock type or pattern of rock types have been traced across the countryside, formal rock unit names, such as Nantahala Formation, Knox Group, or Stone Mountain Granite, are proposed. These formal rock units are what appear on geologic maps, with assigned colors and patterns.

When you see a geologic map for the first time, you may experience a flash of recognition, because many of the landscape's topographical features are strongly influenced by bedrock type. Though no geologist was giving advice when Georgia's largest cities were founded, their locations were determined by geology. Savannah began on the high ground of an ancient beach, through which the Savannah River had eroded a deep valley during the ice ages of Pleistocene time when sea level was low. As the continental glaciers melted and sea level once again rose, the inundated valley became a natural ship channel, which helped Savannah become a major port. Atlanta sits on a relatively high and dry drainage divide just 30 miles southeast of a natural gateway through the mountains, located at a bend in a great, long-dead fault. Both of these geological factors made the location an ideal spot for a rail crossroads, which is how the city got its start. Columbus, Macon, Milledgeville, and Augusta all were sited where rivers cross the Fall Line, the state's most prominent geological boundary. The term *Fall Line* refers to the waterfalls and rocky shoals that developed at this geologic boundary, which separates the ancient crystalline rocks of the Piedmont to the north from the young sedimentary layers of the Coastal Plain. These cities were built at the Fall Line because large riverboats could not navigate past it, and industries arose due to the abundant waterpower provided by the falls.

PLATE TECTONICS

While geologists map rock at the surface, geophysicists learn about Earth's interior by measuring differences in gravity, magnetism, and the time it takes for shock waves from earthquakes to pass through the Earth. At its center, Earth has an iron-nickel metal core, with a radius of some 2,160 miles, surrounded by a rocky mantle that is about 1,800 miles thick. Above the mantle is the crust, ranging from less than 4 miles thick beneath some of the oceans to as much as 40 miles thick under Earth's highest mountains. Compared to the mantle, the crust is rich in aluminum and silicon and poor in iron and magnesium.

Rocks of the crust and uppermost mantle are brittle: under enough stress, they break, sending out earthquake waves. This brittle zone, also known as the lithosphere, varies from roughly 25 to 125 miles in thickness. Beneath the lithosphere, higher temperatures make the rock ductile, meaning it is able to bend and flow like warm taffy. This ductile zone is also known as the asthenosphere. Our recognition of the existence of the lithosphere and asthenosphere was critical to the development of geology's unifying concept of plate tectonics.

During the first century and a half of the development of the modern science of geology, earth scientists recognized that mountains rose and eroded and oceans appeared and disappeared over millions of years, but it was not until the 1960s that they formulated the unifying theory of plate tectonics that explains why. Studies of earthquakes and the ocean floor revealed that Earth's lithosphere is a mosaic of about twenty tectonic plates that are constantly moving relative to each other, creeping along at rates of inches per year. Plates move because, underneath the lithosphere, the mantle's taffy-like rock in the asthenosphere is in constant, extremely slow motion as hot rock rises and cold rock sinks, like a pot of boiling soup.

Most earthquakes and volcanoes are concentrated along plate boundaries. For example, earthquake epicenters and volcanoes dot the Mid-Atlantic Ridge, an underwater feature in the middle of the Atlantic Ocean that roughly parallels the western coastline of Africa and the eastern coastlines of the Americas. This ridge marks the location along which the supercontinent Pangaea rifted apart about 170 million years ago. Since that time, the continents on either side of the rift (North America, South America, Europe, and Africa) have gradually separated as the Atlantic Ocean formed and has grown larger. Magma has risen from Earth's mantle to fill the gap at the rift, creating hot, new seafloor. Since the younger seafloor is hotter, less dense, and more buoyant than the older, cooler seafloor on either side, it "floats" higher on the Earth's mantle, forming the ridge. Rifts like the Mid-Atlantic Ridge represent one of the major tectonic plate boundaries.

Because Earth stays the same size, geologists deduced that seafloor is being recycled elsewhere on the planet at the same rate new seafloor is created at rifts. Around the rim of the Pacific Ocean, and in a few other places, geologists have located places where older, colder, and denser seafloor is being pulled into Earth's mantle. These tectonic plate boundaries, called subduction zones, are marked by deep-sea trenches that develop where seafloor is being subducted (meaning "pulled under") beneath another tectonic plate. As a slab of seafloor descends into the mantle at an angle, water expelled from it lowers the melting point of hot mantle rocks above, causing magma to form and rise. The magma rises to the surface more than a hundred miles from the trench and creates a chain of volcanoes that roughly parallels the plate boundary. Volcanic chains can develop on land or in the sea, depending on where the subduction zone develops. Subduction accounts for the Ring of Fire, a horseshoe-shaped region of volcanism and tsunami-causing earthquakes that encircles the Pacific Ocean. In the United States, the Cascades volcanoes are part of the ring.

Elsewhere, tectonic plates slip past one another along vertical faults, such as California's famous San Andreas Fault, which is part of the boundary between the North American and Pacific plates. These are called transform boundaries.

Ocean basins open and ocean basins close, and when they close, continents collide. Unlike seafloor crust, continental crust is not recycled in the mantle; it is made of low-density rocks, such as granite, which are too buoyant and

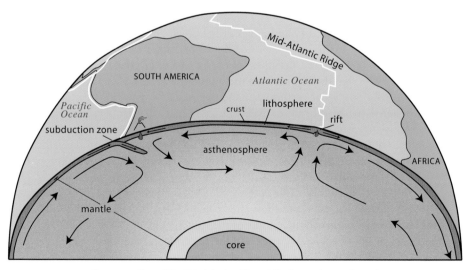

Cross section of Earth's interior. The white lines show a few of Earth's plate boundaries, including the Mid-Atlantic Ridge.

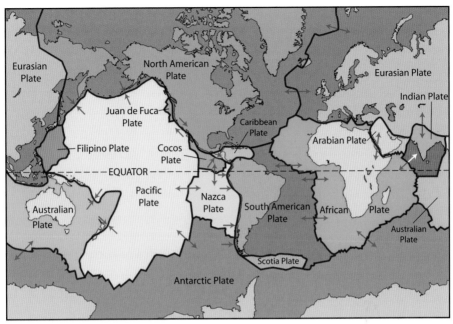

The major tectonic plates of the Earth. The arrows show the relative movement of the plates at their boundaries. —Modified from the U.S. Geological Survey

light to descend into a subduction zone. Instead, when continents collide, the subduction process that closed the ocean basin grinds to a halt. The force of the collision fractures the continental crust and causes it to pile up and overlap, forming mountains and high plateaus. A modern example of such a boundary is the Himalayas and Tibetan Plateau, where the formerly separate Indian and Asian continents are actively colliding.

All of these tectonic scenarios—continents ripping apart, oceans opening and closing, and continents colliding—occurred in Georgia and left their imprint. More than 600 million years ago, a supercontinent that existed long before Pangaea began to rift apart. Northwest Georgia had a ringside seat beside a widening ocean. After that ocean was more than 100 million years old, the regional tectonic processes shifted and subduction began to devour the seafloor. Chains of volcanic islands sprang up above the subduction zones. As subduction zones consumed the seafloor, the volcanic island chains collided with Georgia's coast, adding new territory to North America. About 265 million years ago, Africa—which was part of the supercontinent Gondwana, along with South America and Florida—collided with Georgia. The continental collision formed the Appalachian Mountains, much as the collision of India and Asia is forming the Himalayas. With this collision all of the continents had come together once again, forming the supercontinent Pangaea. For 65 million years, Georgia lay near the center of this vast landmass. Then Pangaea began

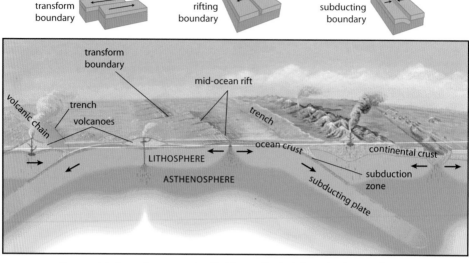

Some plate boundary scenarios that have affected Georgia. Arrows denote relative direction of movement. —Modified from U.S. Geological Survey

to rip apart. The crust of Pangaea cracked at hundreds of scattered locations, and magma welled up to fill each crack. Eventually one set of cracks linked up and became a continuous rift zone that separated Pangaea into the continents we recognize today. As the region continued to pull apart, the rift widened and more magma rose into it. The magma solidified into the Atlantic seafloor, and the rift zone became the Mid-Atlantic Ridge.

GEORGIA'S FIVE LANDSCAPE PROVINCES

Georgia's landscape is divided into five distinct geographic regions called physiographic provinces. The southern half of Georgia lies in the Coastal Plain, the inland edge of which is the Fall Line. The central part of the state is the Piedmont, which gradually ascends northward from the Fall Line to elevations around 1,000 feet. It is cut by steep-sided stream valleys and has a few isolated summits of less than 2,000 feet, such as Stone Mountain outside of Atlanta. *Piedmont* means "foot of the mountains," and to the north lie three distinctive mountain provinces: the Blue Ridge, Valley and Ridge, and Appalachian Plateau. The Blue Ridge has the highest mountains, with some peaks above 4,000 feet. The Valley and Ridge consists of long, parallel ridges separated by flatlands. The Appalachian Plateau is a region of flat-topped mountains, more than 1,700 feet above sea level, interrupted by widely separated, straight valleys.

The provinces played different roles in Georgia's plate tectonic history. The Appalachian Plateau is the eastern edge of a vast sheet of flat-lying sedimentary rocks that blanketed North America and were situated beside the ocean that preceded the development of Pangaea. The Valley and Ridge is the eastward continuation of those sedimentary rocks, which became folded and faulted when North America and Gondwana collided to form Pangaea. The Blue Ridge and Piedmont consist of metamorphic and igneous rocks that are the product of intense deformation, heating, and melting associated with the long series of events that led up to that final collision. The Coastal Plain is composed of sediments that settled on top of the southeastern portion of that deformed region after erosion had planed it down, Pangaea had been pulled apart, and the Atlantic Ocean had started forming.

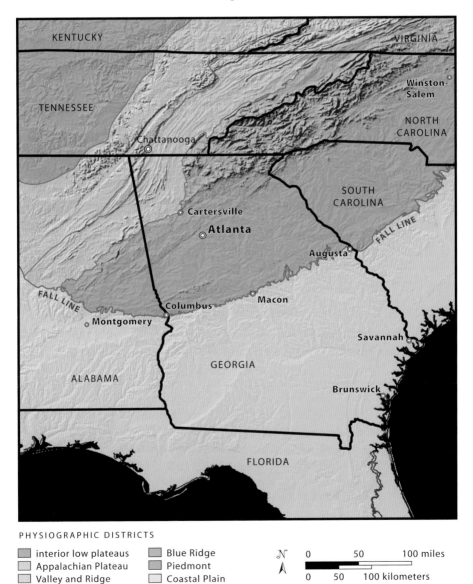

PHYSIOGRAPHIC DISTRICTS

- interior low plateaus
- Appalachian Plateau
- Valley and Ridge
- Blue Ridge
- Piedmont
- Coastal Plain

0   50   100 miles

0   50   100 kilometers

Map of Georgia and the surrounding region showing the five physiographic provinces of the eastern United States. —Landscape image courtesy of U.S. Geological Survey; boundaries added from several sources

Georgia's variation in topography is related to rock type and the history of erosion. The Coastal Plain has gentle topography because it is mainly composed of soft, easily eroded, unconsolidated sediments—that is, it lacks resistant rock to make steep slopes that can stand up to erosion. Rocks rich in the hard and insoluble mineral quartz, including sandstone, form the ridges of the Valley and Ridge and cap the Appalachian Plateau's tabletop mountains, such as Lookout Mountain. Because of its relative resistance to erosion, it has protected these high regions as those around it have been lowered by the elements.

The quartz content of metamorphic and igneous rocks plays a role in the Piedmont's topography, as well. The Piedmont's few isolated mountains, which are called monadnocks, are built mainly of quartz-rich rocks, such as granite, granitic gneiss, and quartzite. The higher, more-rugged terrain of the Blue Ridge may relate, in part, to a greater proportion of quartz-dominated metamorphic rocks, such as metagraywacke and quartzite. Quartz endowed the rocks of both regions with strength that allowed them to remain as other rocks succumbed to erosion.

The distinct and partly random sequence of events by which erosion has worn this landscape down must be part of the explanation for the Blue Ridge and Piedmont topography, because the topography cannot be explained by differences in quartz content alone. In the Piedmont, for example, the isolated monadnocks are only small portions of larger regions composed of granite or granitic gneiss. For example, Stone Mountain represents only 10 percent of the mapped area of the Stone Mountain Granite; the other 90 percent is buried or appears at the surface as broad, flat pavement outcrops. Monadnocks are recognized as a kind of erosional remnant. They are what happen to be left at this particular moment in a long, ongoing process that has already planed off the surrounding landscape.

The abrupt breaks in slope, called escarpments, where the high country of the Blue Ridge steps down to the Piedmont, are other striking landscape features that reflect a snapshot in the continuous process of erosion. They represent the ongoing competition between networks of streams, in which one stream or river system takes over territory formerly drained by another.

Georgia's landscape has been shaped mainly by stream erosion. Networks of streams can steadily reduce elevations as they cut down into underlying rock or sediment. Tributaries lengthen upstream through a process called headward erosion, in which the water gnaws away at sediment and rock at its headwaters. As stream networks compete for territory, the divides between watersheds gradually migrate. If one watershed is at a much lower elevation than its neighbor, the streams at its headwaters will periodically capture tributaries from the higher elevation watershed. This process, called stream capture, occurs when one stream nibbles through a topographical divide and diverts the stream on the other side, capturing its flow. There is evidence for recent or pending stream capture along divides all over Georgia, but it is especially dramatic along GA 52 a few miles west of Amicalola Falls, where three stream networks compete for the flows of the others. (See the GA 52: Chatsworth—Lula road guide in the Blue Ridge–Piedmont chapter for more information about Amicalola Falls and stream capture.)

Because faster-moving water has greater erosive power, a stream flowing down a steeper slope is more likely to erode the land. The stream may erode through a drainage divide and capture the flow of a tributary to another stream.

If two streams have headwaters in roughly the same area at about the same elevation, and one flows a shorter distance to the Atlantic Ocean (perhaps 400

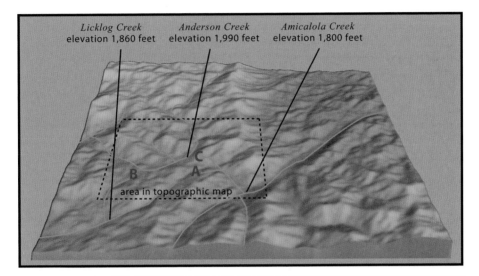

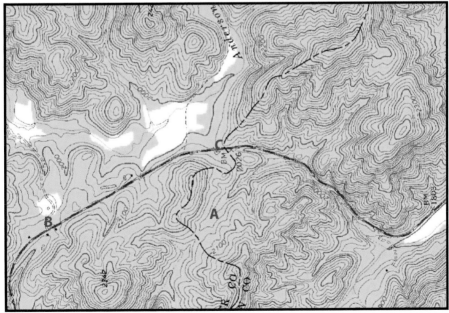

Stream capture in progress. A 3D perspective shows three drainage systems competing for territory near Amicalola Falls. An Amicalola Creek tributary has already captured the upper valley of a Licklog Creek tributary at "A." Anderson Creek, as the highest elevation stream, is vulnerable to capture, either by a Licklog Creek tributary at "B" or an Amicalola Creek tributary at "C." GA 52 passes through both potential capture sites.

miles) and the other flows a much longer distance to the Gulf of Mexico (perhaps 1,500 miles or more), the stream that travels the shorter distance will have a steeper slope and be more erosive (simple geometry). This is the case with two river systems that have headwaters in northeast Georgia. The Savannah River and its tributaries flow about 400 miles from the Blue Ridge Mountains to the Atlantic Ocean. The Tennessee River and its tributaries flow more than 1,500 miles from the same area of the Blue Ridge to the Gulf of Mexico. The Savannah River is more erosive and can therefore capture the tributaries of the Tennessee River and other similarly long, low-slope rivers in the area, such as the Chattahoochee and Coosa rivers. Much of the rugged landscape of the Blue Ridge, with its waterfalls, whitewater, and deep gorges (Amicalola Falls, Tallulah Gorge), can be linked to erosion and stream capture.

Sedimentation

Sediment is at the beginning of the geologic story in most of Georgia, whether you are looking at unconsolidated sediments in the Coastal Plain, sedimentary rocks in northwest Georgia, or metamorphosed sedimentary rocks in the Blue Ridge and Piedmont. Weathering and erosion break down rocks to produce sediment of many sizes, from boulders, gravel, sand, and silt, to clay. Sediment is transported by wind or water and deposited in low areas or places where Earth's crust is sinking, such as today's Atlantic coast. This parade of particles gets buried, compacted, and cemented (by circulating fluids) into sedimentary rock. Gravel, sand, silt, and clay become, respectively, the sedimentary rocks conglomerate, sandstone, siltstone, and mudstone, or shale.

Many marine creatures, such as coral and mollusks, use the carbon dioxide and calcium dissolved in seawater to construct their skeletons, producing the calcium carbonate minerals calcite or aragonite. These minerals can also precipitate directly from seawater and settle on the seafloor. In arid climates, magnesium in seawater can combine with calcium carbonate to make the mineral dolomite. Calcite, aragonite, and dolomite are called carbonate minerals. Over time carbonate minerals collect on the seafloor, sometimes to great depths. They harden into sedimentary rocks called carbonates. Limestone is a carbonate sedimentary rock composed of calcite or aragonite, and dolostone is a carbonate sedimentary rock composed of dolomite.

Other marine organisms, including some sponges and microbes, make their hard parts from submicroscopic crystals of quartz, forming the sedimentary rock chert, better known as agate or flint. Groundwater may dissolve quartz as it flows through silica-rich rocks or layers of volcanic ash. The quartz (or silica) may be precipitated within limestone and other carbonate rocks, forming nodules and replacing fossils. When a carbonate rock weathers, the carbonate minerals dissolve, leaving behind a reddish soil called residuum. It is composed of any insoluble clay and iron oxide impurities from the limestone, any chert nodules, or both (also known as "chert and dirt").

A layer of any kind of sediment deposited by a single geologic event, such as sediment-laden water breaking through a natural levee during a spring flood, is called a bed. The surface of a bed, called a bedding plane, is usually nearly

Two layers of crossbeds atop Tortoise Shell Rock at Rock City Gardens (see the GA 157: Tennessee State Line—Cloudland road guide in the Valley and Ridge and Appalachian Plateau chapter). Each crossbed layer was laid down as a sandbar that migrated from right to left. The top of each layer was partially eroded away before deposition resumed.

Although most sedimentary layers are deposited in nearly horizontal beds, strong currents of water or wind can rework sediment and create a pattern of sloping beds called crossbedding. If you have ever watched strong wind moving sand grains over a sand dune, you have seen crossbeds forming. The wind moves sand up the windward slope of the dune, piling it at the top and in turn causing it to cascade down the steeper leeward slope. Each successive cascade makes a new sloping layer (crossbed), and over time the dune migrates. You might expect the migration of the dune to erase the crossbeds. Indeed, the top of each crossbed generally erodes, but if the area happens to be gradually sinking, at least some of the pattern will be preserved as the sand hardens to rock. As it happens, many coastal regions (such as Georgia's coast) are continually sinking because of sea level rise; because of the subsidence of the Earth's continental crust near the coast under the weight of thousands of feet of deposited sediment; because of the compaction of older sediments; and because adjacent oceanic crust is aging, becoming cooler and denser, and exerting downward pressure on the continental shelf to which it is attached.

The largest and steepest crossbeds are found in rocks formed in windblown sand dunes, but crossbeds made by flowing water, such as those formed by migrating sandbars in rivers or by migrating beach ripples left by tidal currents, are far more common. Crossbeds can have slopes between about 10 and 40 degrees to the horizontal. A layer of crossbeds is bounded above and below by nearly horizontal surfaces, unless it was tilted after having been deposited. In Georgia there are crossbeds in windblown sand deposits in the Coastal Plain, in sandstone in northwest Georgia, and in metamorphosed sandstone in the western Blue Ridge.

horizontal when the bed is deposited. Beds are tilted and folded by tectonic movements, such as those associated with tectonic plate collisions. Tilted beds are said to be *dipping*—gently, steeply, or so many degrees measured from the horizontal (for example, 45 degrees). When geologists refer to the direction of dip, they mean the downslope direction, or the direction that water would run off a tilted bed.

GEOLOGIC TIMESCALE

A volcanic eruption or hurricane can change a landscape overnight, but most changes to the face of our planet happen over unimaginably long spans of time. The best evidence reveals the Earth to be 4.6 billion years old. Intuitively grasping this number is beyond most of us, even as we speak daily of "billions of dollars" or "gigabytes." Understanding the age of Earth is conceivable only by scaling the 4.6 billion years down. For example, if the whole of Earth history were a twenty-four-hour day, modern humans, who evolved about 200,000 years ago, would not have appeared until four seconds before midnight.

In the eighteenth century, geologists began to divide the vastness of Earth time into smaller units. They recognized that, from the bottom to the top of a stack of sedimentary rock layers, fossil life-forms changed, recording the history of life through time. Time boundaries were chosen primarily at major events in the history of life, such as large-scale extinctions. Fossils have allowed geologists to correlate rocks of similar age across the globe, establishing that rocks of the same age formed under a wide range of conditions. The initial timescale that geologists developed from fossils helped them develop a pretty good picture of the part of Earth's past in which fossils existed. Each time slice revealed a distinctive mosaic of landscapes and climates as varied as the surface of the Earth is today.

The first hard body parts of organisms, such as shell and bone, appeared about 542 million years ago. As a result, only rocks formed during the most recent 13 percent of Earth history contain fossils that can be used to correlate rock strata. Moreover, fossils are present almost exclusively in sedimentary rocks, not in metamorphic and igneous rocks. That left geologists with a lot of rock they couldn't assign ages to.

In the early twentieth century scientists developed a new age dating technique called radiometric dating, which employs radioactive elements that occur naturally in rocks, such as uranium and some forms of potassium. The known rate at which a radioactive element decays (changes to another substance) is used to determine how long ago the rock formed, allowing geologists to assign numerical ages to rock bodies.

Fossils and radiometric dating complement each other, because most sedimentary rocks cannot be dated radiometrically. This is because, with few exceptions, any minerals that might contain radioactive elements did not form during sedimentation. Numerical ages can be assigned to fossil-bearing sedimentary rocks when they are interbedded with igneous rocks (such as a lava flow or volcanic ash bed) that can be dated radiometrically. Decades of research and fieldwork by a multitude of individuals around the globe have led to the

GEOLOGIC TIME SCALE

EON	ERA	PERIOD	EPOCH	AGE (millions of years)	MAJOR EVENTS IN THE HISTORY OF LIFE	MAJOR GEOLOGIC EVENTS AFFECTING GEORGIA
PHANEROZOIC	CENOZOIC	QUATERNARY	HOLOCENE	0.012		Present barrier islands form
			PLEISTOCENE		*Homo sapiens* appears; Extinction ends mammoths	Continental shelf exposed during glacial periods
						Older barrier island/marsh shoreline complexes formed
						Windblown sand dunes form on east sides of rivers
		NEOGENE	PLIOCENE	2.59		
				5.3		
			MIOCENE			Global cooling and the growth of the polar ice cap in Antarctica lowers sea level
		PALEOGENE	OLIGOCENE	23		High sea levels
						Carbonate rocks form in southwest Georgia
			EOCENE	33.9	Whales with legs	Tektites from Chesapeake Bay meteorite impact land in Georgia
			PALEOCENE	55.8		
	MESOZOIC	CRETACEOUS		65.5	Great extinction; end of the dinosaurs First flowering plants evolve	First Coastal Plain sediments deposited along shoreline at Fall Line
		JURASSIC		146		Atlantic Ocean opens between North America and Africa
		TRIASSIC		200	First mammals and dinosaurs evolve	Rifting of Pangaea begins Diabase dikes intrude into Piedmont
	PALEOZOIC	PERMIAN		251	Greatest extinction; end of the trilobites	Collision with Africa concludes, resulting in Pangaea Granite intrudes into Piedmont Thrust faulting reaches all of Georgia Strike-slip faulting ends in Piedmont
		PENNSYLVANIAN		299	First reptiles evolve Coal swamps exist	Sandstone and coal form in northwest Georgia; thrust and strike-slip faulting in Piedmont
		MISSISSIPPIAN		318	Abundant crinoids	Collision with Africa begins Carbonate rocks and then shale form in northwest Georgia
		DEVONIAN		359	First nonflowering trees and land vertebrates evolve	Black shale forms in northwest Georgia and much of the continental interior
						Metamorphism in Piedmont
		SILURIAN		416	First land plants evolve	Age of most sandstone underlying prominent ridges in northwest Georgia
		ORDOVICIAN		444		Land rises in Blue Ridge after collision with island chain, sending eroded sand and clay to northwest Georgia
		CAMBRIAN		488	First fossil shells Many marine groups evolve	Oldest rocks in northwest Georgia form in shallow sea: first sandstone and shale, then carbonate rocks
PRECAMBRIAN	PROTEROZOIC	NEOPROTEROZOIC		542	First plants and animals (algae, jellyfish) evolve	Rocks of Blue Ridge–Piedmont begin forming in ocean basin that preceded today's Atlantic
		MESOPROTEROZOIC		1,000		Tectonic collision makes Georgia's oldest rocks (igneous and metamorphic)
		PALEOPROTEROZOIC		1,600		
	ARCHEAN			2,500	First photosynthesis First life (bacteria)	No rocks of these ages exposed in Georgia
	HADEAN			~4,000	Earth forms	
				~4,600		

Events in Georgia through geologic time. Numbers in millions of years.

modern geologic timescale, which is continually refined as new evidence is uncovered.

The major units of geologic time are eons, eras, periods, and epochs. The beginning of the Cambrian period represents the first fossil hard parts in the geologic record. The Cambrian is the first period in the Paleozoic era, which occurs at the beginning of the Phanerozoic eon. The vast amount of time before that—about 87 percent of all Earth history—is informally referred to as Precambrian time. In Georgia, the oldest exposed rocks belong to the most recent eon of the Precambrian, which is called the Proterozoic.

The Phanerozoic eon, of which we know the most, is divided into three eras separated by two great extinction events. The extinction at the end of the Paleozoic era, 251 million years ago, resulted in a loss of 95 percent of marine invertebrate genera (the plural of genus), including the well-known trilobites. The Mesozoic era ended about 65 million years ago with the extinction of many groups, including the dinosaurs. We live in the Cenozoic era.

A WORD ABOUT COLLECTING

The fossils in Georgia's sedimentary layers document how plants and animals appeared in the oceans, colonized the land, and diversified into the life we see today. Georgia's rocks, especially its metamorphic and igneous rocks, host mineral crystals that formed under nearly every possible condition. Much geology can be learned from observing fossils and minerals, and many people delight in collecting them. Invertebrate fossils and shark teeth generally can be collected without problem, but vertebrate bone discoveries may be scientifically significant, so *please check with qualified museum or university personnel before disturbing any you find.* Otherwise, important scientific information could be lost.

The best mineral and fossil localities come and go, with new localities uncovered by construction and old localities succumbing to erosion and collectors. It is easy to observe and collect at many roadside exposures, but elsewhere landowner permission must be obtained, and some private mines and collecting sites charge a fee. Attending collecting trips organized by museums or local mineral clubs associated with the Southeastern Federation of Mineralogical Societies is generally the best way to visit private sites. There are at least ten mineral clubs in Georgia, which also collect fossils (see http://www.amfed.org/sfms/club-ga-ky.html for more information about these clubs), and at least one gold-panning group.

SEA ISLANDS

Eight large barrier islands, or island complexes consisting of a main island associated with one or more smaller islands, line Georgia's 100-mile-long coast. Sounds or tidal inlets at the mouths of rivers separate the islands from each other, and islands are separated from the mainland by a grassy salt marsh several miles wide. From north to south, the major islands (or island complexes) are Tybee (Tybee and Little Tybee), Wassaw, Ossabaw, St. Catherines, Sapelo (Sapelo and Blackbeard), St. Simons (Little St. Simons, Sea, and St. Simons), Jekyll, and Cumberland (Little Cumberland and Cumberland). Only Tybee, St. Simons, and Jekyll islands are accessible from the mainland by road.

How did these islands form? Where large rivers (such as the Savannah and Altamaha) enter the ocean, deltas form because the amount of sediment the rivers supply to the coast exceeds the rate of sea level rise; therefore, the coastline builds seaward as the rivers drop their sediment load. Because of Georgia's large tidal range, meaning the elevational difference between its high and low tides, the deltas along its coast have a different shape than the delta at the mouth of the Mississippi River, where the tidal range is low. Rather than the bird's-foot shape of the Mississippi delta, Georgia's tide-dominated deltas tend to develop into islands. The action of the waves on the seaward side of the islands winnows away the mud, leaving sand grains and forming a beach. Sea breezes across the beach blow the sand into dune ridges. Rising sea levels and high tides have flooded the lower ground behind the dune ridges, forming salt marshes on the inland side of the islands.

LONGSHORE DRIFT AND TIDES

A combination of waves, longshore drift, tides, and sea level change over time shape the Sea Islands. Their beaches are relatively flat, and the waves are small, typically 1 foot or less during the summer. Generally, waves approach the shore at a slight angle. They break in shallow water, and the swash flows up onto the beach and then rushes straight back to sea. Sand grains in the surf zone (where the waves break) tend to move along the beach following this same pattern, moving up onto the beach at an angle and flowing straight back to sea. This is called longshore drift. Over time, sand and shells in the surf zone are transported from north to south along the Georgia coast. As a result, the northern ends of the islands tend to be eroding, whereas the southern ends are building out as sand is transported and deposited. This effect may be more pronounced where rivers or tidal inlets between islands are dredged to deepen them for ship passage. Dredging removes sand that would otherwise be transported southward across tidal inlets to the northern end of the islands.

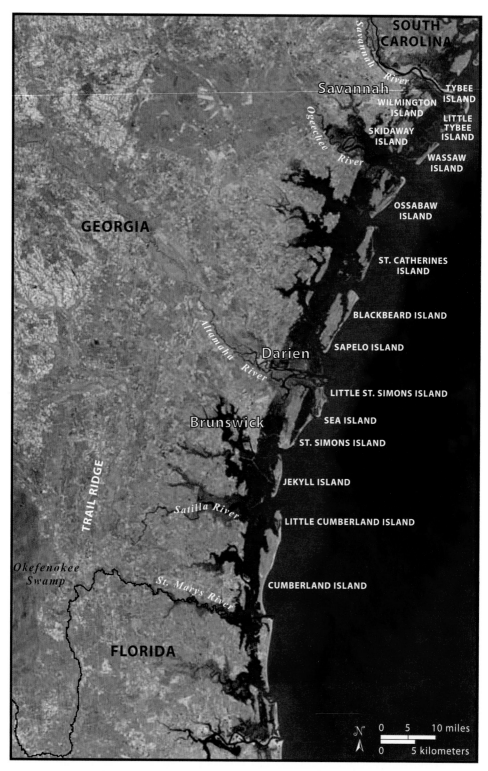

Satellite image of the Georgia coast and Sea Islands. —Created by Clark
Alexander from Landsat imagery provided by the U.S. Geological Survey

The Georgia coast experiences two high tides and two low tides each day. The twice daily tidal flooding creates an extensive intertidal zone composed of beaches and salt marsh that is alternately submerged and exposed with the tides. The water rises and falls between 6 and 10 feet, which is the highest tidal range south of New England. The highest high tides occur at the new moon phase, when the gravitational pull of the sun and the moon are aligned in the same direction.

BEACHES

In general, beaches are flat, wave energy is low, and tidal range is high on the Georgia coast. Because of the flatness of the beaches and the high tidal range, the intertidal zone—that portion of a beach that is submerged during high tide and exposed during low tide—may be 300 or more feet wide. At low tide, the intertidal zone consists of wet, hard-packed sand. Above the high tide line, the upper beach tends to be dry and sandy and contains vegetated dunes. This is the area where sea turtles tend to nest. You can recognize the high tide line by the change from wet to dry sand, but it is also marked by a wrack line consisting of broken pieces of salt marsh grass washed up by waves.

The quartz sand of the beaches is derived from the weathering of granite and gneiss in the Piedmont and Blue Ridge. Low wave-energy removes the finer materials (silt and clay) and leaves sand on the beach. The sand grains are angular. When you walk on dry sand, it sometimes squeaks as the angular grains rub together. The sand is not white like the sand on the Florida panhandle beaches. Instead, there are definite bands and layers of black heavy mineral sand consisting of high-density minerals such as ilmenite, rutile, zircon, magnetite, staurolite, kyanite, tourmaline, and others. Heavy minerals become concentrated in beach sands because waves sort the sand grains by washing away small, lightweight grains and leaving denser and heavier grains behind.

Some beaches have a ridge and runnel system, which is a series of sandbars, or ridges, alternating with lower swales, or runnels, where the water runs. These form where longshore drift deposits a series of curved sandbars with one end anchored on the beach and the other submerged. They run nearly parallel with the shoreline. The runnels between the sandbars fill with water several feet deep as the tide rises and then drain with swift currents as the tide falls.

A distinctive pattern of ripple marks forms in the beach sand in a ridge and runnel system. Straight-crested, symmetrical wave ripples are oriented roughly parallel to the water's edge. They form as waves move back and forth across the ridges and runnels during high tide, sculpting the sand. They may remain on the ridges after the tide has gone out. Asymmetrical current ripples form in the runnels as the tide drops and water drains to the sea in a direction roughly parallel to the shoreline. The crests of asymmetrical ripples are roughly perpendicular to those of the wave ripples and trace wavy lines across the runnels. Asymmetrical ripples have a long side with a low slope and a short side with a steep slope that faces the direction the water flowed. If the current is not strong

A ridge and runnel system on the southern end of Jekyll Island at low tide. The ridges are separated by a water-filled runnel. As the tide falls, current ripples will form in the runnel as the water drains out of it.

Symmetrical wave ripples typically have straight crests, but some of the ripple crests split into two crests.

wave motion

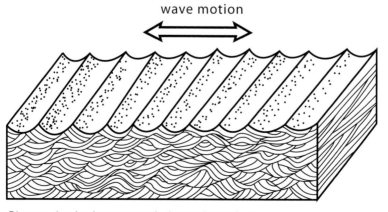

Diagram showing how symmetrical wave ripples form.

enough to completely destroy the wave ripples that formed in the runnels, the result is a set of interference ripples, in which current ripples are superimposed on wave ripples at about 90 degrees.

Few types of organisms can tolerate beach life, where they spend half the time underwater and the other half exposed to dry air and hot sun. Most beach dwellers are burrowers. At low tide, the burrows of ghost shrimp (*Callichirus major*) are recognizable by the short rod-shaped fecal pellets (resembling chocolate sprinkles) surrounding them. Holes in the dry sand of the upper beach that are 1 to 2 inches in diameter are burrows of the ghost crab (*Ocypode quadrata*), named because its pale color makes it nearly invisible as it moves across the sand.

Asymmetrical current ripples with wavy crests that formed in the bottom of a runnel. Each ripple mark has a long side with a low slope (on the left) and a short side with a steep slope (on the right). The current flowed from left to right.

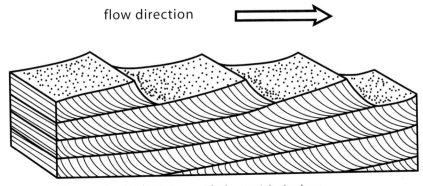

Diagram showing how asymmetrical current ripples form.

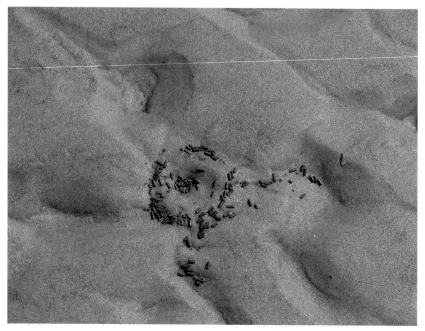

Ghost shrimp burrow surrounded by rod-shaped fecal pellets.

MARSHES

Georgia has more than 700 square miles of coastal marshlands, including salt marsh, tidal channels, and mudflats. This is approximately one-third of all coastal marshland along the east coast of the United States. Salt marshes are the low, muddy areas between the islands and the mainland, and they are threaded with meandering tidal channels. Smooth cordgrass (*Spartina alterniflora*) is the dominant vegetation.

The salt marsh is a critical ecosystem for many life-forms, including humans. It filters pollutants, protects coastal areas from flooding by absorbing excess flow, and provides food and shelter for fish and shellfish, such as oysters, blue crabs, and shrimp, during their juvenile stages. Many organisms spend at least part of their lives in salt marshes along the coast, feeding on organic detritus. If there were no salt marshes offering food and protection, there would be no seafood.

Small forested islands called hammocks (or back-barrier islands) dot salt marshes, and most are accessible only by boat. Some 1,500 of them are present in Georgia. Most are natural islands, but others, near the edges of marshes, were formed by ballast rocks dumped by ships, or by sediment dumped from coastal dredging performed by the U.S. Army Corps of Engineers.

ANCIENT AND MODERN SHORELINES

Sea level rose and fell at least eleven times during the Pleistocene epoch, between about 2.6 million and 12,000 years ago, as the climate fluctuated and glaciers retreated and advanced across North America. During the periods when the glaciers were large and widespread, forming continental-sized ice sheets, enough water was frozen in the ice that sea level was lowered by more than 300 feet, leaving the continental shelf high and dry and the shoreline some 80 miles east of the current one. When the glaciers melted, sea level was often higher than it is today, perhaps more than 100 feet higher, and the waves lapped 50 miles or more inland, near Statesboro, Hazlehurst, and Waycross.

A series of sand ridges, once coastal barrier islands, outlines the high-water marks of the Pleistocene sea. Each shoreline consists of a relict (former) sandy barrier island backed by a low area with muddy salt marsh or lagoon sediments. Swamps, marshes, and river valleys now occupy these back-barrier marshes and lagoons. Each set of former island and salt marsh is called a terrace or shoreline complex. Erosional features called escarpments, or scarps, bound some of the older terraces. Waves cut these scarps into older sediments when sea level was high, leaving a cliff-like face. The Pleistocene-age islands have rich soils that support dense subtropical vegetation. The largest and most prominent relict barrier island is Trail Ridge, located about 40 miles inland, just east of the Okefenokee Swamp. When Trail Ridge was a coastal barrier island, a large lagoon, connected by inlets to the sea, lay to the west, where the Okefenokee Swamp is now. (See also Barrier Island Sequence District in the Coastal Plain chapter.)

Sea level has been rising for much of the last 15,000 years, much of it associated with the melting of the continental ice sheets. Today, it is rising at a rate of about 1 foot per 100 years. In some places, the seas are once again lapping at the shores of the most seaward of the Pleistocene-age barrier islands. Where sea level rise is greater than the rate that sediment is being supplied to the coast by rivers, Holocene-age (modern) barrier islands are eroding back to Pleistocene barrier islands, forming composite islands. Many of the Sea Islands are a composite of a present-day beach attached to (or reoccupying) the seaward side of a Pleistocene-age barrier island (for example, St. Catherines, St. Simons, Jekyll, and Cumberland islands).

US 80
Savannah—Tybee Island
18 miles

Savannah sits on a sandy bluff, which was the barrier island of the former Pamlico shoreline. This shoreline formed during the Sangamon interglacial, a warm period of glacial melting during which sea level stood about 25 feet higher than it does today. When glaciers began developing again sea level dropped, forming a series of islands seaward of the Pamlico shoreline: the Princess Anne shoreline (when sea level was 13 to 15 feet higher than today), Silver Bluff shoreline (when sea level was 5 feet higher than today), and other, younger

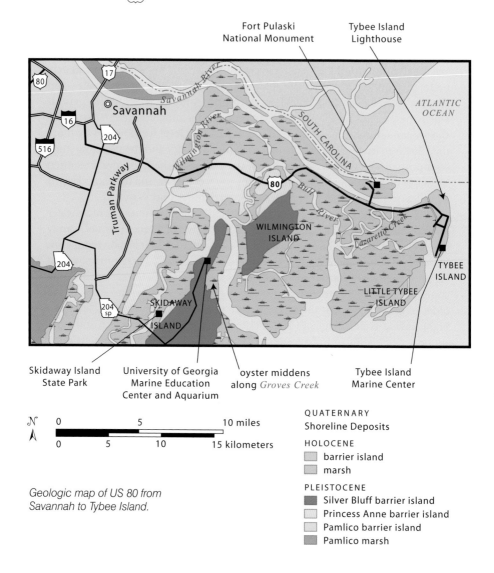

Geologic map of US 80 from Savannah to Tybee Island.

Fort Pulaski National Monument

Tybee Island Lighthouse

Skidaway Island State Park

University of Georgia Marine Education Center and Aquarium

oyster middens along *Groves Creek*

Tybee Island Marine Center

QUATERNARY
Shoreline Deposits

HOLOCENE
barrier island
marsh

PLEISTOCENE
Silver Bluff barrier island
Princess Anne barrier island
Pamlico barrier island
Pamlico marsh

shorelines—now submerged—farther out on the continental shelf. (For more information on Savannah, see the I-16: Macon—Savannah road guide in the Coastal Plain chapter.)

As you leave Savannah on US 80, you cross the Wilmington River, leaving the mainland and entering the salt marsh, a sea of grass with meandering channels called tidal creeks. For about 5 miles the causeway crosses over and north of the remains of the Pleistocene-age Princess Anne and Silver Bluff shoreline complexes. The barrier islands of these complexes are forested and separated by salt marsh.

At the Bull River, a tidal creek, you leave the cluster of back-barrier islands. US 80 turns southeast, following a causeway across about 10 miles of uninterrupted salt marsh on the south side of the Savannah River. The road parallels

McQueens Island Historic Trail, a biking trail that follows the bed of the old Savannah-Tybee Railroad, which carried wealthy Savannahians to the Tybee beach from 1887 until the 1930s. The trail is partially paved with crushed stone and oyster shells.

To the south you will catch a glimpse of some of the meandering tidal channels in the salt marsh. If you can see muddy channel banks covered with oyster shells, you'll know it is low tide. At high tide you will see only cordgrass (*Spartina*) sticking out of the water.

About 4 miles past the Bull River a road to the north leads to Fort Pulaski National Monument on Cockspur Island, part of the delta at the mouth of the Savannah River. Fort Pulaski was commissioned in 1816 by President James Madison to protect the coastal harbor and the city of Savannah following the War of 1812. A series of dikes was constructed to produce a moat and keep the fort and island dry. Robert E. Lee worked on this project after he graduated from West Point. The upland areas of the island are mostly made of material dredged from the Savannah River.

About 1 mile east of the turnoff to Fort Pulaski, US 80 crosses Lazaretto Creek, a wide meandering tidal channel that flows northward into the Savannah River. Just east of the creek, US 80 leaves the marsh and passes onto Tybee Island.

Tybee Island is the northernmost, and most developed, of the Sea Islands. It is a Holocene-age shoreline complex formed of sediment transported southward by longshore drift from the mouth of the Savannah River. About 10 miles of salt marsh separate the beach at Tybee Island from the Pleistocene-age barrier islands. This distance is much greater than that of the other islands along

Salt marsh on the south side of US 80 between Savannah and Tybee Island at high tide. The forested area along the horizon is a Pleistocene-age barrier island.

Satellite image of Sea Islands at the mouth of the Savannah River. The modern, Holocene-age Tybee and Wassaw islands are separated from the Pleistocene-age Wilmington and Skidaway islands to the west by a wide expanse of salt marsh. —Created by Clark Alexander from Landsat imagery provided by the U.S. Geological Survey

the Georgia coast due to the large sediment load the Savannah River carries and deposits at the river's mouth—sediment that has allowed the modern shoreline to build out a greater distance from the older shorelines.

Since the 1880s, the northern end of Tybee Island has been eroding as a result of human interference with the natural system. The river has been dredged since the late 1800s to keep a 42-foot-deep channel open to the Port of Savannah, the fourth-largest container port in the nation. You can see huge container ships passing Tybee Island on their way to Savannah. The river channel traps sand moving southward from South Carolina, and before dredging began the sand would then move southward to nourish the Tybee beaches. Today the sediment dredged from the river (totaling 7 million cubic yards per year) is dumped in upland and offshore disposal areas because it would be too costly to dump it on the beaches.

The U.S. Army Corps of Engineers occasionally pumps sand from offshore to renourish the beaches. The last renourishment, in 2009, cost $10 million. You can find a mixture of modern and fossil mollusk shells on the beaches if sand was pumped from offshore Miocene-age deposits. The fossil shells are dark gray. Although renourishment provides sand to the beaches, it also buries and kills burrowing marine life.

North Beach Area

From US 80, follow the brown signs north to Tybee Lighthouse. The Tybee Lighthouse at the north end of the island is the tallest (154 feet) and oldest in Georgia. The lighthouse survived the Charleston earthquake of August 31, 1886 (approximately 7.5 on the Richter scale), but was cracked near the middle, where the walls were 6 feet thick. The earthquake also moved the structure's 1-ton lens about 1.5 inches to the northeast. Tybee Island is only about 80 miles from Charleston. The quake was one of the relatively rare historical examples of a destructive earthquake centered far from a plate boundary. Its tectonic causes are not well understood.

Across from the lighthouse is historic Fort Screven (an early twentieth-century coastal defense station) and beachfront public parking (for a fee). Follow the boardwalk from the parking lot across the dunes to reach North Beach. Walking on the dunes destroys the vegetation that stabilizes them, so boardwalks have been built to protect the dunes and provide beach access. Look for the groin at the northern end of the beach. It is a line of rock perpendicular to the beach at the northeastern corner of the island. The groin's purpose is to prevent localized longshore currents from eroding the beach.

If you carefully examine the sands at North Beach, you may find small, dark gray to black, fossilized shark teeth. Visitors occasionally find other Pleistocene-age and older fossils that were carried here by wave action or were dredged from older deposits.

South Beach Area

Near the south end of the island are the Tybee pavilion and pier. Snow fences have been erected on the beach in this area to slow the wind, encouraging the deposition of sand and the building of dunes. The Tybee Island Marine Science

A rock groin on North Beach at Tybee Island, built in 1975.

Fences on the beach slow the wind and promote sand deposition and dune building. Note the Tybee pavilion and pier and the boardwalk over the dunes in the background.

Center, located near the pavilion and pier, is designed for the public, with touch tanks and saltwater aquarium displays. The center also offers beach discovery walks and marsh tours.

Near the southern tip of the island, look for the south groin, constructed of rocks in 1985, running perpendicular to the shore, along with the Tybee seawall, which parallels the shore and was built by the Works Progress Administration in 1938. Additional concrete groins were constructed farther south in 1995 to decrease erosion.

Wassaw Island and Sapelo Island

South of Tybee Island, the next four major barrier islands—Wassaw, Ossabaw, St. Catherines, and Sapelo—are accessible only by boat. Skidaway Island, behind Wassaw Island, is accessible by road and is worth visiting. It is home to Skidaway Island State Park, the Skidaway Institute of Oceanography, and the University of Georgia Marine Extension Service Marine Education Center and Aquarium (MECA).

Skidaway Island is a Pleistocene-age barrier island of the Silver Bluff shoreline complex. It is a 15-mile drive southeast of downtown Savannah. Follow Truman Parkway to GA Spur 204, which becomes the Diamond Causeway leading to Skidaway Island. To reach MECA, turn left onto McWhorter Drive and follow it to the north end of the island (about 4.2 miles). MECA is open to the public, with many educational exhibits (see the appendix for more information).

On the road south of MECA, at the edge of the marsh on the northeastern side of the island, there are oyster middens: piles and rings of oyster shells left by the Indians who hunted and fished on Skidaway 3,000 to 4,000 years ago. Archaeologists have found evidence of Indian middens on many of Georgia's Pleistocene-age barrier islands.

Wassaw Island is similar to Tybee Island in that its present-day Holocene-age beach is separated from the Pleistocene shoreline by a broad expanse of salt marsh, about 4 miles wide in this case. Wassaw Island (and the marsh behind it) is a national wildlife refuge. It is unique among the islands in that it has a virgin forest that has never been cut, and it has never been farmed.

Ossabaw Island lies to the south, across the Ogeechee River and Ossabaw Sound. It is a state heritage preserve. Ossabaw is a V-shaped, composite barrier island. The present-day Holocene-age barrier island has migrated landward with rising sea level and has come into contact with the Pleistocene-age shoreline at the southern end of the island, giving the island its V shape. The northern end of Ossabaw Island's modern beach has retreated less than the southern end due to sediment supplied by the Ogeechee and Savannah rivers.

St. Catherines Island lies south of Ossabaw Island, separated from it by several short, meandering tidal creeks that come together at St. Catherines Sound. It is owned by a private foundation and is used for research. St. Catherines is

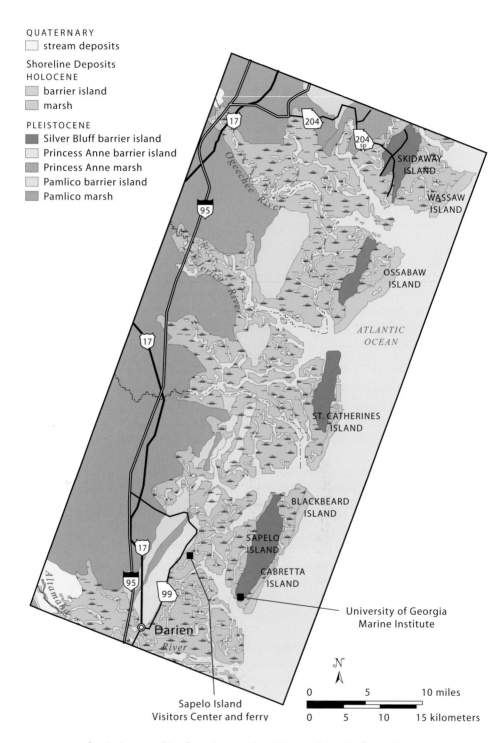

SKIDAWAY
ISLAND

WASSAW
ISLAND

OSSABAW
ISLAND

Ogeechee River

ATLANTIC
OCEAN

Vernon River

ST. CATHERINES
ISLAND

BLACKBEARD
ISLAND

SAPELO
ISLAND

CABRETTA
ISLAND

University of Georgia
Marine Institute

Darien

Altamaha

River

Sapelo Island
Visitors Center and ferry

N

0   5   10 miles

0   5   10   15 kilometers

Geologic map of the Georgia coast from Wassaw Island to Sapelo Island.

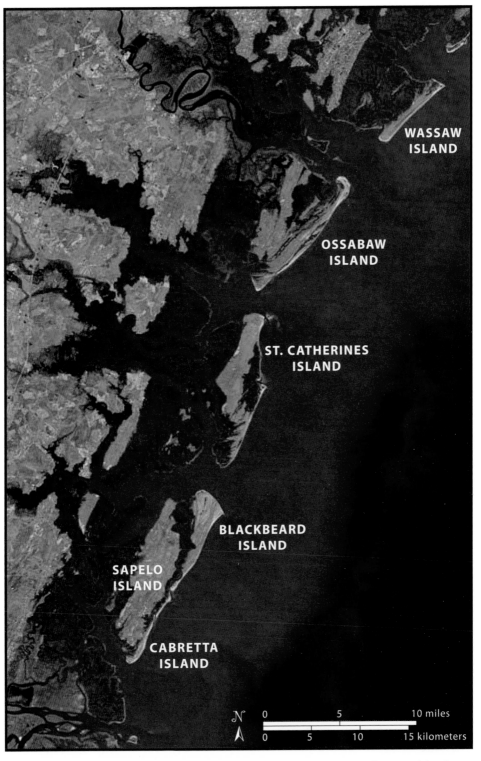

WASSAW
ISLAND

OSSABAW
ISLAND

ST. CATHERINES
ISLAND

BLACKBEARD
ISLAND

SAPELO
ISLAND

CABRETTA
ISLAND

N

| 0 | | 5 | | 10 miles |
| 0 | 5 | | 10 | 15 kilometers |

Satellite image of the Georgia coast from Wassaw Island to Sapelo Island. Ossabaw Island's V shape is very distinct. —Created by Clark Alexander from Landsat imagery provided by the U.S. Geological Survey

also a composite island made up of an eroding Pleistocene-age core on the northern end, with modern Holocene-age beaches on the northeastern and southeastern ends.

The next major island to the south is Sapelo Island. Most of it is owned by the state. Sapelo is home of the Sapelo Island National Estuarine Research Reserve and the University of Georgia Marine Institute (a scientific research center). You can take a thirty-minute ride on a state-run ferry to Sapelo Island from the Sapelo Island Visitors Center on Landing Road, near Meridian and north of Darien. Reservations are required. The Georgia Department of Natural Resources also leads guided island tours that include marsh and beach walks.

Sapelo Island is a Pleistocene-age barrier island of the Silver Bluff shoreline complex. As at Skidaway and other coastal islands, prehistoric people left oyster middens on Sapelo, some as high as 12 feet tall and 300 feet in diameter. Indian relics have been found with the shell mounds. Radiocarbon dates indicate the mounds were constructed between 3,000 and 4,600 years ago.

Darien is the only sizable town between Savannah and Brunswick. It sits on the edge of a salt marsh at the mouth of the Altamaha River and was a major lumber port in the nineteenth century. The Altamaha is Georgia's largest river, formed from the confluence of the Oconee and the Ocmulgee rivers 137 miles upstream near Hazlehurst. The river is largely undisturbed by humans. It is

Meridian, Georgia. —Courtesy of Bob Tolford

not dammed (except on tributaries), roads cross it in only five places, and it is not dredged. As a result, sediment collected from the entire length of the Altamaha travels to the coast, where it is deposited on or around a number of small marshy islands near its mouth. Longshore drift carries some of the sediment south, where it is deposited on Little St. Simons, Sea, and St. Simons islands.

Gray's Reef National Marine Sanctuary

About 20 miles off the coast of Sapelo Island, and 60 feet below sea level, lies the largest live-bottom reef in the southeastern United States. It's a little hard to visit, but Gray's Reef National Marine Sanctuary is a popular site for fishing, scuba diving, and scientific research. The reef is not a tropical carbonate reef. Instead it is an area of sandstone with dolomite minerals (called dolomitic sandstone). Overhanging rock ledges and sandy, flat-bottomed areas characterize the reef.

The reef hosts a rich assortment of marine life. The dolomitic sandstone provides a firm seafloor for bottom-dwelling invertebrates to colonize. Rising and falling sea level shaped the reef during the last several million years. The dolomitic sandstone started as sand and lime mud deposited in the sea during Pliocene time, between 5 and 2 million years ago. At some point the sediment was cemented to form the hard dolomitic sandstone when it was exposed to fresh air and sun during a time of low sea level. Fossil bones of mastodons, horses, and other mammals found under the sea near Gray's Reef indicate that the area was dry land during the last glaciation, some 25,000 years ago. At that time the shoreline was about 80 miles east of where it is today. Seawater flooded the Gray's Reef area about 7,000 years ago as the ice melted.

St. Simons Island and Its Neighbors

St. Simons Island, readily accessible by road, is a popular vacation destination and the most populated of Georgia's Sea Islands. The causeway that connects the island with US 17 in Brunswick is a narrow strip of land built above the high tide line. It passes through a roughly 5-mile expanse of salt marsh—a sea of *Spartina* grass. You cross several tidal channels, the last and largest of which is the Frederica River, part of the Intracoastal Waterway. Pleistocene mammal skeletons, including giant ground sloth (*Eremotherium mirabile*), elephant (*Elephas columbi*), mastodon (*Mammut americanum*), and several species of horse and bison, have been excavated from salt marshes and recovered by divers in the Intracoastal Waterway on the western side of some of the Sea Islands.

Look north along the western edge of the island and you will see a slightly higher area called Gascoigne Bluff. In 1794, a number of old live oak trees were cut here and sent to shipyards in Boston to build the first vessel of the U.S. Navy, the famed warship USS *Constitution*, or "Old Ironsides." You will see old live oak trees with their sprawling branches as you drive around the island because they thrive in the sandy soils of the coastal region. The live oak retains many of its leaves year-round, hence it appears to be "live" during the winter.

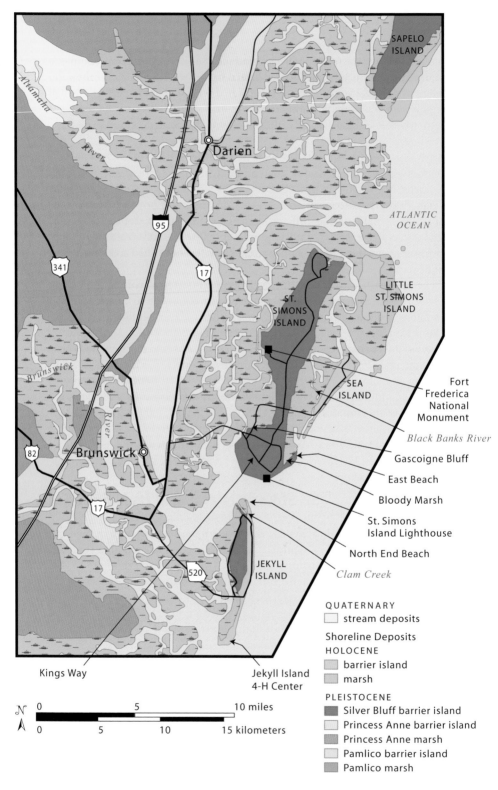

SAPELO
ISLAND

Altamaha

River

Darien

95

17

341

17

ATLANTIC
OCEAN

LITTLE
ST. SIMONS
ISLAND

ST.
SIMONS
ISLAND

Brunswick

River

SEA
ISLAND

Fort
Frederica
National
Monument

Black Banks River

82

Brunswick

Gascoigne Bluff

East Beach

Bloody Marsh

St. Simons
Island Lighthouse

17

North End Beach

520

JEKYLL
ISLAND

Clam Creek

QUATERNARY
☐ stream deposits

Shoreline Deposits
HOLOCENE
☐ barrier island
☐ marsh

Kings Way

Jekyll Island
4-H Center

PLEISTOCENE
☐ Silver Bluff barrier island
☐ Princess Anne barrier island
☐ Princess Anne marsh
☐ Pamlico barrier island
☐ Pamlico marsh

N

0   5   10 miles

0   5   10   15 kilometers

Geologic map of St. Simons and Jekyll Islands and vicinity, including the delta of the Altamaha River.

St. Simons is a Pleistocene-age barrier island (Silver Bluff shoreline complex) with rich soil and lush subtropical vegetation. It lies just south of the mouth of the Altamaha River in a complex of smaller, mostly marshy islands, including the Holocene-age Little St. Simons and Sea islands. Little St. Simons Island has enlarged tremendously since the 1860s due to the sediment-rich nature of Altamaha's delta. Little St. Simons, like Sapelo and Skidaway islands, was inhabited by prehistoric Indians who constructed oyster middens; however, you can only reach the island by boat. Sea Island, a private, narrow barrier island, lies along the northeastern edge of St. Simons.

One of the major attractions of St. Simons Island is the beach. There are two main public beaches: one at the southern end of the island near the lighthouse, which is only accessible at low tide, and East Beach, near the Coast Guard Station along the island's southeastern coast. Both beaches have been affected by erosion.

To reach the southern beach, travel south on Kings Way about 2.5 miles from where the causeway crosses onto the island. There is a visitor center at the southern end of the island near the pier, as well as Neptune Park and the lighthouse. As you cross one of the wooden stairways leading down to the beach, or go out on the pier, you'll notice large boulders beneath the stairs rimming the southern edge of the island. In 1964 Hurricane Dora eroded large portions of the beach and washed numerous homes out to sea. The boulders are part of the large seawall that President Lyndon B. Johnson had constructed along the shoreline of St. Simons Island (and also part of Jekyll Island, to the south) to help prevent the coastal erosion caused by pounding waves, strong currents, and longshore drift. The seawall, which came to be known as the "Johnson Rocks" or the "LBJ Wall," was installed before the beach had recovered from the storm, so in places large parts of the wall were soon covered by dune sand. In some places, such as the north end of the island, the boulders are being exposed again as the beach in front of them grows narrower due to erosion. Scientists agree that armoring shorelines with rocks promotes the scouring of beach sand as waves are reflected off the rocks. Armoring also makes it difficult for sea turtles to nest and for people to access the beach.

To reach East Beach, travel northeast on Ocean Boulevard. As you travel northward you will see Bloody Marsh on your left, the site of a battle between Spanish and British forces in 1742. East Beach is a modern, Holocene-age shoreline. Near the King and Prince resort, the beach is armored with a seawall of boulders and is eroding. Farther north, near the historic Coast Guard Station (1st Street), the beach is building outward.

Why is the southern portion of East Beach eroding while it is growing outward at its northern end, near the Coast Guard Station? Unlike the Altamaha River to the north, the Brunswick River south of St. Simons Island is dredged to accommodate shipping. This dredging is partially responsible for the erosion at the south end of St. Simons Island and the north end of Jekyll Island. Without dredged sand to work with, longshore currents instead remove sand from the islands. Near the Coast Guard Station, however, the beach is building outward as longshore drift carries and deposits sand from the mouth of the Altamaha River, Sea Island, and Little St. Simons Island. The Coast Guard Station and

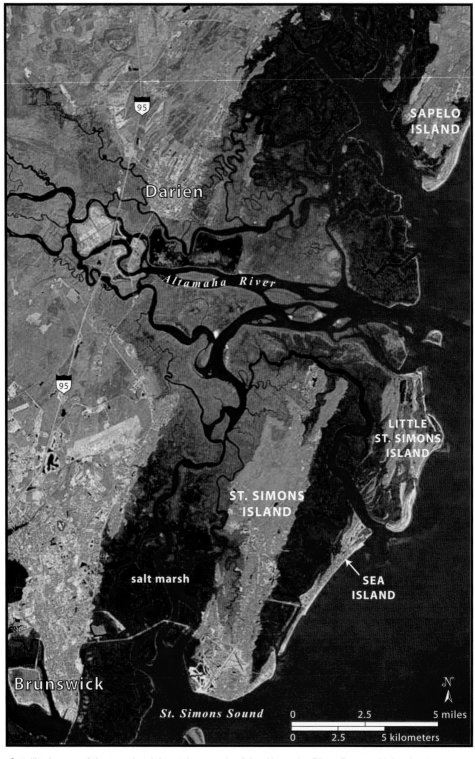

Satellite image of the marshy delta at the mouth of the Altamaha River. Forested islands are green, salt marsh is red, and Brunswick is light pink. —Created by Clark Alexander from Landsat imagery provided by the U.S. Geological Survey

St. Simons Island Lighthouse at the southern end of the island at high tide. Note the seawall of boulders between the submerged beach and the lighthouse, and the wooden stairway built for public access to the beach.

houses, which were once beachfront property, are now about 0.25 mile inland. Seawalls are also partially responsible for the erosion. Although very little of Georgia's coast has been developed, a large percentage of the developed areas have been stabilized by seawalls. Erosion is a particularly acute problem on a populated island like St. Simons. Man-made structures hamper the natural landward migration of beaches as sea level rises. There have been numerous government and public discussions about beach renourishment and the construction of a concrete groin to hold the sand in place near the King and Prince resort. If improperly designed or constructed, a concrete groin would accelerate erosion rates on the down-current side of the groin, an area that includes the lighthouse and a number of homes.

In addition, there are negative aspects to beach renourishment that are not immediately obvious. Renourishment can change the quality of the sand. At present, Georgia's beaches contain well-sorted sand with a low percentage of shell fragments. Sand dredged from offshore to renourish beaches may have a high concentration of broken shell fragments (as much as 40 percent according to one assessment), which would make it difficult to walk barefoot on the renourished beach. In addition, sand dumped on the beach buries and smothers organisms that inhabit the beach and are food for shorebirds. The dredged areas suffer as well. After dredging and renourishment there is a significant decrease in marine life and an interruption of the food chain for about a year. Fisheries also suffer for about the same length of time. Waves, currents, and tides will ultimately erode the sand added to the beach. Renourished beaches typically last several years, sometimes as long as six or eight years, before renourishment is needed again, and renourishment comes with a price tag of millions of dollars.

Cross Bloody Marsh on the East Beach Causeway and head north to Fort Frederica National Monument, on the western side of the island and overlooking the Frederica River. The ruins of the old fort and buildings, established in 1736 by English settlers, are made of tabby, a mixture of equal volumes of oyster shells, lime (burned oyster shells), sand, and water. The mixture was poured into wooden forms about 12 inches wide and 18 inches deep, layer by layer, to build up the walls. A few other tabby buildings still stand around the island.

From the fort follow Frederica Road south to Sea Island Road. Turn right (west) onto Sea Island Road and follow it for about 2.7 miles. Turn right onto Frederica Road and right again onto the causeway, which leads back to US 17. Follow US 17 south for about 4 miles and turn left (east) onto Georgia 520, or Jekyll Island Road. Along the way you'll see a number of small, forested islands sticking up out of the golden sea of *Spartina* grass in the salt marsh. These are hammocks.

One of the tabby buildings at Fort Frederica.

Close-up of tabby showing the oyster shells.

Jekyll Island

The causeway that leads to Jekyll Island from US 17, just south of Brunswick, crosses 6 miles of salt marsh with *Spartina* grass. The causeway was built from sediment dredged from local waterways. Jekyll Island is shaped like a turkey leg, with a central core of a Pleistocene-age barrier island of the Silver Bluff shoreline complex, which has richer soils and a greater diversity of plant species. The northern and southern tips of the island are younger—less than 2,000 years old.

The island is slowly migrating to the south. More than 1,000 feet of beach has eroded from the northern end of the island since the early 1900s, when dredging began in the Brunswick River to allow ships to reach Brunswick, the sixth-largest port for automobile imports in the United States and a significant center for grain export. At the same time, more than 1,000 feet of sediment has been deposited along the southern end of the island. Longshore drift transports the sand along the shore and forms a spit at the southern end of the island. Because the waves tend to curve around the tip of the island as they approach the shallow water of the shore at an angle, the spit curves toward the mainland. Marshy wetlands have formed in the quiet area behind the sand spit, where they are shielded from the waves. The northern end of the island is eroding because sand that longshore drift would normally transport to the island is dropped in the dredged St. Simons Sound at the mouth of the Brunswick River. The

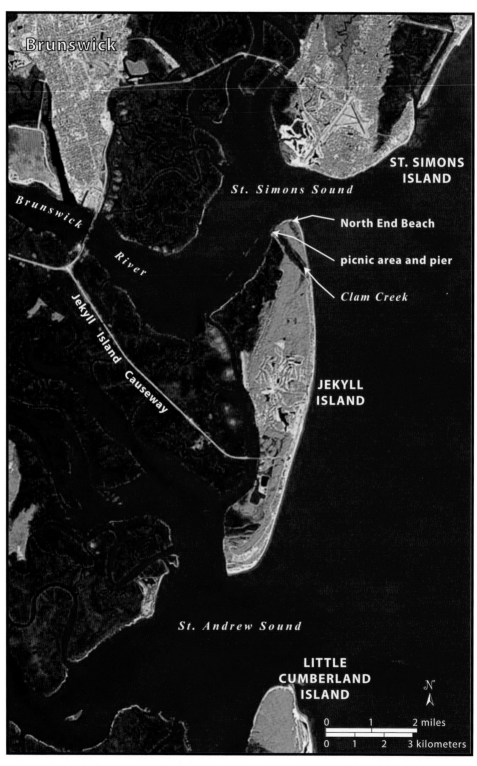

Brunswick

ST. SIMONS
ISLAND

St. Simons Sound

Brunswick

River

North End Beach

picnic area and pier

Clam Creek

Jekyll Island Causeway

JEKYLL
ISLAND

St. Andrew Sound

LITTLE
CUMBERLAND
ISLAND

N

0		1		2 miles
0	1	2		3 kilometers

Satellite image of Jekyll Island and the southern tip of St. Simons Island.
—Created by Clark Alexander from Landsat imagery provided by the U.S. Geological Survey

dredged material is dumped elsewhere, so it never reaches the shores of Jekyll Island. Meanwhile, the currents and wave energy are still acting on the north end of the island, picking up and carrying beach sand to the south.

The best place to see the effects of erosion is on North End Beach (at the northern end of the island) near the Clam Creek picnic area and fishing pier. The northern end of Jekyll Island consists of a narrow Pleistocene-age dune ridge (Silver Bluff shoreline complex), which Clam Creek Road follows to the picnic area. West of the road is the salt marsh behind the Silver Bluff shoreline, where you can see clusters of living oysters on the brown marsh mud at low tide. To the east is the salt marsh behind the modern Holocene-age beach and dune ridges, and Clam Creek runs down the middle of it. Clam Creek's meandering over the years has eroded the narrow dune ridge. In several places, the meanders come nearly to the edge of the road. (Erosion is typically greatest on the outside of the curve of a meandering stream.) One day the tidal creek may erode through the dune ridge, although concrete and rubble have been added to try to prevent this from happening anytime soon. This scene will give you an idea of how hammocks form as meandering tidal creeks cut through narrow remnants of beaches, leaving small, isolated islands surrounded by marsh. The pier is attached to the Pleistocene-age dune ridge, and a footbridge leads across Clam Creek to the Holocene-age shoreline. The best time to visit this area is at low tide, when you can see "boneyard beach." Instead of real bones, what you see are the carcasses of trees that were once part of the island's forest.

West of the Clam Creek Road you can see oyster shells on the marsh mud at low tide.

North End Beach at low tide. St. Simons Sound and St. Simons Island are visible on the horizon. Remains of dead trees litter the beach. The sand forms a thin veneer over the dark brown mud (foreground) of the salt marsh.

North End Beach at high tide.

Salt water killed the trees and erosion undercut the roots, leaving them lying in repose on the beach. At high tide, the water comes up to the edge of the remaining forest.

Along most of the eastern side of the island, the Holocene-age beaches have migrated back onto the Pleistocene-age island as a result of rising sea level and reoccupied the ancient shoreline. There are eroding beaches lined with boulders near the north end of the island and flat, sandy beaches farther south where longshore drift is depositing sediment.

The Jekyll Island 4-H Center, located at the southern end of the island between Jekyll Point and St. Andrews Beach, off South Beachview Drive, is a good place to visualize how fast the southern end of the island is growing. Buildings come and go as an area is redeveloped, but older structures are good points of reference to see the widening of a beach over time. The 4-H Center occupies a building constructed in 1956, during the segregation era, as a motel for African-American beachgoers. The one-time beachside motel is now more than 1,000 feet inland. Blocks of homes just west of the 4-H Center were beach-front property in the early 1970s, but some are now more than 2,000 feet from the sea because of sediment deposition. A boardwalk from the 4-H Center to the beach crosses an expanse of dune ridges (with low shrubs) separated by swales or depressions, some of which contain freshwater marshes, or sloughs. This stretch of beach is also a good place to see ridge and runnel systems, with asymmetrical and symmetrical ripples.

Before you leave the island, the Georgia Sea Turtle Center, located in the Historic District of Jekyll Island near the early twentieth-century millionaire

The fossil sea turtle Archelon *at the Georgia Sea Turtle Center.*
—Courtesy of the Georgia Sea Turtle Center

"cottages" and historic hotel, is worth a visit. From the causeway, turn left onto Riverview Road and then right onto Stable Road (the center is at 214 Stable Road). The center is open to the public with exhibits on sea turtles in Georgia and also serves as a rehabilitation center for ill and injured sea turtles. A highlight of the center is the giant fossil sea turtle *Archelon*, which is hanging from the ceiling. It was recovered from the Cretaceous-age Pierre Shale in South Dakota.

Cumberland Island National Seashore

Cumberland Island, the largest and southernmost of the Georgia Sea Islands, lies just north of the Florida border, near the town of St. Marys and the Kings Bay Naval Submarine Base. You can access the island by passenger ferry (no cars) that departs from St. Marys. The number of visitors each day is limited, so reservations are recommended. The ferry winds down the meandering St. Marys River through several miles of salt marsh to Cumberland Sound, part of the Intracoastal Waterway, and heads to a dock on the west side of Cumberland Island. Once at the dock you are on foot, unless you decide to rent a bicycle to traverse the shell and dirt roads and paths of this protected natural area. From the dock, the beach is about 0.5 mile due east.

The island has an interesting history, going back at least 4,500 years to when Indians built villages and mound- and ring-shaped oyster shell middens. Atop one of these mounds a four-story tabby mansion, called Dungeness, was built in the 1790s. The walls of the mansion were 6 feet thick at their base. The mansion burned in 1866. In the 1880s, Thomas Carnegie acquired the property and built a castle on the ruins of Dungeness, but it too burned and lies in ruins.

Erosion has impacted Cumberland Island quite differently than the other Sea Islands, due in part to the presence of wild horses brought to the island by the Spanish, English, and millionaire industrialist Thomas Carnegie. The horses graze on vegetation that would help hold the soil in place, and they cut paths that cause rainwater runoff to be channelized, increasing erosion. The western side of the island is eroding roughly 2 to 3 feet per year due to tides, natural meandering of the channels, and wakes from boats and submarines traveling the waterway. A seawall helps protect the shore, but erosion at the end of the seawall has created a steep bank and undercuts trees. Erosion of the shore has exposed old middens in places.

The bulk of the island is composed of a series of forested Pleistocene-age (Silver Bluff shoreline) beach ridges backed by extensive salt marshes. About 10 percent of the island is Holocene in age, less than 2,000 years old. Sediments of this age comprise a thin veneer of sand that lies upon the Pleistocene-age beach primarily along the eastern side of the island. Sand dunes have begun to bury trees and shrubs southeast of Dungeness as wind transports beach sand inland. Typically, vegetation stabilizes them, but in places feral horses and other grazing animals have destroyed the vegetation. In some places on the island, a set of secondary, large parabolic dunes has formed inland of the original dunes along

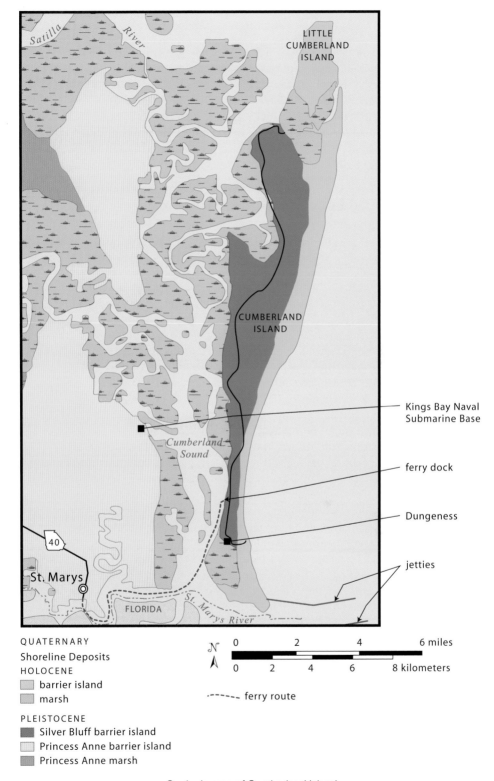

Geologic map of Cumberland Island.

Oyster shell midden exposed by erosion. —Courtesy of C. J. Jackson

A wild horse on Cumberland Island near a sand dune that is encroaching upon trees and shrubs. —Courtesy of C. J. Jackson

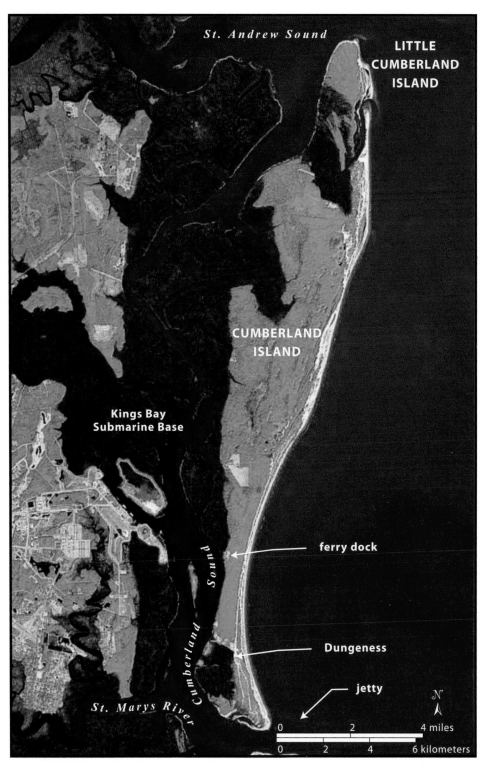

Satellite image of Cumberland Island. Note the jetty extending eastward from the south-eastern tip of the island. —Created by Clark Alexander from Landsat imagery provided by the U.S. Geological Survey

the beach. Driven by prevailing winds, these secondary dunes are migrating into the maritime forest and burying trees.

Most of the drinking water on Cumberland Island comes from artesian (free-flowing) wells drilled into the porous limestone that is hundreds of feet below the surface. This limestone is part of the Floridan aquifer and can be traced in the subsurface to Albany, about 180 miles to the west. Impermeable clay lies over the limestone, which causes the water to be at considerable pressure near the coast. A 680-foot-deep well at Dungeness drilled by the Carnegies in 1887 flowed at nearly 800,000 gallons per day, and the water rose 51 feet above the ground surface. The water gets its taste and slight odor from the minerals and sulfur—in the form of hydrogen sulfide gas—that it picks up in its subsurface journey.

At the southeastern tip of Cumberland Island, there is a 2.5-mile-long jetty, or wall of stone, built out into the water. Another jetty is present at the north end of Amelia Island in Florida, south of the inlet separating the two islands. The jetties line the sides of the inlet to protect it from drifting sand, allowing submarines access to the submarine base.

Both Cumberland Sound, behind the island, and the inlet between Cumberland and Amelia islands are dredged to accommodate submarines. Dredging, in combination with the strong tidal currents, causes fine-grained sediments and organic debris from the marsh behind Cumberland Island to erode. These sediments settle into the sound and channel, so both must be dredged frequently. Dredged sediment is dumped in mounds called spoil piles along the western edge of the island. Fossils have been collected from the spoil piles, including shark teeth, ray plates, pieces of turtle shell, and pieces of blackened fossil bones, the color of which is due to phosphate minerals. Grand Avenue, the main north-south road down the center of the island, has been paved with dredged sediment, and if you look carefully you can find small black or dark gray shark teeth in the road.

COASTAL PLAIN

SEDIMENTARY FORMATIONS

The Georgia Coastal Plain comprises the southern half of the state, south of the Fall Line cities of Columbus, Macon, Milledgeville, and Augusta. It is dominated by relatively flat land dissected by rivers, and a series of sandy ridges separated by swampy valleys, both of which parallel the coast. The portion of the Coastal Plain in which rivers flow to the Atlantic Ocean contains sediments dominated by sand and clay, whereas the portion with rivers flowing to the Gulf of Mexico contains large areas of limestone.

The oldest sediments exposed in the Coastal Plain are Late Cretaceous in age, between 100 and 80 million years old. Cretaceous deposits extend southeastward from the Fall Line for up to 35 miles and continue in the subsurface. Progressively younger sediments are exposed to the south and east of the Cretaceous deposits. Paleocene- to Pliocene-age sediments cover most of the Coastal Plain, and younger Pleistocene-age sediments cover a 40-mile-wide belt along the coast.

The Coastal Plain sediments form a wedge that thickens southeastward from the Fall Line toward the coast and continues out beneath the Atlantic continental shelf. At the coast, sedimentary deposits 1 to 2 miles thick overlie bedrock. About 150 miles offshore the deposits reach their maximum thickness of about 9 miles, beyond which they taper to a few hundred feet thick about 300 miles offshore, beyond the outer edge of the continental shelf.

The sediment making up this great wedge is derived from the weathering of the Appalachian Mountains, which once towered like the Himalayas in the Piedmont and Blue Ridge regions. At the Fall Line, Late Cretaceous–age sediments overlie Paleozoic-age and older metamorphic and igneous rocks (more than about 250 million years old) of the Piedmont.

Sea level was much higher during the Late Cretaceous, and instead of a Gulf of Mexico there was a seaway that extended far inland across the Great Plains region of the central United States and northward into Canada. A part of the coastline of this great seaway passed through southwestern Georgia, roughly from Columbus to Albany. Rivers flowed west and southwest across central Georgia toward this coastline. Today we find seashells and shark teeth southwest of this old coastline, where Late Cretaceous sand and clay were deposited

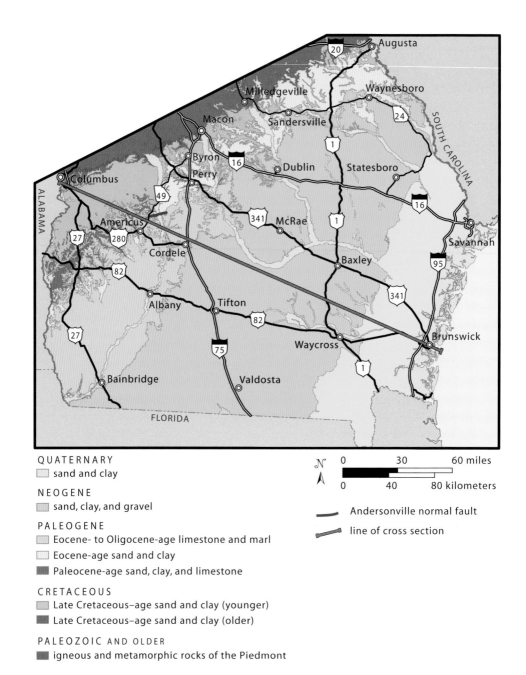

Geologic map of the Georgia Coastal Plain.

QUATERNARY
sand and clay

NEOGENE
sand, clay, and gravel

PALEOGENE
Eocene- to Oligocene-age limestone and marl
Eocene-age sand and clay
Paleocene-age sand, clay, and limestone

CRETACEOUS
Late Cretaceous–age sand and clay (younger)
Late Cretaceous–age sand and clay (older)

PALEOZOIC AND OLDER
igneous and metamorphic rocks of the Piedmont

0   30   60 miles
0   40   80 kilometers

Andersonville normal fault
line of cross section

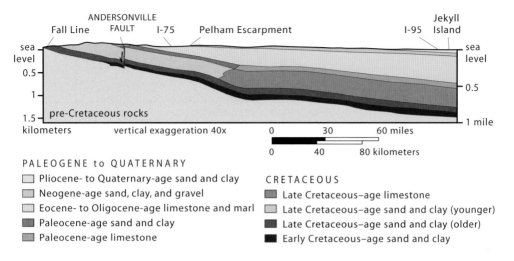

Cross section from Columbus to Jekyll Island showing the wedge of Coastal Plain sediments. (Modified after American Association of Petroleum Geologists, 1995.)

in shallow marine and nearshore environments. Farther east, in central and eastern Georgia, sediments of Late Cretaceous age were deposited in river and floodplain environments.

Sea level was high in Eocene to Oligocene time as well, and limestone was deposited across much of the Georgia Coastal Plain. Fossil shells of oysters and other marine organisms can be found in limestone of the Dougherty Plain of southwest Georgia and at Clinchfield (south of Macon), Sandersville, and Shell Bluff along the Savannah River south of Augusta.

The sea retreated during the Miocene epoch (associated with global cooling and the growth of glacial ice in Antarctica), and the former sea bottom was exposed. Eocene- and Oligocene-age limestone began to weather and dissolve, leaving behind a residuum of red clay. Rivers flowed across the land, eroding and redistributing the red soil. Reddish orange residuum and river deposits of the Miocene-age Altamaha Formation blanket large areas of the Coastal Plain and are practically indistinguishable since their origins are so intertwined. There is a lot of reddish orange soil in south Georgia.

Sea level rose and fell several times during Pliocene and Pleistocene time, and waves cut into the Miocene-age river deposits that blanket south Georgia, forming a series of terraces—and an escarpment—across the region. Pleistocene sea level changes produced a series of barrier island shoreline complexes—sandy beach ridges alternating with muddy, back-barrier marsh and lagoon sediments—paralleling the coastline.

STRATIGRAPHIC COLUMN FOR THE KAOLIN BELT AREA OF CENTRAL AND EASTERN GEORGIA

ERA	PERIOD	EPOCH	GROUP	FORMATION	MEMBER	DESCRIPTION AND DEPOSITIONAL ENVIRONMENT
CENOZOIC	NEOGENE	PLIOCENE				Strata eroded away or not deposited.
		MIOCENE		Altamaha Formation		Reddish orange clayey sandstone and conglomerate. Deposited in rivers.
	PALEOGENE	OLIGOCENE				Strata eroded away or not deposited.
		LATE EOCENE	Barnwell Group (same age as Ocala Limestone)	Tobacco Road Sand		Burrowed pebbly sand; in places comprises well-layered and fine-grained chert and montmorillonite clay. Fossils include oysters replaced by silica. Deposited in a nearshore environment.
					Sandersville Limestone	White limestone with clay and chert nodules. Fossils include clams, snails, sand dollars, oysters, bryozoans, manatee bones, and shark teeth. Replaced by chert near Louisville. Deposited in a continental shelf environment with deepening water.
				Dry Branch Formation	Irwinton Sand	Quartz sand. Fossils include clams and oysters. Deposited in a nearshore environment.
					Twiggs Clay	Gray fuller's earth or montmorillonite clay with glauconite; commonly weathered to a brownish red, mottled residuum. Fossils include mollusk shells, carbonized wood, whale bones, and fish teeth and scales. Deposited in salt marshes behind barrier islands and includes windblown volcanic ash.
				Tivola Limestone		Tan, porous, highly fossiliferous limestone with scallop shells, sand dollars, bryozoans, shark teeth, ray teeth, and molds of clams and snails. Deposited in seas when they were at their highest level in the area.
				Clinchfield Formation		Fine-grained sand with fossil shark teeth, manatee bones, and scallops and other mollusks. Deposited on a beach.
		MIDDLE EOCENE	Oconee Group	Huber Formation		Fining-upward sand sequences capped by hard white kaolin beds. Also includes gray to black unoxidized kaolin and carbonized wood. The upper part contains burrows of intertidal shrimp. Deposited in rivers and floodplains to a brackish to shoreline environment.
		EARLY EOCENE				
		LATE PALEOCENE				
		MIDDLE PALEOCENE				Strata eroded away or not deposited.
		EARLY PALEOCENE				
MESOZOIC	LATEST CRETACEOUS			Buffalo Creek Formation or Gaillard Formation		Fining-upward sand sequences capped by soft white kaolin beds. Deposited in rivers and floodplains.
	LATE CRETACEOUS			Pio Nono Formation		Feldspar-rich clayey sand with gravel. No fossils (grades into the Eutaw Formation). Deposited in a river.
				Piedmont igneous and metamorphic rocks		

Stratigraphic column for the central and northeastern Georgia Coastal Plain, compiled from various sources. Wavy black lines represent unconformities.

STRATIGRAPHIC COLUMN FOR THE WESTERN GEORGIA COASTAL PLAIN

ERA	PERIOD	EPOCH	GROUP	FORMATION	DESCRIPTION AND DEPOSITIONAL ENVIRONMENT
CENOZOIC	PALEOGENE	LATE OLIGOCENE	Undifferentiated residuum		Reddish orange clayey sandstone and conglomerate, locally with chert. Rare fossils. Overlies erosional surfaces that cut down through underlying sediments, including the Providence Sand. In some places it forms from the dissolution of limestone, and in others it may be a fluvial deposit.
		EARLY OLIGOCENE		Bridgeboro and Suwannee limestones	Limestone with chert, weathered to pale orange granular residuum. Fossils include scallops, echinoids, and foraminifera. Deposited in an offshore continental shelf environment.
		LATE EOCENE		Ocala Limestone	Fossil-bearing limestone with clay and chert. Weathers to reddish orange clayey residuum with chert. Fossils include corals, sand dollars, and clams. Deposited in an offshore continental shelf environment.
			Barnwell Group	Clinchfield Formation	Limey sand and sandy limestone. Deposited in a shallow marine, nearshore environment.
		MIDDLE EOCENE	Claiborne Group		White, yellow, tan, or red crossbedded sand and sandstone with some clay. Deposited in a shallow marine environment.
		EARLY EOCENE		Tuscahoma Sand	Dark greenish gray clay and silt interlayered with gray sandstone, with glauconite and phosphate pebbles. Fossils include fish teeth and foraminifera. Deposited in a marine environment.
		LATE PALEOCENE		Nanafalia Formation	Sandstone deposited in a river environment, and kaolinite and bauxite-rich clay (near Andersonville). Deposited in coastal estuary. Fossils include clams, oysters, snails, and shark teeth.
		PALEOCENE		Clayton Formation	Grayish orange to reddish brown clayey sand with pebbles and ironstone at Providence Canyon. Elsewhere, fossil-bearing limestone with some chert and dark gray clay. Forms chert-bearing weathered residuum. Fossils include oysters. Deposited in a coastal to nearshore shelf environment.
	LATEST CRETACEOUS			Providence Sand	White to tan, clayey, kaolin-bearing crossbedded sand. Fossils include oysters, snails, ammonites, and echinoids. Deposited in a shallow marine environment.
MESOZOIC	LATE CRETACEOUS			Ripley Formation	Light gray to orange, massive to crossbedded sand and clay. Fossils include shark teeth, snails, clams, crabs in concretions and cemented ledges, and burrows. Deposited in the inner continental shelf.
				Cusseta Sand	Coarse sand with large-scale crossbeds. Contains fossil bivalves. Deposited in estuary, barrier island, tidal marsh, lagoon, and inner shelf environments.
				Blufftown Formation	Crossbedded sandstone with mudstone, silt, shale, and carbonaceous clay. Some glauconite. Contains abundant fossils of oysters, clams, snails, ammonites, crabs, and bryozoans. Vertebrate fossils include shark teeth, fish fossils, reptile bones and teeth, turtles, mosasaurs, crocodiles, and dinosaurs. Deposited in a coastal marine environment with a broad estuary, sand shoals, and muddy tidal flats.
				Eutaw Formation	Crossbedded sandstone with fine carbonaceous sand and montmorillonite clay. Contains fossils of oysters, clams, snails, ammonites, shark teeth, teeth of rays and skates, and bones of flying reptiles. Deposited in a range of shoreline, nearshore, marsh, and lagoon environments associated with a barrier island complex.
				Tuscaloosa Formation	Interbedded sandstone and mudstone with petrified wood. Poorly sorted kaolinitic, feldspar-rich clayey sand with white kaolinite clay. Deposited in a river to delta environment.
				Piedmont igneous and metamorphic rocks	

Stratigraphic column for the western Georgia Coastal Plain, compiled from various sources. Wavy black lines represent unconformities.

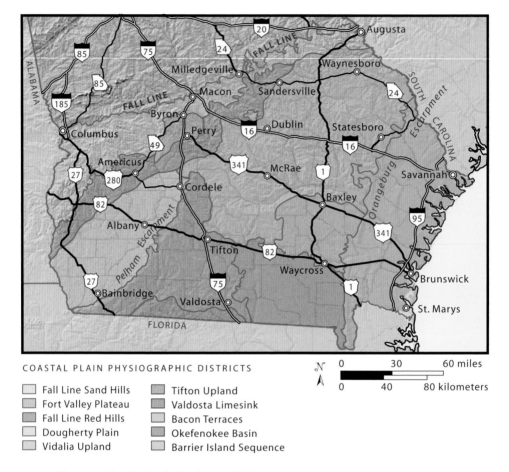

COASTAL PLAIN PHYSIOGRAPHIC DISTRICTS

☐ Fall Line Sand Hills
☐ Fort Valley Plateau
☐ Fall Line Red Hills
☐ Dougherty Plain
☐ Vidalia Upland
☐ Tifton Upland
☐ Valdosta Limesink
☐ Bacon Terraces
☐ Okefenokee Basin
☐ Barrier Island Sequence

𝒩
A

| 0 | 30 | 60 miles |
| 0 | 40 | 80 kilometers |

Physiographic districts in the Coastal Plain. —Landscape image courtesy of U.S. Geological Survey; boundaries modified after Griffith et al., 2001, and Clark and Zisa, 1976

PHYSIOGRAPHIC DISTRICTS

The Georgia Coastal Plain is divided into eight physiographic districts based on topography and types and ages of rocks and sediments.

Fall Line Hills

The Fall Line Hills District is adjacent to the Fall Line and is underlain by loose sand, gravel, and clay. The district is subdivided into the Sand Hills, which are directly adjacent to the Piedmont and composed of Cretaceous- to Eocene-age sand, and the Red Hills, composed of Eocene-age sand and clay formations. The sediments of the Fall Line Hills are Late Cretaceous river and floodplain deposits across much of Georgia, but south of Columbus they are coastal and marine deposits. The economically significant white clay of the Kaolin Belt lies

Bluffs of sand with clay in it (probably the Pio Nono Formation) north of Huber along the east side of US 23.

within this district. Rainwater percolates into Cretaceous-age sand and gravel of the Fall Line Hills and recharges subsurface aquifers that supply groundwater to south Georgia.

The sediments in the Fall Line Hills erode easily, so rivers and streams have produced highly dissected topography with very little level land other than floodplains and stream terraces. The characteristics of rivers change as they flow from the Piedmont to the Fall Line Hills. In the Piedmont, where there is bedrock at the surface, channels are narrow and rocky, but in the Fall Line Hills river channels meander through wide, swampy floodplains. South of the Fall Line Hills the floodplains narrow and become less swampy, and the river channels become straighter where the sediments have higher clay content and are harder to erode.

Fort Valley Plateau

The Fort Valley Plateau, south of Macon, has few streams and is unusually flat because it is capped by residuum from Eocene-age sand containing clay and silt, which is less easily eroded than the sand of the Fall Line Hills. In the eastern part of the Fort Valley Plateau, the sediments are mainly river and floodplain

Flat agricultural land of the Fort Valley Plateau with reddish brown soil, north of Montezuma.

deposits of the Oconee Group, but along the western side of the plateau the sediments were deposited in coastal marine environments and comprise the Fort Valley Group, which grades westward into the Ripley and Clayton formations south of Columbus.

The Fort Valley Plateau is bounded by the Ocmulgee River on the east, the Flint River on the west, and Hogcrawl and Big Indian creeks on the south. The plateau slopes gently to the southeast, toward the Perry Escarpment, a steep slope where the land rises southward to the higher elevations of the Fall Line Hills.

Vidalia Upland

The Vidalia Upland covers a broad area in east-central Georgia, southeastward of the Fall Line Hills. The terrain is lower and flatter than that of the Fall Line Hills. It is sometimes called the Atlantic Southern Loam Plains because of the loamy (sand plus clay) soils it contains. The Vidalia Upland is underlain by the reddish brown, mottled, Miocene-age Altamaha Formation, which is mainly composed of massive sandy clay and clayey sand. The Altamaha Formation is the most widespread sedimentary rock unit exposed in the state, covering about half of the Coastal Plain in the Vidalia Upland and Tifton Upland districts. The formation is interpreted to be a complex unit of river deposits, about 20 million years old, that grade southeastward into marine deposits in the subsurface. The lowermost beds are more clay rich, consisting primarily of kaolinite. The upper beds contain crossbedded sandstone containing pebbles and gravel, which is called Altamaha Grit.

Mottled, clay-rich sediment of the Altamaha Formation exposed along US 341 in the Vidalia Upland between McRae and Lumber City. The mottling develops as clay and iron oxides, which form from the weathering of iron-bearing minerals, migrate downward in the formation.

Tifton Upland and Valdosta Limesink Districts

The Tifton Upland extends northward from the Florida border through Valdosta and Tifton to Cordele. It stands as much as 200 feet higher than the Dougherty Plain to the west. The steep west-facing slope marking the edge of the upland is known as the Pelham Escarpment. The 150-to-200-foot-high bluffs along the southern side of the Flint River arm of Lake Seminole, in the southwestern corner of the state, are good exposures of the escarpment and the sediments of the Tifton Upland. Erosion-resistant, clay-bearing sand, silt, and clay of Miocene to Holocene age, including the Altamaha Formation and several other units, compose the Tifton Upland. A small part of this district—south and west of Valdosta near the Florida state line—is called the Valdosta Limesink District. It is composed of limestone containing sinkholes, or lime sinks.

Dougherty Plain

The southeastern border of the Dougherty Plain is the Pelham Escarpment, extending from the southwestern corner of Georgia to Cordele, about 10 miles east of the Flint River. To the northwest it butts up against the Fall Line Red Hills. The plain is underlain by 30 to 60 feet of weathered residuum overlying Eocene- to Oligocene-age limestone. Slightly acidic rainwater percolates into the ground and enters cracks in the limestone, dissolving the rock and forming

A small lime sink within a larger, shallow lime sink (margin lined by pine trees) in the Dougherty Plain west of Albany.

underground holes or caverns in the bedrock. When the water table is high, the buoyancy of the water supports the roof of an underground cavity, but when water levels drop, the roof and residuum collapse into the cavity in the bedrock, forming lime sinks. The limestone is part of the Floridan aquifer, and water flowing through its cavities and cracks provides groundwater for commercial, agricultural, and domestic use throughout south Georgia and Florida.

Bacon Terraces

The Bacon Terraces District lies 40 to 100 miles inland of Georgia's south-central coastline, west of the Orangeburg Escarpment, which is a major Coastal Plain boundary and former coastline. The Bacon Terraces are a peculiar series of east-facing, steplike terraces cut by waves during former highstands of sea level. Although almost imperceptible from the ground, the terraces are relatively flat areas separated by east-facing slopes. The oldest, most weathered terraces form the highest-elevation portion of the district at its western edge, and they grow younger and lower toward the coast, recording a relative lowering of sea level over time. They are composed of Pliocene- to Pleistocene-age sands and gravels.

The terraces are difficult to date because they are erosional features, but the oldest terrace, the Hazlehurst Terrace, may mark the worldwide lowering of sea level about 5 million years ago, at the end of the Miocene epoch. This drop in sea level is associated with the first growth of glacial ice in Antarctica, which heralded global cooling and the development of ice sheets that locked up a great deal of water in ice.

Barrier Island Sequence District

The Barrier Island Sequence District dominates the lower Coastal Plain and stretches from the coastline to 40 miles inland. This is one of the most fascinating areas of the Coastal Plain because it contains a 2-million-year record of cyclic climatic changes associated with the Pleistocene glaciations. The area is dominated by the remnants of six barrier island shoreline complexes. Each complex consists of a sandy ridge (the former barrier island beach) with a low, clayey area on its western side. The low areas, now occupied by rivers, marshes, and swamps, are the remnants of the salt marshes and lagoons that existed behind the barrier islands. The six shoreline complexes extend for hundreds of miles, paralleling the shoreline from Florida to Maryland. (See Ancient and Modern Shorelines in the Sea Islands chapter.)

How did these shoreline complexes form, and why are they located so far inland? Sea level shifted as the climate changed. During glaciations, sea level stood much lower than today because much of the world's water was tied up in continental ice sheets. During the interglacial times, those periods between the glaciations, the ice sheets melted and sea level stood much higher than today. As a result, the coastline shifted back and forth over approximately 100 miles during the Pleistocene epoch. A sea level high-stand about 2.1 million years ago formed Trail Ridge, a former barrier island along the east side of the Okefenokee Swamp that is now 40 miles inland. At sea level low-stands, the shoreline was as much as 80 miles east of its present position, far out on today's continental shelf. Like the Bacon Terraces, the barrier island shoreline complexes become younger toward the coastline, with the youngest complex (called Silver Bluff) forming the core from which most of today's Sea Islands have built seaward.

In southwestern Georgia, there are several terraces west of the Okefenokee Swamp—west of Trail Ridge, the inland boundary of the Barrier Island Sequence District in Georgia. These terraces indicate that in the past the shoreline was as much as 150 miles inland of its present location. These terraces are difficult to date. The oldest of these, the shorelines farthest west and at the highest elevations, may be of Miocene or Pliocene age. They probably mark the worldwide sea level high-stand that occurred about 5 million years ago, during Miocene time, just before glacial ice began forming in Antarctica, which lowered sea level worldwide. As with the Bacon Terraces, the barrier island shoreline complexes grow younger to the east.

Okefenokee Basin

The Okefenokee Basin, the low area west of Trail Ridge, contains the Okefenokee Swamp and several smaller swamps. Trail Ridge dams the eastern edge of the Okefenokee, and without it there would be no swamp. When Trail Ridge was a barrier island, a large saltwater lagoon lay to the west. The lagoon was connected to the sea by inlets, located where the St. Marys and Satilla rivers now flow, though now most of the Okefenokee Swamp drains slowly southwestward into the renowned Suwannee River of Florida. The swamp formed as sea level dropped and rainwater filled the depression. Peat has been accumulating in the swamp for at least the last 6,500 to 7,000 years, indicating that the water has been fresh for that amount of time.

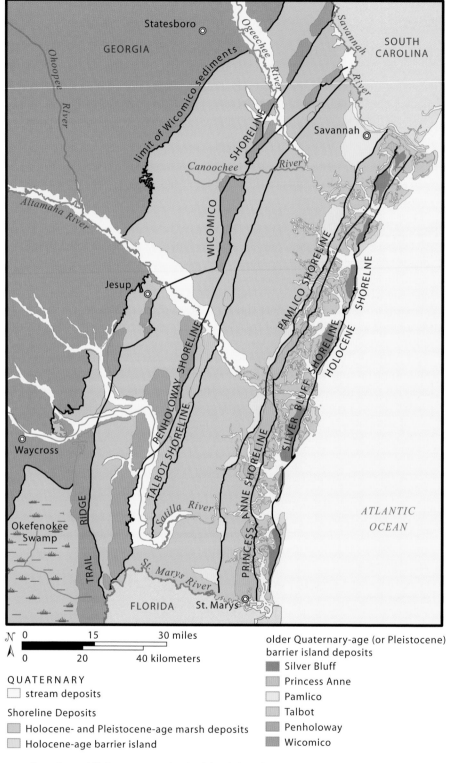

Locations of Pleistocene-age barrier island shoreline complexes in the Georgia
Coastal Plain. (Based on an original by Clark Alexander; modified from Hoyt, 1968.)

Pleistocene Shoreline Complexes in the Barrier Island Sequence District		Elevation (in feet above sea level)	Probable Age
1	Silver Bluff shoreline (found in central part of today's Sea Islands)	5 feet	Pleistocene
2	Princess Anne shoreline	13–15 feet	Pleistocene
3	Pamlico shoreline	25 feet	Pleistocene (Sangamon interglacial)
4	Talbot shoreline	40–46 feet	Pleistocene
5	Penholoway shoreline	70–76 feet	Pleistocene
6	Wicomico ("wye-COM-uh-co") shoreline (Trail Ridge)	95–105 feet	early Pleistocene
Erosional Marine Terraces in the Bacon Terraces District		Elevation (in feet above sea level)	Probable Age
7	Okefenokee (Sunderland)	125 feet	late Pliocene or early Pleistocene
8	Waycross terrace (Sunderland)	130–150 feet	late Pliocene or early Pleistocene
9	Argyle terrace (Sunderland)	170–175 feet	late Pliocene or early Pleistocene
10	Claxton terrace	200 feet	late Pliocene or early Pleistocene
11	Pearson terrace (Coharie)	216–225 feet	Pliocene or early Pleistocene
12	Hazlehurst terrace (Brandywine)	225–275 feet	Miocene or Pliocene or early Pleistocene

Table summarizing the six Bacon Terraces and the six barrier island shoreline complexes, from youngest to oldest.

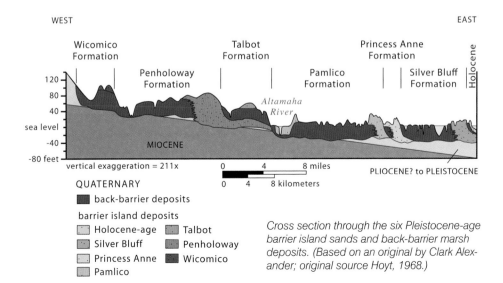

Cross section through the six Pleistocene-age barrier island sands and back-barrier marsh deposits. (Based on an original by Clark Alexander; original source Hoyt, 1968.)

LANDFORMS

A number of unusual landforms can be found on the Coastal Plain. All are fairly young—Quaternary in age. Some are relatively small, and others form boundaries between the physiographic districts.

Pleistocene Sand Dune Fields

One of the most surprising features of the Coastal Plain is the Pleistocene-age dune fields—relatively large areas composed of a multitude of sand dunes. You can see the white sand of the dune fields along I-16 on the east side of the Ohoopee and Canoochee rivers near Metter, and along Georgia Spur 204 west of I-95 beside the Ogeechee River near Savannah. Dune fields are also present along the east side of the Flint, Little Ocmulgee, and Altamaha rivers. There are several distinct bands of sand dunes along the Altamaha River, particularly near Jesup, ranging in age from 15,000 to 45,000 years old. The dunes nearest the river are youngest, and they become older to the northeast.

Today, these rivers are mostly meandering streams on broad, swampy, muddy floodplains, due in part to the warm, humid climate with abundant rainfall and relatively high sea level, which causes the stream gradient, or slope, to be low and ultimately results in sediment deposition within the floodplain rather than erosion. But during the Pleistocene epoch the situation was much different. North America was in the grip of an ice age, and most of the land north of what is now Kentucky was buried beneath ice. In Georgia, the climate was cold and dry. The coast was tens of miles farther offshore than it is today, and much of the continental shelf was exposed, with rivers flowing across it to the sea. Because of lower sea level, riverbeds were steeper and the rivers themselves eroded their channels more deeply. Steeper gradients lead to faster flowing water. Unlike today's rivers, many of these streams and rivers had braided channels broken up by elongated bars of sand and gravel. In some cases, rivers may have dried up due to the cold, dry climate.

The end result of these climatic changes was a great deal of unconsolidated sandy sediment within the river channels. Strong winds from the west and southwest blew the river sediment into huge sand dunes as tall as 50 to 75 feet. Some of the first satellite imagery of the inland Coastal Plain, taken in the 1970s, revealed the curved, parabolic shape of them. Stunted trees, such as scrub oak (turkey oak) and longleaf pine, evergreen shrubs (rosemary and woody mints), and cactus have colonized the dune fields. The trees are stunted because the sandy soil doesn't hold moisture or retain soil nutrients. Several rare and endangered plants and animals, including the gopher tortoise and the eastern indigo snake, which commonly lives in the tortoise burrows, are adapted to the harsh, dry conditions of the dune fields.

The Mysterious Carolina Bays

Probably the most mysterious and hotly debated features in the Coastal Plain are the Carolina bays, oval lakes or wetlands. Some people call them *pocosins*, a Native American word meaning "swamp on a hill." Others call them "round ponds." Carolina bays range in size from a few hundred yards to several miles across. Hundreds of thousands of them are present on the Coastal Plain in a belt stretching from New Jersey to northern Florida, but they are most abundant in the Carolinas.

More than a thousand Carolina bays have been identified in Georgia. There is a large concentration northeast of Grand Bay, near Lakeland, and along the Alapaha River. Those along the Alapaha are some of the largest bays in the state, with Banks Lake the largest at nearly 4 miles across. They are called "bays" for the bay tree, an evergreen, not because they are filled with water. Various species of bay tree grow in or around them. Carolina bays also provide habitat for rare insect-eating plants, such as the Venus flytrap and pitcher plant.

Carolina bays are underlain with organic-rich, clay-rich sediment, which contrasts with the surrounding sandy soils. Fossilized pollen in the clay-rich sediments suggests that the bays formed between 110,000 and 40,000 years ago, during cold, dry glacial episodes. They are some of the youngest features of the Coastal Plain. Both the sediments and pollen in the bays provide scientists with information about the plant communities and climate of the Coastal Plain during the Pleistocene epoch.

Oddly, the longer axes of the oval bays are oriented in a north-south direction, parallel with one another, or slightly overlapping. Many of the bays have a rim of sand on their southeastern edge, which geologists interpret to be dunes formed by windblown sand. Because of their regular geometric shape and parallel orientation, Carolina bays have drawn much interest and curiosity over the years from scientists and laypeople alike.

A number of possible origins have been suggested for the bays, some supported by scientific evidence and others far-fetched. Their origin is likely related to one or more of the following scenarios, for which there is scientific evidence: they are (1) depressions scoured by currents or tidal eddies when the sea covered the area; (2) places where peat bogs burned during drought with windblown sand along the resulting rims; (3) basins formed by springs near sand dunes; (4) valleys or depressions dammed by sandbars or giant sand ripples; (5) low areas between sand dunes or at the foot of a marine terrace; (6) lakes or low areas that were elongated by the erosive power of wind blowing in a particular direction for a long period of time; or (7) lime sinks with windblown sand along their rims.

The scenarios from the far-fetched list include: (1) craters formed by a swarm of meteorites striking the Earth; (2) craters formed when a black hole struck the Hudson Bay area of Canada, throwing ice chunks that made impact craters in Coastal Plain sediments; (3) craters formed by shock waves from a comet exploding in the atmosphere; (4) nests made by giant schools of spawning fish; (5) whale wallows. The extraterrestrial hypotheses are the least likely explanations from a scientific perspective as no celestial fragments have been found associated with the bays. Another argument against the extraterrestrial theories is that the bays are shallow surficial features that don't appear to be sitting on deeply disturbed bedrock. If a meteorite or comet slammed into Earth, the bedrock should show signs of the collision. The association of the bays with parabolic dunes along their rims suggests that the bays are natural depressions, shaped by wind.

Orangeburg Escarpment

The Orangeburg Escarpment, named for a town in South Carolina, was once a cliff that erosion has reduced to a gentle slope. Roughly parallel to the coast, it extends from near Fayetteville, North Carolina, to Jesup, Georgia, just south of the Altamaha River. The escarpment is one of the oldest landforms in the Coastal Plain and serves as the dividing line between the upper Coastal Plain (Vidalia Upland and Bacon Terraces) and the lower Coastal Plain (Barrier Island Sequence District). The escarpment marks a change in elevation of 50 to 100 feet or so over a distance of 1 to 2 miles.

The escarpment marks a change in depositional environment and age of sediments, too. West of the escarpment is the Miocene-age Altamaha Formation (at least 4.3 million years old), with sediments deposited in rivers or along the coast. To the east, at the base of the slope, sediments are younger (late Pliocene to early Pleistocene, 4.3 to 1.8 million years old) and were deposited in the ocean. So how did it form? Waves cut the escarpment between 3.5 and 3 million years ago, in the late Pliocene, when sea level was about 150 feet higher. The escarpment was a significant boundary for several million years, suggesting that it may be the surface expression of a deeply buried fault.

The elevation of the Orangeburg Escarpment decreases from north to south. Near the South Carolina border it tops out at 230 to 250 feet, but farther south, near the Altamaha River, the elevation is only 140 feet. Farther south it merges with one of the Bacon Terraces (Waycross terrace), which merges southward with Trail Ridge.

Pelham Escarpment

The north-south-trending Pelham Escarpment separates the Dougherty Plain from the higher elevations of the Tifton Upland to the east. The Dougherty Plain is mainly Eocene- and Oligocene-age limestone, whereas the Tifton Upland has younger, Neogene-age sand and clay that is more resistant to erosion. Streams flow northwest down the escarpment and into caves and sinkholes along the eastern edge of the Dougherty Plain.

THE GNAT LINE

If you get out of the car in south Georgia on a hot summer day and are assaulted by a swarm of tiny gnats buzzing in your ears and flying into your eyes, you have crossed the gnat line. This invisible line separates areas inhabited by sand gnats from the inland regions that lack them. There are about sixty species of these tiny insects.

You can blame geologic factors for their distribution, because they inhabit wetland areas such as the Okefenokee Swamp or coastal salt marshes. Only the female bites because she must feed on blood before laying eggs. She is not picky, biting birds, amphibians, and reptiles in addition to humans and any other mammals that come her way. The larvae are aquatic or semiaquatic, thus requiring the damp, organic-rich sediment that occurs near the edges of salt marshes and swamps. You may think that nothing loves a gnat but another gnat; however, in the salt marsh, tiny killifish enjoy gnat larvae as part of a food chain that includes larger fish and dolphins.

It Came from Outer Space

You probably know that in the past asteroids struck Earth, even contributing to the extinction of the dinosaurs. But did you know that an asteroid struck near Cape Charles, Virginia, at the mouth of the Chesapeake Bay? It left a crater about 54 miles wide and 1 mile deep while pulverizing and, in some cases, melting Coastal Plain sediments and underlying granitic rocks. The Chesapeake Bay impact crater is the largest impact feature in the United States. Buried under about 1,000 feet of sediment, it was detected by geologists studying well cores and other data.

Although the asteroid impact was more than 600 miles away, its effects were felt in Georgia. Debris from the blast, including water, sediment, shattered rock, and droplets of molten rock, was thrown high into the atmosphere, falling back to Earth hundreds to thousands of miles away—as far as Texas. The impact generated a tsunami, which probably inundated parts of the Piedmont and Coastal Plain, perhaps even reaching the Blue Ridge Mountains.

Evidence of the impact in Georgia consists of green, glassy, pitted objects called tektites, which formed from molten droplets of Earth's crust that cooled as they were flung through the air. Georgia's tektites (called georgiaites), typically 1 to 2 inches in diameter, are most commonly found in freshly plowed soil. Approximately two thousand Georgia tektites have been found, most of them collected in Bleckley and Dodge counties, which are part of a larger area in which tektites are distributed. The tektites have been radiometrically dated to

Georgia tektites. —Courtesy of Robert "Bobby" Strange

about 35 million years and have been found in areas with Eocene- to Oligocene-age sediment. Both the radiometric age and the age of the surrounding sediment indicate the impact occurred around the Eocene-Oligocene time boundary.

Evidence of earlier asteroid impacts has been found at other Coastal Plain sites. A layer of what's called shocked quartz has been found near the base of the Twiggs Clay in Warren County, about 40 miles west of Augusta. The microscopic crystal structure of these quartz grains, which gives them their name, was created by the intense pressures associated with an asteroid impact. South of Columbus, at the top of the Cretaceous-age sediment in the region, geologists have found breccia and conglomerate containing rock fragments not typical of Georgia. Some interpret the breccia and conglomerate to be impact breccia—pulverized bedrock fragments that were flung into the air and fell far from their place of origin—resulting from the asteroid impact that ended the reign of the dinosaurs at the end of the Cretaceous period, about 65 million years ago. This impact occurred near the Yucatan Peninsula in the Gulf of Mexico, creating a huge crater and sending rock debris and water flying in all directions.

FOSSILS

The fossils in the Coastal Plain include both invertebrates and vertebrates, along with plant fossils and single-celled organisms such as foraminifera. The invertebrates include various types of mollusks (clams, oysters, scallops, snails, scaphopods), corals, bryozoans, arthropods (crabs, barnacles, and ostracodes), and echinoderms (echinoids and sand dollars). Most foraminifera (or forams) are tiny planktonic organisms, 1 millimeter or less in diameter, but the *Lepido-cyclina* forams, which lived on the seafloor, grew to the size and shape of a coin. Plant fossils include carbonized wood and microscopic pollen grains.

Fossils are preserved in different ways. When deposited in porous and sandy sediment, most calcium carbonate shells are dissolved by groundwater, leaving only impressions or internal molds. Impressions are exactly what they sound like: the impression of a shell's outer surface left behind in soft sediment that hardens into rock. Internal molds form when sediment fills the inside of a shell and is left behind after the shell is gone. Shells are more likely to be preserved in clay-rich sediment, where groundwater cannot penetrate easily. Generally, aragonite shells are fossilized less often, whereas calcite shells (scallops, oysters, and sand dollars) are more likely to be preserved. Some fossils, such as agatized coral found along the Withlacoochee River, form when silica minerals replace the original mineral matter of the fossilized organism. Bones of vertebrates are commonly preserved through permineralization, the filling of the pore spaces in the bone with minerals precipitated from solution, and by replacement, in which phosphate and/or carbonate minerals replace the bone.

Sediment deposited in marine environments is more likely to have fossil seashells. Sediment deposited in rivers is less likely to contain fossils, other than petrified wood or chert cobbles with shell impressions. Since weathered residuum and river deposits mantle much of the Coastal Plain, the best places to collect fossils are in man-made excavations, in bluffs, or in streambeds where

Large Lepidocyclina *foraminifera from the Tivola Limestone.*

Internal molds of clams from the Tivola Limestone.

*Fossil scallop (*Aequipecten spillmani*) from the Tivola Limestone.*

*Internal mold of a snail (*Turritella*) from the Tivola Limestone.*

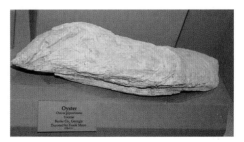

Clam and snail fossils in the clay-rich Nanafalia Formation near Fort Gaines. —Courtesy of Alan Cressler

*Fossil Eocene-age oyster (*Ostrea gigantissima*) on display at the Tellus Science Museum in Cartersville.*

*Fossil sand dollar (*Periarchus quinquefarius*) from the Sandersville Limestone.*

Agatized coral from the Withlacoochee River.

fossil-bearing sediment has been exposed by erosion. Spoil piles of dredged sediment along the Intracoastal Waterway commonly contain fossils as well.

Sharks and Other Fish

Shark teeth, the official state fossil of Georgia, are relatively common in marine deposits of the Coastal Plain. They are widely distributed, having been found in Cretaceous-age sediments near Columbus; in Eocene limestone in the Sandersville area; in the Intracoastal Waterway, which passes through the back-barrier salt marshes; and on beaches of the Sea Islands. The largest shark teeth come from *Carcharodon megalodon*. Unlike recently shed shark teeth, which are white, fossil shark teeth are gray, black, or brown. Shark skeletons are rarely preserved because they are made of cartilage rather than bone. Cartilage decomposes relatively easily.

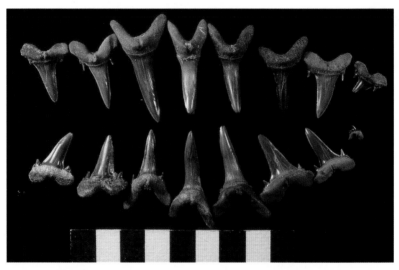

A set of shark teeth of Scapanorhynchus texanus *from the Cretaceous-age Blufftown Formation. Scale is in centimeters.*

Bony fish fossils are present at a number of places in Cretaceous formations of Georgia. They include vertebrae and teeth of *Xiphactinus*, the largest bony fish known from the Late Cretaceous of North America. Complete specimens of this genus from other localities are up to 18 feet long.

Bones of a giant coelacanth fish have also been found in Georgia. The 80-million-year-old fossils found in Georgia are among the youngest coelacanths known. They range up to about 15 feet long. Scientists thought the coelacanth lineage was extinct until one was caught in 1938 in the Indian Ocean near Madagascar, hence its common epithet as the "living fossil."

Marine Reptiles

A 30-foot long, 78-million-year-old fossil mosasaur (*Tylosaurus proriger*) resides in the Georgia Southern University Museum in Statesboro, not far from I-16.

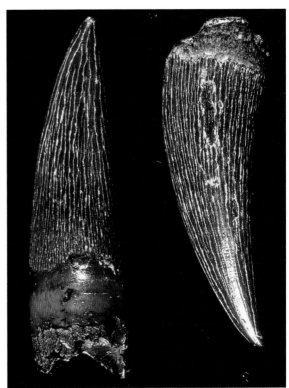

Ferocious, swimming meat eaters, mosasaurs ruled the Cretaceous seas covering south Georgia during Cretaceous time, when dinosaurs dominated the land. Mosasaurs, however, were not dinosaurs but marine reptiles more closely related to monitor lizards. Although the complete skeleton in the Statesboro museum was found in South Dakota, vertebrae and teeth of mosasaurs have been found in Georgia. Mosasaurs were primarily fish eaters, but they also ate mollusks. Interestingly, a mosasaur bone found in the Blufftown Formation south of Columbus had a shark tooth imbedded in it, indicating that the mosasaur was not necessarily at the top of the food chain.

Plesiosaurs, the long-necked sea monsters of the Cretaceous, are another type of marine reptile that inhabited the seas in Georgia. Plesiosaur teeth are present in the Blufftown Formation just south of Columbus.

Dinosaurs and Other Reptiles

Some of the most exciting fossils in the Coastal Plain are the remains of dinosaurs. Fossils of two major groups of dinosaur are present: theropods, carnivorous dinosaurs that walked on two legs, and ornithischians, plant-eating dinosaurs. Although no complete skeletons have been found in Georgia so far, remains of both groups, including leg bones, vertebrae, and teeth, have been found in the Blufftown Formation near Columbus. In addition to bones, a layer of shale with feathers, possibly from dinosaurs, has been found in the Eutaw Formation near Columbus. All of the Georgia dinosaur remains from the Columbus area were collected and described through the efforts of Dr. David Schwimmer at Columbus State University.

The theropods of the Southeast include *Appalachiosaurus* (related to *Tyrannosaurus rex*) and a dinosaur related to *Ornithomimus*. Fossil hunters can recognize carnivorous dinosaurs by examining their teeth, which are serrated like a blade for cutting. The ornithischian dinosaurs in the Southeast were plant-eating hadrosaurs (duck-billed dinosaurs), such as *Lophorhothon*.

Georgia dinosaur bones are found in coastal marine deposits, and they are typically dismembered. Dinosaurs lived on land, but raging rivers and flood-waters carried their bloated carcasses out to sea. As the floating carcasses decayed, bones, teeth, and other parts dropped to the seafloor, where they were entombed in sediment with a host of marine fossils. Sharks also helped to dismember dinosaur carcasses. We know this because shark teeth have been found embedded in dinosaur bones. Bones of *Deinosuchus*, a huge reptile (up to 35 feet long) distantly related to modern alligators, have been found in the Blufftown Formation. This enormous predator probably also ate dinosaurs.

Not all reptiles lived on the land or in the sea. Long, thin, hollow bones of flying reptiles called pterosaurs have been found in the Eutaw Formation in Georgia.

Pleistocene-Age Vertebrates

An astonishing variety of Pleistocene-age vertebrate fossils, many belonging to animals now extinct, have been found in coastal Georgia near Savannah and Brunswick. Most of the fossils come from the marsh mud between the Pamlico and Princess Ann shoreline complexes.

Fossils from more than one hundred types of vertebrate animals have been found near Savannah, including twelve species of sharks and rays, twenty-five of bony ray-finned fish, seven of amphibians, twenty of reptiles (such as turtles, tortoises, snakes, and alligators), four of birds, and thirty-five of mammals (including giant ground sloths, elephants, mastodons, bison, horses, boars, bears, rodents, lynx, and a small leopard). Most were collected from marsh mud just above a layer of fossil seashells, near Skidaway Island, and range in age between 75,000 and 125,000 years old. In the 1830s, a nearly complete mastodon skeleton was discovered in the salt marsh near Skidaway. Complete skeletons are relatively rare.

In the 1830s, a canal was built near Brunswick to connect the Altamaha and Turtle rivers. The abandoned canal lies about 1 mile east of I-95. The fossilized bones of several large mammals were discovered during its construction, including the giant ground sloth, elephant, mastodon, and several species of horse and bison. The vertebrate fossils were found, along with petrified wood, 4 to 6 feet below the surface just above a layer of sand with fossil seashells. Dredging operations around Brunswick occasionally bring up other vertebrate fossils, including tapir, deer, turtle, crocodile, and giant beaver.

Giant ground sloths (*Eremotherium laurillardi*) roamed North America between 700,000 and 8,000 years ago. North America's first fossil specimen was

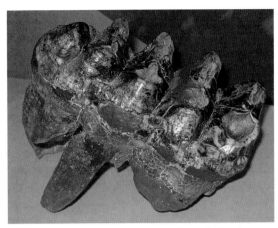

A mastodon tooth, on display in the Georgia Southern University Museum in Statesboro.

Giant ground sloth on display in the Tellus Science Museum in Cartersville.

found at Skidaway Island in 1822. About 350 bones from at least three individuals were discovered during the construction of I-95 near Brunswick in 1970. The animals died after getting stuck in the salt marsh mud. A partial skeleton from this find is on display in the Science Library at the University of Georgia in Athens.

In 1992 amateur fossil hunters diving in the Intracoastal Waterway near St. Simons Island discovered a nearly complete fossil skeleton of a giant ground sloth. When alive this sloth was 22 feet long and weighed 6 to 7 tons.

KAOLIN DEPOSITS

Georgia is the nation's leading producer of the white clay known as kaolin. More than half of the world's kaolin resources—an estimated 5 to 10 billion tons—are located in the Georgia Coastal Plain. There are about 110 kaolin mines in central Georgia in a belt stretching between Macon and Augusta, and kaolin contributes roughly $900 million to Georgia's economy.

Kaolin has been used since ancient times for making porcelain. It's named for a hill called Kao-Ling in China, where kaolin was mined. The commercial production of kaolin in Georgia began in the 1700s, when it was shipped to England for use in Wedgwood china.

Today, the demand for kaolin is driven more by reading than fine dining. Most of the kaolin mined in Georgia is used as filler and coatings for paper. About 30 percent of the weight of glossy paper used in magazines and books is kaolin. Kaolin coating provides a smooth surface for printing, ensuring crisp, readable type and sharp graphics. Chances are that you are holding Georgia kaolin in your hand at this moment. Kaolin is also an important ingredient in toothpaste, plastics, rubber, paint, ceramics (including toilets), fiberglass insulation, and pharmaceuticals. Pigment-grade kaolin is mined between Macon and South Carolina. Ceramic-grade clay with associated bauxite (an aluminum ore) is mined near Andersonville, south of the Fort Valley Plateau.

Kaolin is mined in open pits up to 200 feet deep and covering 2 to 12 acres. In order to reach the kaolin, a considerable amount of overlying sediment must be removed. Kaolin is excavated using backhoes, scrapers, and draglines, a large machine with a bucket that can scrape up and remove 20 cubic yards

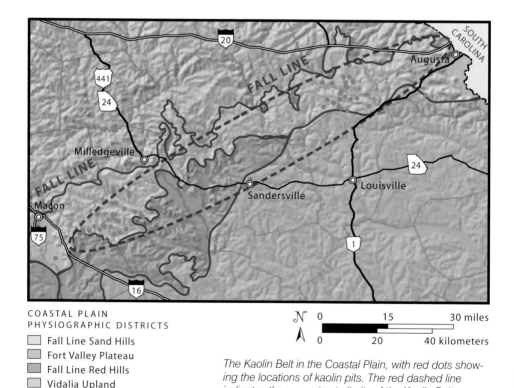

COASTAL PLAIN
PHYSIOGRAPHIC DISTRICTS

- Fall Line Sand Hills
- Fort Valley Plateau
- Fall Line Red Hills
- Vidalia Upland

The Kaolin Belt in the Coastal Plain, with red dots showing the locations of kaolin pits. The red dashed line indicates the approximate limits of the Kaolin Belt.

of sediment at a time. A kaolin pit migrates over time as the previously mined area is filled back in.

Kaolin pits are stunning. Reddish orange clays overlie white, cream, bluish gray, and tan clays and sands that frame brilliant turquoise blue water, the color of which is related to fine clay particles suspended within the water that scatter sunlight. The reddish orange soil overlies the Twiggs Clay, which tends to be bluish gray where it has not oxidized. The Twiggs Clay is a type of clay commonly called fuller's earth but known to geologists as montmorillonite. This unusual clay most likely developed from the alteration of windblown volcanic ash. The ash, deposited in salt marshes behind barrier islands, probably came from volcanoes in the western United States or the Caribbean. A sharply defined contact separates the Twiggs Clay from the underlying light-colored sands and clays of the Oconee Group, which contains the kaolin.

Georgia's kaolin deposits lie southeast of large weathered granite bodies in the Piedmont. When it weathers, the feldspar in granite is chemically altered to white clay. Over time granite bodies weather to a soft, crumbly residuum called saprolite (meaning "rotten rock"). The hot and humid climate 100 to 45 million years ago, during the Late Cretaceous and Paleogene time, caused an intense period of rock weathering that produced thick saprolite in the Piedmont. Heavy rains eroded the saprolite, transporting white clay particles, clumps of clay, and quartz sand grains to the sea, which at that time was near the Fall Line. The sediments were deposited in shallow marshes and deltas that may have been similar to the present-day salt marshes and deltas along the Georgia coast.

Bedding planes exposed in the walls of kaolin mines are not always flat. Sometimes they are uneven or irregular and cut by scoop-shaped channels, which are indicative of times of low sea level when streams and rivers eroded down into the sediment. These periods of low sea level were critical to the formation of kaolin. During these periods marsh sediments were exposed to the atmosphere, oxidized, burrowed by organisms, subjected to soil-forming processes, and leached by acidic rainfall, which removed organic matter and altered or removed certain minerals as it percolated downward through the sediments. As a result, the marsh mud that contained a diverse assemblage of clay minerals and organic matter, like mud deposited near the coast today, was altered to almost pure kaolinite, the mineral that makes up kaolin.

Georgia's richest concentration of kaolin lies within the Buffalo Creek and Huber formations of the Oconee Group, which is of Late Cretaceous to middle Eocene age. These units contain sedimentary layers that become finer in grain size as you move upward through the rock strata, from crossbedded sandstone to lenses and layers of nearly pure kaolin up to 90 feet thick. Geologists can use this sequence of sedimentary rock types to interpret the environment in which the kaolin was deposited. They can then use this information to explore for kaolin deposits elsewhere.

The presence or absence of the Twiggs Clay affects the quality of the underlying kaolin. Where the Twiggs Clay overlies the kaolin in a continuous layer that has not been breached by erosion, it provides a seal that blocks oxidizing surface waters from reaching the kaolin. In these areas the kaolin is gray,

In the side of a colorful kaolin mine near Sandersville, notice the curved, scoop-shaped traces of bedding planes, seen here in cross section. These are the remains of ancient erosion surfaces.

A natural exposure of kaolin along the Oconee River near GA 24.

CLAY EATING

White kaolin may not look very appetizing, but people in the South have been known to eat it. The medical term for this is *geophagia* (literally, "earth eating"), and it is a type of pica—an appetite for nonfood substances. Those who eat the clay, which some call "chalk" (a misnomer), talk of craving it and say they like the taste. Kaolin was, at one time, a principal ingredient in an antidiarrheal medicine. Not unexpectedly, those who eat kaolin often suffer from constipation. Eating kaolin is not recommended because the clay tends to absorb nutrients from food in the digestive tract, particularly iron, which can lead to anemia. Clay eating also decreases hunger, leading to malnutrition. You can find "snack kaolin" for sale in plastic bags in some convenience stores and in the produce department of some grocery stores.

because it contains pyrite (fool's gold) and organic carbon, and has less commercial value. In areas where the Twiggs Clay has been breached by erosion, acidic waters percolated from the surface and leached out organic material and pyrite, producing the highest-quality white kaolin.

The kaolin-bearing Huber Formation parallels the Fall Line in a belt about 25 miles wide, extending from Andersonville, in west-central Georgia, nearly to Augusta. It is also present in a small fault-bounded basin in the Piedmont about 29 miles inland of the Fall Line, near Warm Springs. About 140 miles to the northwest, near Cartersville, there are lime sinks filled by bauxite and kaolin of similar age and type. The Huber Formation may have originally extended far to the north of the present Coastal Plain, having been deposited during a time of extraordinarily high sea level.

HEAVY MINERAL SAND

The black sand you see on Georgia beaches is composed of dark mineral grains that have a high specific gravity, which is a ratio of the density (mass of a specific volume) of the mineral against the density of water. Called heavy minerals, they are concentrated in some beach sand because of differential sorting, a process by which waves wash away lightweight grains of sand with low specific gravity and leave behind the heavier grains with high specific gravity. Heavy mineral sand is present at Trail Ridge and in other former beach ridges of the Barrier Island Sequence District. The heavy minerals at Trail Ridge include zircon, ilmenite, rutile, staurolite, kyanite, sillimanite, tourmaline, topaz, corundum, and monazite. By definition, heavy minerals have a specific gravity greater than 2.85, meaning they are 2.85 times more dense than water. For comparison, quartz (not a heavy mineral) has a specific gravity of 2.65, whereas rutile, zircon, and magnetite have a specific gravity over 4.

Heavy minerals are mined for commercial use or because they contain valuable elements. Zircon is used in glazes for ceramic tiles and china. Ilmenite and rutile are a source of titanium, which is a super-white pigment in paint, plastic, paper, and hundreds of other products (including the white powder on donuts, the white filling of sandwich cookies, and in the white letters on M&M's candy). Monazite is a source of neodymium and other rare earth elements, which are used for magnets in hybrid cars, wind turbines, electronics, and solar cells.

Companies have shown interest in mining heavy minerals from Trail Ridge because it is the largest known deposit of heavy minerals in the Coastal Plain. There are fears, however, that strip-mining Trail Ridge might damage the hydrology of the Okefenokee Swamp. In 2003, in response to citizens' concerns for the environment, a company that had planned on mining Trail Ridge retired its mining rights and donated a 16,000-acre tract of land for preservation—the biggest land conservation gift in Georgia's history. However, there are mining operations removing heavy minerals from the Trail Ridge area, such as from the Penholoway shoreline complex east of Nahunta (on US 82).

US 27
Columbus—Florida State Line
147 miles

US 27 travels over sediments that are progressively younger southward, from Cretaceous at Columbus to Miocene at the Florida state line. Between Columbus and Cusseta, US 27 passes through Fort Benning, a U.S. Army training center established in 1918 on an old plantation. The rolling hills covered by pine trees and grasslands are underlain by sand and clay of the Tuscaloosa, Eutaw, and Blufftown formations, and deep ravines and valleys dissect the unconsolidated sediments. The sediment is exposed in several roadside embankments along US 27, such as just south of Upatoi Creek and at Cusseta. The Tuscaloosa Formation is mainly north of Upatoi Creek. It is the oldest and lowermost of the Coastal Plain sediments in this part of Georgia, lying directly on top of the eroded surface of the Piedmont's metamorphic and igneous rocks. The Tuscaloosa is composed of poorly sorted sand and white kaolinite clay. Its beds are arranged in sequences in which the grain size becomes finer upward, a characteristic of meandering-river deposits. It also contains petrified wood, but few other fossils. The top of the Tuscaloosa Formation is marked by an erosional surface, which probably formed during a drop in sea level.

Sea level rose and ocean waters deposited the Eutaw Formation over Tuscaloosa Formation river deposits. The Eutaw Formation is sandstone with varying amounts of clay, mudstone, and marl, all deposited in environments associated with a barrier island complex. Its fossils are 86 to 83 million years old and include oysters, clams, snails, ammonites, shark teeth, teeth of rays and skates, and the bones of flying reptiles.

South of Upatoi Creek, US 27 crosses the Blufftown Formation, consisting of gray to olive green, mica-rich fine sand, silt, and marly clay deposited in a coastal marine environment. The lower part of the unit has crossbedded sand containing the branching burrows of a shrimplike arthropod called *Ophiomorpha*, which inhabits shallow, nearshore marine environments. The upper part of the Blufftown contains carbon-rich clay and silt, crossbedded sand,

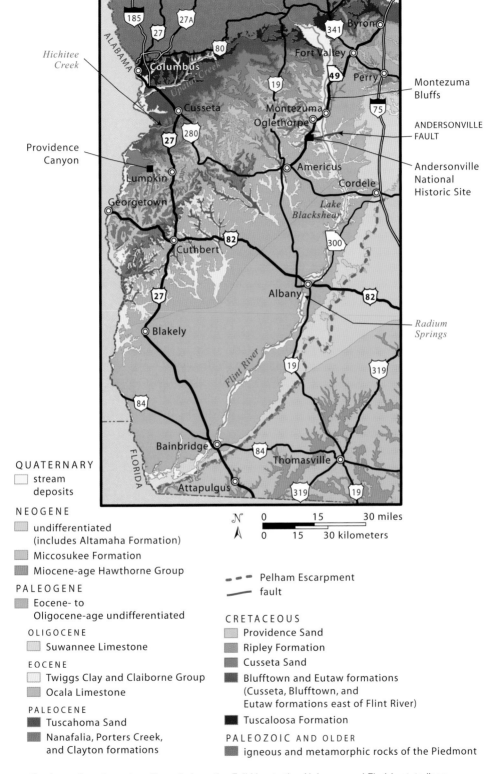

QUATERNARY
☐ stream
   deposits

NEOGENE
☐ undifferentiated
   (includes Altamaha Formation)
☐ Miccosukee Formation
☐ Miocene-age Hawthorne Group

PALEOGENE
☐ Eocene- to
   Oligocene-age undifferentiated

OLIGOCENE
☐ Suwannee Limestone

EOCENE
☐ Twiggs Clay and Claiborne Group
☐ Ocala Limestone

PALEOCENE
☐ Tuscahoma Sand
☐ Nanafalia, Porters Creek,
   and Clayton formations

- - - Pelham Escarpment
—— fault

CRETACEOUS
☐ Providence Sand
☐ Ripley Formation
☐ Cusseta Sand
☐ Blufftown and Eutaw formations
   (Cusseta, Blufftown, and
   Eutaw formations east of Flint River)
☐ Tuscaloosa Formation

PALEOZOIC AND OLDER
☐ igneous and metamorphic rocks of the Piedmont

Geology of southwestern Georgia from the Fall Line to the Alabama and Florida state lines.

fossiliferous clay, and glauconite-bearing sand. It was deposited in a lagoon behind a barrier island complex. Blufftown Formation fossils are 83 to 70 million years old and include a variety of oysters and other bivalves, gastropods (snails), ammonites, crabs, and bryozoans, in addition to vertebrate fossils and calcareous nannofossils.

The Cusseta Sand, exposed around Cusseta, overlies the Blufftown Formation and consists of coarse sand with large crossbeds and locally abundant *Ophiomorpha* burrows. These beds were deposited in a barrier island and lagoon environment. The Ripley Formation overlies the Cusseta Sand. US 27 crosses the Ripley Formation between Hichitee Creek and Lumpkin. The Ripley Formation was deposited on the inner continental shelf. Using the nannofossils it contains, geologists have determined it is between 70 and 65 million years old, the Late Cretaceous.

US 27 passes Lumpkin, which sits on a hill topped by clay residuum weathered from Eocene- and Oligocene-age limestone. The upland surfaces are relatively flat, but the region has been highly dissected by tributaries of the Chattahoochee River, which lies about 15 miles to the west.

Providence Canyon State Outdoor Recreation Area

Providence Canyon, known as Georgia's Little Grand Canyon, is one of the Seven Natural Wonders of Georgia. It is located about 6.5 miles west of Lumpkin on GA 39C. The area contains sixteen spectacular canyons, or erosional gullies, as much as 200 feet deep. The recreation area includes a fenced viewing area along the upper rim, picnic areas, and hiking trails that descend into the canyons.

The canyons were not here when the first settlers arrived in the early 1800s. They formed as the result of poor farming practices that led to runaway soil erosion. When settlers first cleared the land for agriculture, they tended to plow straight up and down the hills rather than along the contours. The plowed furrows were good conduits for rainwater runoff, and by 1850 gullies ranging from 3 to 5 feet deep appeared in the fields. Once the gullies cut through the erosion-resistant clay cap of the Clayton Formation and into loose Providence Sand below, erosion rates accelerated, and the land became useless for farming. The gullies deepened and widened into canyons, which continue to expand.

The geologic formations exposed in Providence Canyon are readily distinguished by color. The uppermost formation is the orange residuum of the Clayton Formation, an iron-rich, sandy clay of Paleocene age (65 to 55 million years old). It overlies the white to tan Providence Sand, a kaolin-bearing sand of Late Cretaceous age (70 to 65 million years old). The contact between the two formations is an erosional surface that represents the change from the Mesozoic era, when the dinosaurs lived, to the Cenozoic era, after dinosaurs became extinct. The Ripley Formation forms the floor of the canyon, though it is poorly exposed and mostly overgrown. It is dark gray, locally weathering to orange, and ranges in texture from clay to sand.

Crossbeds in the Providence Sand indicate it was deposited by currents and probably was part of a barrier island complex. The overlying Clayton

Formation, perhaps deposited in a shallow marine environment, has been highly altered by weathering, producing the orange residuum.

You can have a close look at the contact between the Clayton Formation and Providence Sand in a road cut about 2 miles east of the recreation area's entrance, along GA 39C. If you rub the white, clayey Providence Sand between your fingers, you can feel the quartz sand grains it contains, and a white residue that feels like talcum powder will remain on your fingers. This is the clay mineral kaolinite. The sand is crossbedded and contains small rounded pebbles of white clay, indicating that clumps of the clay were rolled along in the current that formed the crossbeds.

A layer of sandy brown iron ore, called goethite, rests on the uneven erosional surface of the Providence Sand. Goethite, also known as limonite, occurs in layers, lenses, nodules, concretions, and hollow geodes near the base of the Clayton Formation. Some of the iron geodes are more than 1 foot in diameter and when broken open may have powdered ochre inside. These are known as "Indian paint

Iron nodules cover the ground in some places at Providence Canyon.

White Providence Sand overlain by the orange Clayton Formation in Providence Canyon. The contact between the Clayton Formation and the Providence Sand is undulatory; before the Clayton Formation was deposited, rivers cut channels in the Providence Sand, which were then filled with the Clayton sediments.

A layer of iron ore separates the orange Clayton Formation from the white Providence Sand in this road cut on GA 39C.

pots." The iron ore has been mined near Lumpkin and is suitable for steel manufacturing, although some of it is too sandy to be of commercial use.

Lumpkin to Florida State Line

Between Lumpkin and Cuthbert, US 27 crosses creeks that drain to the Chattahoochee River, about 18 miles to the west. The terrain is hilly, with deep valleys underlain by easily eroded sediments, including the Ripley Formation, Providence Sand, Clayton Formation, and Tuscahoma Sand. Cuthbert, like Lumpkin, sits on clayey residuum from the weathering of Eocene- and Oligocene-age limestone.

About 20 miles south of Cuthbert, US 27 reaches the north edge of the Dougherty Plain, one of Georgia's flattest regions, which is underlain by red clay residuum weathered from the Ocala Limestone. The Ocala contains networks of underground cavities. The ground surface subsides or collapses into some of these, producing lime sinks. The lime sinks are the surface exposure of the Floridan aquifer, which is recharged by rainwater and surface runoff seeping into the ground. The area is good for agriculture, and you can see fields of peanuts and cotton irrigated by huge rolling sprinklers.

At Bainbridge, US 27 crosses the Flint River. Spanish explorer Hernando de Soto reached this spot on March 5, 1540. At that time, the Indian name for the waterway was *Thronateeska*, which translates to "Flint River." Indians used flint (technically, chert) gathered along the river for tools and weapons. The chert, some of which has Oligocene-age fossils in it, is found in the clayey residuum left after the limestone in the Flint River basin weathered. What is the difference between flint and chert? Both are varieties of quartz with extremely tiny crystals, which geologists call cryptocrystalline quartz. Flint is a black variety of chert. However, the chert found along the Flint River is white, yellowish, gray, red, or colorless, but never black, so the name "Flint River" is not entirely accurate, but close enough. Farther downstream the river enters Lake Seminole, where it joins with the Chattahoochee River to form the Apalachicola River, which flows to the Gulf of Mexico.

On the bridge in Bainbridge, US 27 also begins crossing the Pelham Escarpment, the boundary between the Dougherty Plain to the northwest and the Tifton Upland to the southeast. Caverns are present in limestone along the escarpment northeast of Bainbridge. For example, Climax Caverns, in Decatur County, is one of the largest caves in the South. The road continues to climb the Pelham Escarpment for several miles to the southeast. The Tifton Upland is higher than the plain because erosion-resistant, clayey, Neogene-age sediments overlie the Eocene-age limestone exposed at the surface of the Dougherty Plain.

Attapulgus, along with nearby Quincy, Florida, is the world's center of attapulgite production, a light-gray clay that is a type of fuller's earth. In this region attapulgite beds in the Miocene-age Hawthorne Group are up to 10 feet thick. Fuller's earth was named for its use in "fulling," the process of washing, removing oil from, and thickening wool cloth. Attapulgite is absorbent and can soak up several times its own weight in liquids. The petroleum industry values

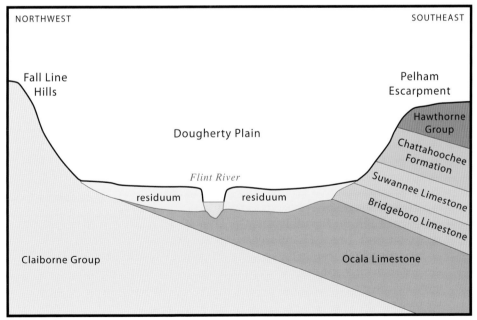

not to scale

Cross section of the Dougherty Plain, showing the surface topography and some of the underlying geologic units. The Dougherty Plain is about 50 miles wide. (Based on Carter and Manker, 1995.)

attapulgite for its ability to stay suspended in salt water (rather than clumping and settling out), which makes it ideal for use as a lubricant (drilling mud) to aid in the drilling of deep boreholes. Attapulgite is also used in oil and grease absorbents, cat litter, adhesives, paint, roofing granules, ceramic tiles, cosmetics, pharmaceuticals, pesticides, and other products. The Florida state line is about 4 miles south of Attapulgus.

I-75
Macon—Florida State Line
166 miles

I-75 enters the Coastal Plain where it diverges from I-16 in Macon and crosses several physiographic districts of the Coastal Plain from here to Florida. The Macon Museum of Arts and Sciences, located at 4182 Forsyth Road, has a 40-million-year-old (Eocene-age) fossil whale excavated from a kaolin mine in Twiggs County. Named "Ziggy," short for the genus *Zygorhiza*, the whale was found with the vertebrae of a small shark inside, remnants from its last meal.

Between Macon and Byron I-75 crosses the Fall Line Sand Hills. This region is highly dissected by streams and dominated by flat-topped hills separated by deep valleys. Near exit 156 (I-475), I-75 crosses the Tobesofkee Creek valley, a broad panoramic valley about 5 miles wide and 200 feet deep. Just south of exit 156 you can see a clay pit on the east side of the road with light orange sediment and white kaolin of Late Cretaceous to Paleocene age. You can also see kaolin on the west side of the interstate, north of mile 154.

Near mile marker 151, I-75 crosses Echeconnee Creek, which occupies a valley about 5 miles wide and 200 feet deep. This valley separates the Fall Line Sand Hills to the north from the Fort Valley Plateau to the south. On the south side of the valley you can see reddish brown Paleocene- to Eocene-age sediment overlying white kaolin in the median.

The Fort Valley Plateau is a raised, flat area of prime agricultural land topped by Eocene-age clayey sand of the Fort Valley Group and dark red residuum. This is the major peach-growing region in Georgia, as you can probably guess from the signs. Peach trees grow better in the sandy loam soils of this area than in Piedmont clay or Coastal Plain sand, and this area is far enough south to avoid late-season frosts. The flat topography is ideal for orchards.

Near exit 144, the Russell Parkway, you can see a forested area on the east side of the road. This is a Carolina bay. It is not farmed because of the clayey, poorly drained soil. Trees in the Carolina bay differ from those nearby. In well-drained soils, thick stands of loblolly pines are common, but the pines do not grow in Carolina bays. Several similar Carolina bays are present farther from the road near here.

The Fort Valley Plateau ends where I-75 crosses Big Indian Creek, south of Perry. South of the creek, I-75 climbs the Perry Escarpment, the northern edge of the Fall Line Hills, onto younger sediments of Oligocene age. They consist of reddish brown residuum—a mixture of sand, clay, and chert weathered from

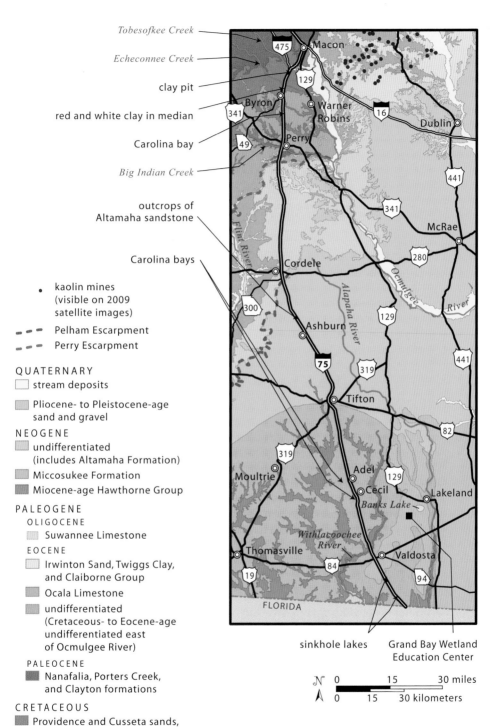

Tobesofkee Creek

Echeconnee Creek

clay pit

red and white clay in median

Carolina bay

Big Indian Creek

outcrops of
Altamaha sandstone

Carolina bays

- kaolin mines
  (visible on 2009
  satellite images)

- - - Pelham Escarpment

- - - Perry Escarpment

QUATERNARY
□ stream deposits

▨ Pliocene- to Pleistocene-age
  sand and gravel

NEOGENE
▨ undifferentiated
  (includes Altamaha Formation)

▨ Miccosukee Formation

▨ Miocene-age Hawthorne Group

PALEOGENE
  OLIGOCENE
  ▨ Suwannee Limestone

  EOCENE
  ▨ Irwinton Sand, Twiggs Clay,
    and Claiborne Group

  ▨ Ocala Limestone

  ▨ undifferentiated
    (Cretaceous- to Eocene-age
    undifferentiated east
    of Ocmulgee River)

  PALEOCENE
  ▨ Nanafalia, Porters Creek,
    and Clayton formations

CRETACEOUS
▨ Providence and Cusseta sands,
  Ripley, Blufftown, Eutaw,
  and Tuscaloosa formations

Tobesofkee Creek
Macon
475
129
Byron
341
Warner
Robins
16
Dublin
Perry
49
441
Flint River
341
McRae
280
Cordele
Ocmulgee River
300
Alapaha River
Ashburn
129
441
75
319
Tifton
82
319
Adel
129
Moultrie
Cecil
Lakeland
Banks Lake
Withlacoochee River
Thomasville
Valdosta
19
84
94
FLORIDA

sinkhole lakes   Grand Bay Wetland
                     Education Center

PALEOZOIC AND OLDER
▨ igneous and metamorphic rocks of the Piedmont

N   0   15   30 miles
A   0   15   30 kilometers

Coastal Plain geology along I-75 south of Macon.

limestone. South of Perry you can see distant hills to the east. This is the Vidalia Upland, another physiographic district, on the far side of the Ocmulgee River valley.

Around exit 104, north of Cordele, you can see groves of pecan trees on both sides of the interstate. Changes in the types of trees commonly signal a difference in geology. Here the trees indicate that I-75 is skimming along the eastern edge of the Dougherty Plain, which is underlain by the Eocene-age Ocala Limestone and younger Oligocene-age limestone. Pecan trees do best in deep, fertile, well-drained soils. Most Georgia soils tend to be highly acidic and require the addition of lime to keep the pH within the optimal range to support pecan trees, so areas underlain by limestone are well suited for growing pecans. Level topography and lack of standing water are also favorable for orchards.

Cordele bills itself as the Watermelon Capital of the World. The loose, sandy soil, hot, humid climate, and availability of water for irrigation promote the growth of watermelons, which are said to be some of the sweetest in the nation.

I-75 begins ascending the Pelham Escarpment around exit 97. Elevation increases about 125 feet over several miles as you climb to the Tifton Upland. Between exit 97 and the city of Tifton, I-75 crosses the Miocene-age Altamaha Formation, which includes both a loose orangish red, clayey sand and well-cemented sandstone. A belt of erosion-resistant sandstone serves as the drainage divide along the Pelham Escarpment, separating the Flint River watershed to the west from the Alapaha watershed. Where the Tifton Upland is underlain by sandstone, vegetation tends toward stunted oak trees and plants you would normally associate with deserts, such as cactus and yucca. The sandstone crops out in bluffs along streams and as flat surfaces in fields. North of Ashburn, about 2 miles north of exit 84 (GA 159) near a rest area on the northbound side of the road, you can get up-close views of the sandstone behind the rest area parking lot. Sandstone is also exposed about 800 feet north of the rest area near the bridge over the West Fork Deep Creek.

From Tifton southward to the Florida state line, the Miccosukee Formation (Miocene- to Pliocene-age) is present at the surface. The Miccosukee Formation is grayish orange to grayish red, mottled clay, sand, and gravel deposited in a delta environment.

Between exits 41 (Sparks) and 32 (Cecil) there are several Carolina bays. Most are swampy and forested, but west of Adel, No Mans Friend Pond (named because you can easily get lost in the swamp) has open water. Immediately north of the Cecil exit, the forested areas on the west side of I-75 are Carolina bays.

Grand Bay, about 15 miles northeast of Valdosta on Moody Air Force Base, is part of the Grand Bay Wildlife Management Area. This wetland area, one of the larger Carolina bays in the state, is filled with peat and has plant communities similar to those found in the Okefenokee Swamp. At the Grand Bay Wetland Education Center, accessed from Knights Academy Road, a 2,000-foot board-walk and observation tower allow you to see the wetlands—and possibly many waterbirds—without getting your feet wet.

There is a large concentration of Carolina bays northeast of Grand Bay, near Lakeland, and along the Alapaha River. Some of the largest Carolina bays in the state occur in this large grouping, with the largest, Banks Lake, being nearly 4 miles across.

Between Cordele and Valdosta, the southward-flowing Alapaha River, to the east, roughly parallels I-75. The river is peculiar. It is a relatively small, with little to no flow during the dry season; however, it occupies a very wide floodplain—much too wide for such a small river. One hypothesis to explain this disparity is that the Ocmulgee River at one time flowed to the Gulf of Mexico through this area, forming the wide floodplain before the Altamaha River captured its headwaters and diverted it to the Atlantic. The Ocmulgee River makes a sharp bend and begins to flow to the northeast at a point east of Ashburn. This sharp bend, coupled with the oddly sized Alapaha floodplain, is strong evidence in favor of the stream capture hypothesis.

South of Valdosta, near the state line, is the Valdosta Limesink District, which has a large number of ponds, lakes, and swampy depressions that developed in lime sinks. These are the result of the sand, gravel, and clay at the surface collapsing into underlying caves. You can see lime sink lakes south of Valdosta around exit 11 (GA 31) and exit 5 (GA 376). Lake Louise is about 7 miles south of Valdosta. The lake, just east of I-75, is a 13-acre blackwater lake that developed in a lime sink. Blackwater lakes have dark water resembling tea due to tannic acid from decaying vegetation.

About 8 to 10 miles west of I-75, and south of Valdosta, the Withlacoochee River cuts through Miocene-age Hawthorne Group limestone. Mineral collectors have flocked to this stretch of river to look for agatized coral. The minerals composing the coral have been dissolved and replaced by a type

A lime sink west of I-75 near Valdosta. —Courtesy of Don Thieme

An agatized coral geode with botryoidal quartz from the Withlacoochee River, on display at the Tellus Science Museum in Cartersville. The cylindrical projections into the geode are molds of boring clams, probably of the Lithophaga genus.

of silica called chalcedony, or agate. This process formed geodes lined by agate, botryoidal (grape-shaped clusters) quartz in various colors, and tiny quartz crystals. The coral is found along with fossils that indicate the limestone formed in relatively shallow water, far from sources of clay and silt that would have prevented the other organisms from living in the water. The coral probably developed in patch reefs, isolated shallow areas.

Georgia 49
Byron—Americus
55 miles

See the map on page 75.

From Byron GA 49 passes through the Fort Valley Plateau. This is prime agricultural land because of the rich soils and relatively high, flat topography that developed on weathered residuum, sand, and clay of the Fort Valley Group. These sands and clays are the coastal and marine equivalents of kaolin-bearing sediments deposited in river and floodplain environments to the east during Eocene time. You can see some of the region's 617,000 peach trees and 73,000 pecan trees between Fort Valley and Montezuma. In March the peach trees bloom, and in June and July the peaches arrive.

It is hard to understand how the town of Fort Valley got its name, because it never had a fort and it isn't in a valley. The town is located where a trading post was established at the intersection of two Indian trails. As they usually did, the Indian trails followed the high, flat land, avoiding unnecessary stream and river crossings, and subsequently these routes developed into major transportation routes, now US 341, GA 49, and the railroads.

South of the town of Fort Valley, about 0.2 mile south of State University Drive, you can see a deep open pit on the west side of the road that exposes orange residuum and lighter-colored sediment below, which is Eocene-age sand and kaolinite of the Fort Valley Group.

Between Fort Valley and Marshallville, the terrain is flat, and in a few places you can see light-orange residuum. Patches of forest on both sides of the road are Carolina bays, now filled in with trees because their clayey soil is poorly drained and difficult to farm.

In Marshallville, GA 49 turns right (west) and then due south again, paralleling the Flint River (west of the highway) to Montezuma. About 2 miles south of the landfill on the west side of the road (2 miles north of Montezuma), Crooks Landing Road leads west to Montezuma Bluffs Natural Area. The bluffs are steep slopes, about 150 feet high, overlooking the Flint River. Although the bluffs are heavily forested, sand and clay of the Providence Sand and Clayton and Marshallville formations, and the overlying Fort Valley Group sand topped by reddish residuum, are exposed. Fossil collectors have removed giant oysters (*Ostrea crenulimarginata*) from gray clay and thin limestone layers of the Clayton Formation in these bluffs.

GA 49 crosses the Flint River between the sister cities of Montezuma and Oglethorpe. In July 1994, Tropical Storm Alberto dumped more than 20 inches

An open pit along GA 49, south of Fort Valley, exposing orange residuum and Fort Valley Group sand with kaolinite.

of rain over this area in three days, causing a 500-year flood. The normal channel width here is less than 0.1 mile, but during the flood the entire 1.5-mile-wide floodplain was under water up to 35 feet deep—more than the height of a three-story building. Roads, railroads, the water filtration plant, and many buildings were underwater. More than forty thousand people were displaced during the flooding, and thirty-two people were killed. Even the dead were displaced as floodwaters unearthed caskets, some of which floated downstream.

Montezuma and Oglethorpe sit on opposite sides of the Flint River on old river terraces. These terraces, which are 40 to 50 feet above the current floodplain, represent the position of the floodplain in the past, when sea level was higher. Sea level dropped and caused the river to erode down to its present level. At the Flint River, GA 49 leaves the Fort Valley Plateau and heads into the Fall Line Hills District, which in this area is underlain by the Providence Sand with the slightly younger Nanafalia Formation on the hilltops.

About 3 miles south of Oglethorpe, there is a road to the east with a sign for a mine named Mulcoa Plant #5. You have now entered the Andersonville Bauxite Mining District, from which both kaolin and bauxite are mined. There is a larger mine south of Andersonville, on the west side of the road.

Andersonville and Bauxite

Bauxite was discovered near Andersonville in 1912, and the mining of bauxite for aluminum began in 1914. Bauxite is an unusual-looking rock that consists of small, rounded nodules ranging from cream colored to reddish brown or gray. It contains the mineral gibbsite (an aluminum oxide), along with kaolinite. There are about sixteen active mines in the Andersonville area mining bauxite, kaolin, and other clays. Bauxite is mined for aluminum, and the clays are used to manufacture refractory materials (high-temperature ceramics), in metal casting, in the production of semiconductors, as abrasives, and for other industrial uses.

Bauxite layers range from a few inches thick up to 25 feet thick within thicker and more extensive layers of kaolin in the late Paleocene to early Eocene Nanafalia Formation. Like kaolin (see Kaolin Deposits in the chapter introduction), bauxite developed from the clay that weathered from the granite of the Piedmont and was carried by streams to this area. Once deposited, certain elements were leached from the clay minerals in a tropical climate. Clay contains both aluminum and silicon, so in order for the aluminum-rich rock bauxite to form, the leaching had to be extensive enough to remove the silicon, meaning the clay had to be above sea level for some time.

Andersonville is best known as the site of a Civil War prison camp, where more than forty-five thousand Union soldiers were held in an open stockade that is now a national historic site. As the Confederates ran out of supplies, the prisoners were not adequately fed and nearly thirteen thousand died. If you visit this site off GA 49 in Andersonville, look carefully at the stone directional posts. They are made of fossiliferous Oligocene-age chert.

Structural features like folds and faults are not common in the Coastal Plain. One major exception is the Andersonville fault, which extends in an east-west

Fossiliferous Oligocene-age chert with scallops and clams at Andersonville National Historic Site.

direction along Sweetwater Creek, about 1 mile south of Andersonville. This high-angle normal fault cuts Coastal Plain sediments, and the ground on the north side of the fault has dropped down about 100 feet relative to that on the south side. The fault is at least 20 miles long, and the sediment on either side of the fault appears to be deformed for about 0.5 mile in either direction, the result of movement along the fault. Movement along the fault probably occurred intermittently in the Paleogene and Neogene periods, and perhaps even the Holocene. Stick around a while. Maybe there will be an earthquake!

GA 49 continues south to Americus. This area is underlain by residuum created by the intense tropical weathering and dissolution of Eocene- and Oligocene-age limestone. The residuum consists of maroon, brown, orange, and white clay, sandy clay, sandstone, and chert with Oligocene-age fossils.

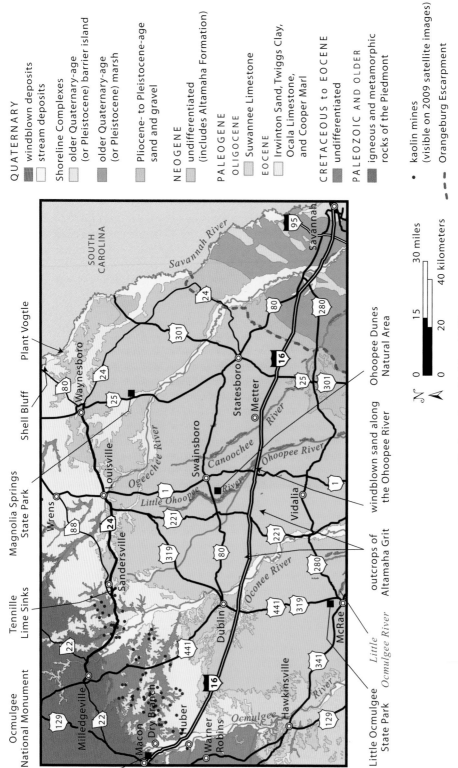

QUATERNARY

Shoreline Complexes

windblown deposits

stream deposits

older Quaternary-age
(or Pleistocene) barrier island

older Quaternary-age
(or Pleistocene) marsh

Pliocene- to Pleistocene-age
sand and gravel

NEOGENE

undifferentiated
(includes Altamaha Formation)

PALEOGENE

OLIGOCENE

Suwannee Limestone

EOCENE

Irwinton Sand, Twiggs Clay,
Ocala Limestone,
and Cooper Marl

CRETACEOUS to EOCENE

undifferentiated

PALEOZOIC AND OLDER

igneous and metamorphic
rocks of the Piedmont

• kaolin mines
(visible on 2009 satellite images)

- - - Orangeburg Escarpment

Ocmulgee
National Monument

Magnolia Springs
State Park

Tennille
Lime Sinks

Shell Bluff

Plant Vogtle

SOUTH
CAROLINA

Savannah River

Savannah

95

24

301

80

280

16

25

301

Metter

Statesboro

Canoochee River

Ohoopee River

Swainsboro

Ogeechee River

Little Ohoopee River

Louisville

Wrens

88

24

Sandersville

319

221

80

Dublin

441

319

280

Vidalia

221

1

Oconee River

McRae

341

Hawkinsville

129

Ocmulgee

Little
Ocmulgee River

River

16

441

Warner
Robins

Huber

Dry Branch

Macon

Milledgeville

129

22

22

Little Ocmulgee
State Park

windblown sand along
the Ohoopee River

Ohoopee Dunes
Natural Area

outcrops of
Altamaha Grit

N

0   15   30 miles

0   20   40 kilometers

Geology along GA 24 and I-16.

Georgia 24
Milledgeville—Waynesboro
154 miles

Milledgeville, high on the west bank of the Oconee River, served as the capital of Georgia in the early 1800s. The river marked the boundary of the Indian Territory prior to 1802. In the 1800s, cotton was shipped from Milledgeville along the Oconee River to Darien at the coast. GA 24 crosses the Oconee River on the east side of Milledgeville and descends from the Piedmont, crossing the Fall Line and onto sediments of the Coastal Plain. The road follows an old Indian trail.

The character of the Oconee River varies depending on the type of geology it crosses. In the Piedmont, the river is relatively straight as it cuts through hard igneous and metamorphic rock, but where it crosses the easily eroded Coastal Plain sediments it meanders, winding back and forth in a broad floodplain.

Between 8 and 12 miles east of Milledgeville, flat land with light-colored sandy soil and small scrubby trees distinguishes the Fall Line Sand Hills District. Developed on Cretaceous-age sands, the soils of this physiographic province are poor for agriculture because they don't retain water or soil nutrients.

Although you might think the top of the Piedmont's igneous and metamorphic rocks would slope smoothly beneath the Coastal Plain sediments from the Fall Line toward the coast, that is not the case in this part of Georgia. The depth to this bedrock varies dramatically because of underground faults associated with Triassic-Jurassic-age basins buried beneath the sediments. Outliers of Piedmont rock are exposed in some places in the Coastal Plain, while nearby the same bedrock is buried more than 870 feet below the surface. These faults may be responsible for the small earthquakes that rumble through this part of Georgia from time to time.

Faults also seem to control the pattern of the river and stream channels in the Coastal Plain. In soft sediments, like we see in this area, we would expect to see branching, treelike stream patterns (called dendritic) draining southeast toward the coast. Instead, some stream valleys, such as Buffalo Creek west of Sandersville, flow due south, probably following subsurface faults.

Kaolin Mines

About 14 miles east of Milledgeville, near the Baldwin-Washington county line, GA 24 enters the Kaolin Belt, which lies within the Fall Line Red Hills. Kaolin mines are visible to the north near the intersection with GA 272. The white clay is kaolin, which is mined from the Buffalo Creek and Huber formations of Late Cretaceous to middle Eocene age.

Heading east toward Sandersville, the Twiggs Clay forms an impermeable layer over the kaolin, which prevents oxidizing water from reaching the clay. As a result, the kaolin here is gray and contains the mineral pyrite. The higher elevations are capped by the Irwinton Sand and the overlying Tobacco Road Sand, both of Eocene age.

The development of Sandersville, known as the Kaolin Capital of the World, began near White Pond, one of a number of lime sinks in the Eocene-age

A kaolin mine north of GA 24 near the GA 272 intersection. The white kaolin you see here has been mined from deep within the pit and piled on the surface.

Sandersville Limestone that have long since drained. Before the Civil War, a small limestone quarry was opened near Sandersville to obtain limestone for making mortar to build the town. Following the war, the quarry was reopened to obtain lime for rebuilding the courthouse. General Sherman's army burned the courthouse in November 1864 on their March to the Sea, during which they followed the precursor to GA 24.

As you travel through downtown on N. Harris St. (GA 24), look for the courthouse to the west. It has a tower with an eagle on top of it. Local geologists will tell you that at one time, sea level was so high in this area that only the eagle would have been above the waves. But, of course, that was about 50 million years ago.

Attesting to the sea that once covered this area, Eocene-age marine fossils, including oysters (*Crassostrea gigantissima*), scallop shells and other bivalves, internal molds of gastropods (*Turritella* genus), bryozoans, decapods, shark teeth, and stingray tooth plates have been found in the Clinchfield Formation overlying the kaolin in one of the region's mines.

Dinosaur bones have been found in Cretaceous-age sediments in western Georgia near Columbus, but no dinosaur bones have been found in this part of the state, despite the fact that many millions of tons of kaolin have been mined from Cretaceous deposits. This is probably due to the postdepositional alteration and leaching of the sediments by acidic rainwater and groundwater, which could have dissolved the bones.

Stored kaolin in Sandersville.

Kaolin Processing Facility

Just south of downtown Sandersville, west off GA 15 on Kaolin Road, you can see one of the kaolin processing facilities. Most of us expect paper to be white and smooth, so kaolin used for paper coatings has to be processed to meet standards for whiteness, brightness, and particle size. Trains deliver kaolin to the processing facility, where washing, settling, and centrifuging processes remove gritty quartz, sparkling muscovite mica, and dark-colored heavy minerals while separating the kaolin into a variety of product sizes. The clay is then treated with bleach and other chemicals to remove iron minerals, plant matter, and other impurities. Kaolin to be used in paint is baked at approximately 950°F in furnaces, a process called calcining, which drives off water in the mineral structure to produce a very bright white, opaque pigment. The discarded material from these processes is then dumped in large impoundments that can cover many acres. Kaolin plant tours are offered during the Kaolin Festival held in October each year, and tours can be arranged in advance for groups.

Tennille Lime Sinks

You might not associate central Georgia with caves and lime sinks; however, they exist in the Sandersville Limestone, a discontinuous, 40-foot-thick unit. The limestone occurs at the surface in a relatively small area east of the kaolin mines and is exposed in streambeds south of Sandersville. Tennille Lime Sinks is a little place back in the woods, on private property, nestled along a branch of Limestone Creek near the town of Tennille. *Tennille* is pronounced "tunnel," and not like it is in Captain and Tennille.

Some of the best evidence that Washington County once lay under the sea can be found in the Sandersville Limestone. A wide variety of fossils have been collected from it, including shark teeth, skate teeth, stingray spines, manatee ribs, sea turtle vertebrae, crocodile teeth, sand dollars, giant oysters, and bivalve and gastropod mollusks.

Hidden Basins

Along the Fall Line, Late Cretaceous–age Coastal Plain sediments (younger than about 100 million years) lie directly on Paleozoic-age and older igneous and metamorphic rocks of the Piedmont, so you might expect that Piedmont rocks lie immediately beneath the oldest Coastal Plain sediments everywhere. However, oil exploration wells drilled south and east of Sandersville passed through about 1,000 feet of Coastal Plain sediments and then encountered more than 7,000 feet of red conglomerate, sandstone, shale, and dark igneous diabase of Triassic and Jurassic age (about 200 million years old) before hitting Piedmont rocks at 8,350 feet. The 7,000 feet of Triassic-Jurassic sedimentary and igneous rock beneath the Coastal Plain sediments fill an ancient basin, which geologists call the Riddleville Basin. This "hidden" basin lies just south of GA 24 in the Sandersville area far below the surface. Perhaps this hidden basin should not have been so surprising to geologists, because similar rocks of similar age are common in fault-bounded basins from the Piedmont of the Carolinas northward into Canada.

These basins formed during the Triassic period, as the supercontinent Pangaea was pulled apart and the Atlantic formed in the resulting scar. Along the east coast of North America, regions of the crust dropped down along normal faults due to these tectonic forces, forming basins. One of these faults roughly

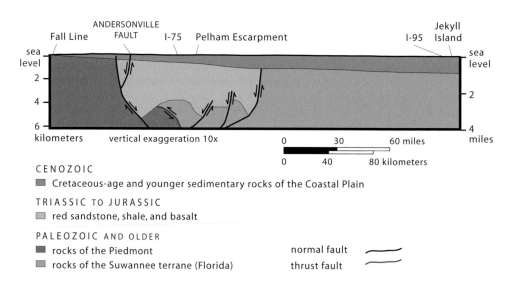

Cross section of Triassic-Jurassic sedimentary rocks beneath the Coastal Plain along a line from Columbus to Jekyll Island. (Modified after American Association of Petroleum Geologists, 1995, with data from Hatcher et al., 2007.)

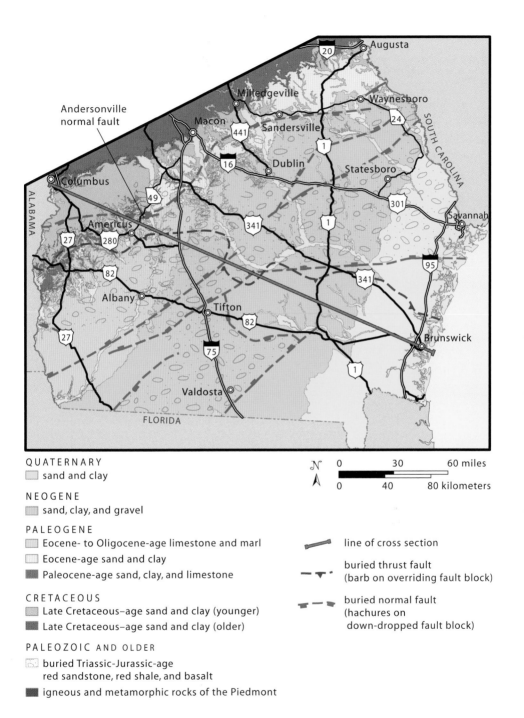

QUATERNARY
sand and clay

NEOGENE
sand, clay, and gravel

PALEOGENE
Eocene- to Oligocene-age limestone and marl
Eocene-age sand and clay
Paleocene-age sand, clay, and limestone

CRETACEOUS
Late Cretaceous–age sand and clay (younger)
Late Cretaceous–age sand and clay (older)

PALEOZOIC AND OLDER
buried Triassic-Jurassic-age
red sandstone, red shale, and basalt
igneous and metamorphic rocks of the Piedmont

line of cross section

buried thrust fault
(barb on overriding fault block)

buried normal fault
(hachures on
down-dropped fault block)

N   0   30   60 miles
      0   40   80 kilometers

Triassic-Jurassic basins hidden beneath Georgia's Coastal Plain.
(Buried fault traces after Hatcher et al.)

Cores of Triassic-age sedimentary rock from beneath the Georgia Coastal Plain, including (from left to right) gray, coarse-grained sandstone with pebbles, red siltstone with calcite nodules, and red and gray siltstone.

parallels GA 24, but does so more than 1,000 feet below the surface. A similar fault-bounded basin, called the Dunbarton Basin, lies about 40 miles to the northeast near Waynesboro, and a much larger basin, called the South Georgia Rift, lies beneath the Statesboro area and may contain as much as 10,500 feet of Triassic-Jurassic sedimentary rock.

Sandersville to Waynesboro

Between Sandersville and Statesboro, GA 24 crosses the Vidalia Upland, which is lower and flatter than the Fall Line Hills. The sediment is orange gravelly, clayey sand and sandy clay of the Miocene-age Altamaha Formation, which was deposited in river and estuary environments. The Altamaha Formation is the most widespread sedimentary unit exposed in Georgia.

About 25 miles east of Sandersville, near Louisville (pronounced "Lewis-ville"), GA 24 crosses the Ogeechee River. Because of its location on the river, Louisville was once an important port and one of colonial Georgia's first inland settlements, serving as Georgia's capital from 1795 to 1805. Sherman's troops burned the town during the Civil War during their March to the Sea.

South of Louisville, the Irwinton Sand (deposited in a nearshore marine environment) grades into finer-grained sediments of the Twiggs Clay, which was deposited farther offshore. Just above the Twiggs Clay is a layer of chert in the Sandersville Limestone. The chert is unusual because it contains sand dollar fossils in which the original calcite was replaced by silica.

East of Waynesboro, near the Savannah River, is Plant Vogtle, one of Georgia's nuclear power plants. Excavations during the construction of the power plant in 1983 uncovered one of the most unusual fossils ever found in Georgia: a whale with legs. This nearly complete, 12-foot-long whale skeleton, now on display at the Georgia Southern University Museum in Statesboro, provides a link between land mammals and whales.

The skeleton was found about 30 feet below the surface with shallow-water marine fossils, including mollusks and tiny calcareous nannoplankton, which helped date the rock to between 41 and 40 million years old, making it the oldest whale skeleton ever found in North America. This early ancestor of the modern whale lived in the Eocene seas. It was also an unknown genus and species, so it was named *Georgiacetus vogtlensis*, the Georgia whale from Vogtle.

The whale had a long snout, a mouth full of teeth, and a long tail. Like modern whales and dolphins, *Georgiacetus* swam by moving its tail up and down. Modern whales have a blowhole on the top of their head, but *Georgiacetus* had a blowhole near the tip of its snout. But its most unusual feature was a pair of hind legs. Even more remarkable, an examination of the hip sockets showed that the legs were functional. The Vogtle whale is an important link between modern whales and their hippopotamus-like, land-dwelling, four-legged mammal ancestors.

The cast of Georgiacetus vogtlensis, *the whale with legs, on display in the Georgia Southern University Museum in Statesboro.*

About 12 miles south of Waynesboro, on US 25 near Perkins, Magnolia Springs State Park is home to a clear spring that produces 7 million gallons of water a day from Ocala Limestone. Because of the ample water supply, it served as a Confederate prisoner of war camp during the Civil War and once held ten thousand Union soldiers.

Shell Bluff

About 16 miles northeast of Waynesboro is Shell Bluff, described by geologists as one of the classic exposures of the Georgia Coastal Plain. The bluffs along the Savannah River near Shell Bluff Landing, and other bluffs several miles downstream, expose more than 100 feet of sandstone, limey sandstone, limestone, clay beds, and thick beds of oyster shells (*Crassostrea gigantissma*) and other marine organisms. Charles Lyell, one of the foremost geologists of his day, collected fossils here in 1842. These rocks are part of the Dry Branch Formation of Late Eocene age. Beneath the Dry Branch Formation is a fossiliferous limestone in the Clinchfield Formation, with caves and lime sinks. Fossils in this formation include molds of gastropods and bivalves, scallop shells, sand dollars (*Periarchus* genus), crab claws, bryozoans, shark teeth, and stingray tooth plates.

To get there, head northeast from Waynesboro on GA 80 for 13.5 miles to the intersection with GA 56 Spur (River Road). Continue straight ahead on the unpaved road until you reach the Savannah River, about 2.2 miles.

I-16
Macon—Savannah
170 miles

See the map on page 88.

I-16 crosses virtually the entire Coastal Plain from the Fall Line almost to the coast. The interstate runs along the Ocmulgee River for about 5 miles at Macon. North of the Fall Line, the river is relatively straight and narrow where it cuts through Piedmont rock, but to the south the floodplain widens and the river begins to meander where it crosses easily eroded Coastal Plain sediments.

About 3 miles east of the I-75/I-16 interchange, at the point where the floodplain widens, Ocmulgee National Monument is a scant 300 yards north of the interstate. The monument is sited on an old Indian settlement that contains a record of 17,000 years of continuous human habitation. Its location indicates that the Fall Line area was not just significant for the location of our modern-day cities, but also for Indian settlements. The mound builders of the Mississippian culture (AD 900 to 1650) left the most conspicuous evidence. The monument consists of two tracts of mounds, one located directly on the Fall Line bluffs along the river, and a smaller area about 2 miles downstream called the Lamar Mounds.

I-16 leaves the Ocmulgee River floodplain and the terrain becomes hilly as it crosses the Fall Line Sand Hills. You may catch a glimpse of sandy sediment or white kaolin clay along the road. About 9 miles from I-75 (at the Twiggs County line) you enter the Fall Line Red Hills District. The district gets its name from the red, pink, and orange color of weathered Twiggs Clay capping the hills. Miners strip off the red clay to get to the kaolin underneath. For the next 10 miles or so, you pass through the Kaolin Belt, and there are numerous strip mines, particularly on the northeast side of the interstate around the town of Dry Branch.

Around exit 12 and southward, the kaolin mines and processing plants near Huber are present on both sides of the road, and you may be able to see them. South of Flat Creek, near exit 18, I-16 leaves the Kaolin Belt. Here the Twiggs Clay overlies the Eocene-age Tivola Limestone instead of kaolin beds.

The landscape becomes much flatter southeast of exit 27 as I-16 enters the Vidalia Upland. A complex unit of river deposits as much as 20 million years old, consisting of massive sandy clay and gravelly, clayey sand of the Miocene-age Altamaha Formation, underlies the upland. The upper part of the unit contains crossbedded, pebbly to gravelly sandstone called the Altamaha Grit.

Around mile 68, near Soperton, the terrain is not flat anymore. There are numerous outcrops of sandstone belonging to the Altamaha Grit in this area, although you probably will not see them from the interstate. The grit surfaces in bluffs near streams or as flat outcrops. Due to the fact that sandstone doesn't retain nutrients or water, areas where it outcrops tend to have stunted oak trees and plants that live in arid climates, such as cactus and yucca.

Altamaha Grit along GA 46, 6.3 miles east of Soperton near exit 84.

Just east of mile marker 87, I-16 crosses the Ohoopee River, which has distinctly different vegetation and soil on either side of the river. On the western side, tall pine and oak trees grow in orange to tan clayey soil. On the east side, however, the sparse vegetation, consisting of small, stunted, turkey oaks and lichens, grows in white sand. The white sand is part of a larger dune complex that formed during Pleistocene time, when the climate was drier and winds blew riverbed sand into huge dunes along the eastern side of the region's rivers. Some of the best exposures of these relict dunes are at Ohoopee Dunes Natural Area, about 10 miles north near Swainsboro (see the US 1: South Carolina State Line—Florida State Line road guide).

By exit 98, I-16 is east of the white dune sand and the soil is orange and pink clay. Near Metter (exit 104, GA 23/GA 121), I-16 crosses the Canoochee River and its tributary Fifteenmile Creek. Windblown white sand is present on the east side of these streams, too.

I-16 leaves the Vidalia Upland south of Statesboro, at exit 127, and descends the Orangeburg Escarpment, a topographic feature that waves eroded about 3.5 to 3 million years ago. The elevation of the roadway drops from about 186 feet to about 130 feet over a couple of miles. East of the escarpment, between Statesboro and Savannah, I-16 crosses the Barrier Island Sequence District.

Just east of the Orangeburg Escarpment, the interstate crosses a 2.5-mile stretch of the Okefenokee terrace, the northward extension of the youngest of the Bacon Terraces District to the south. Pliocene- and Pleistocene-age sand composes the terrace, which was cut by waves when the sea stood about 100 to 120 feet above present sea level.

A 13-mile stretch of somewhat swampy land, lying just east of the terrace, is the former back-barrier marsh of the Wicomico (Trail Ridge) and Penholoway barrier island complexes. The slight rise between exit 143 (GA 30/US 280) and the west bank of the Ogeechee River is the relict Wicomico barrier island, which is about 1.5 million years old. The Ogeechee River has removed the 1-million-year-old Penholoway barrier island that normally lies east of the Wicomico. As you saw along the Ohoopee and Canoochee rivers, stunted oaks grow in riverine dunes along the east side of the Ogeechee River. You can examine these dune sands up close if you take exit 148 south.

Between the Ogeechee River and the Little Ogeechee River to the east, the road crosses the marsh that formed behind the barrier island of the Talbot barrier island complex. The Little Ogeechee River has eroded away the Talbot barrier island, so the Talbot marsh grades eastward into the Pamlico marsh. As you approach the modern coastline, the barrier island complexes become younger.

About 10 miles east of I-95, I-16 delivers you to the heart of Savannah. The city was built on Yamacraw Bluff on the Savannah River, where General James Oglethorpe landed in 1733 to establish the thirteenth colony. The sandy bluff is part of the Pamlico barrier island complex.

The Port of Savannah is the fourth-largest and fastest growing container port in the United States. Kaolin clay, mined in the Georgia Coastal Plain, is the port's dominant mineral export. Nearly 100,000 containers of kaolin are shipped from Savannah each year, bound for destinations including English china factories and Canadian paper mills.

Pleistocene sand dunes east of the Ogeechee River along
the east side of Old River Road, just south of exit 148.

Savannah's City Hall was built in 1901 atop the bluff overlooking the river.
The lower part of the exterior is constructed of Stone Mountain Granite, with
its characteristic white-rimmed patches of the mineral tourmaline.

A container ship on the Savannah River, with Yamacraw Bluff at left and the US 17 bridge in the background. —Courtesy of Fang Guo

The Savannah River is a few blocks north of the end of I-16. When Oglethorpe came to Georgia in the 1700s, the river was only about 12 feet deep. It has since been dredged to a depth of 42 feet and is slated to be deepened to 48 feet to allow larger ships to enter the port.

US 341
Fall Line—Brunswick
174 miles

Between Musella and Roberta US 341 crosses the Fall Line, which separates the Piedmont and Coastal Plain. At the intersection of US 341 and Dixon Road, a road cut exposes a spectacular outcrop of pale grayish orange sediment overlying white to grayish red, deeply weathered metamorphic rock. The two rock units are separated by an unconformity—a sharp, irregular contact marking an ancient erosional surface. Known as the Fall Line unconformity, this is one of the few places in Georgia where you can see the ancient metamorphic rocks of the Piedmont in contact with the much younger, overlying sediments of the Coastal Plain. In most places the contact is obscured by vegetation, thick soil, or water. Although the Piedmont rocks here are now weathered to saprolite, with their feldspar altered to kaolinite, the ancient metamorphic foliation (alignment of sheetlike minerals) is preserved. The overlying sediments belong to the Late Cretaceous–age Pio Nono Formation.

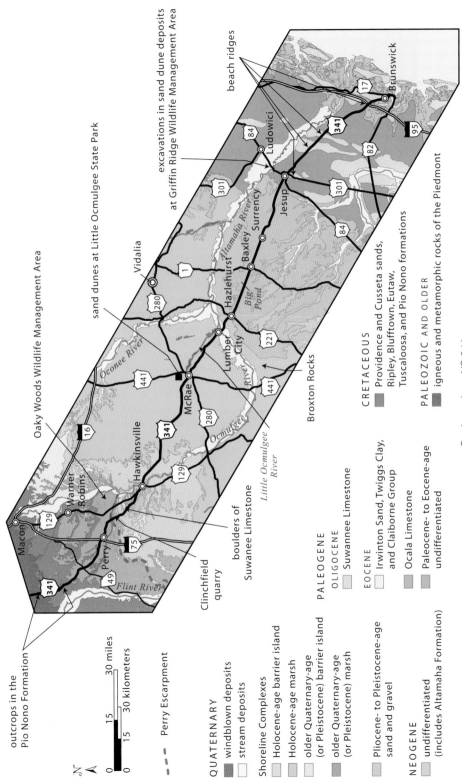

Geology along US 341.

QUATERNARY

windblown deposits

stream deposits

Shoreline Complexes

Holocene-age barrier island

Holocene-age marsh

older Quaternary-age (or Pleistocene) barrier island

older Quaternary-age (or Pleistocene) marsh

Pliocene- to Pleistocene-age sand and gravel

NEOGENE

undifferentiated (includes Altamaha Formation)

PALEOGENE

OLIGOCENE

Suwannee Limestone

EOCENE

Irwinton Sand, Twiggs Clay, and Claiborne Group

Ocala Limestone

Paleocene- to Eocene-age undifferentiated

CRETACEOUS

Providence and Cusseta sands, Ripley, Blufftown, Eutaw, Tuscaloosa, and Pio Nono formations

PALEOZOIC AND OLDER

igneous and metamorphic rocks of the Piedmont

- - - Perry Escarpment

0   15   30 miles

0   15   30 kilometers

outcrops in the Pio Nono Formation

Oaky Woods Wildlife Management Area

sand dunes at Little Ocmulgee State Park

excavations in sand dune deposits at Griffin Ridge Wildlife Management Area

beach ridges

Clinchfield quarry

boulders of Suwanee Limestone

Broxton Rocks

The Fall Line unconformity separates the orange Pio Nono Formation from the deeply weathered, white saprolite of Piedmont metamorphic rocks beneath it at the northwest corner of US 341 and Dixon Road.

Clayey, kaolinite-bearing, crossbedded sand of the Pio Nono Formation south of Roberta.

Five miles south of Roberta, white, crossbedded, clayey, kaolin-bearing sand of the Pio Nono Formation is exposed on both sides of the road. The soil along US 341 is white near this road cut but changes to red near House Road, where the highway leaves the Fall Line Hills and enters the Fort Valley Plateau District, about 8 miles south of Roberta. The area is underlain by weathered Eocene sand and sandy clays. It is heavily farmed due to its flatness, and major crops include peaches and pecans.

The entrance to Atlanta Sand and Supply is 1 mile south of the road cut, at East Sand Pit Road, where sand is mined from several pits in the Gaillard Formation and other units, on the west side of US 341.

Oaky Woods Wildlife Management Area

Oaky Woods Wildlife Management Area, an undeveloped partially state-owned tract on the west side of the Ocmulgee River, has yielded a treasure trove of Eocene-age fossils, including sand dollars, mollusk shells, shark teeth, *Basilosaurus* whale bones, bryozoans, and microscopic foraminifera. Surface collecting has been allowed, but no excavations. Check with a ranger for access.

To get there, in Perry turn left (northeast) onto GA 127, 1.7 miles east of I-75. Continue 8.4 miles until GA 127 ends at US 129. Turn right (south) and continue 0.7 mile. Turn left (east) onto Oaky Woods Road and continue 1.5 miles to the wildlife management area's main gate.

Perry to McRae

The CEMEX quarry in Clinchfield may seem an unlikely place to hunt for fossil seashells, but it has yielded its share. Approximately 7 miles south of Perry, on the east side of US 341, boulders of fossiliferous Tivola Limestone mark the quarry entrance. CEMEX is one of the few cement-making plants in Georgia. Groups may call in advance to arrange a tour of the quarry.

The Clinchfield Formation, the lowermost unit exposed in the quarry, is an ancient beach deposit dominated by sandstone, with fossils of mollusks and vertebrates, including shark teeth and manatee bones. The late-Eocene Tivola Limestone is made up almost entirely of the twiglike remains of tiny bryozoans, animals that lived in colonies, along with scallops, oysters, and other mollusks; sand dollars; foraminifera (*Lepidocyclina* species), the shells of single-celled protozoans; shark and stingray teeth; and fossil whale bones. The limestone records what may be the time of highest sea level in the Coastal Plain.

The dark gray Twiggs Clay overlies the Tivola Limestone and consists of a type of clay called fuller's earth that swells when wet and shrinks when dry. The Twiggs Clay contains microscopic shells of foraminifera, along with mollusk shells composed of aragonite. The clay protected the mollusk shells from downward-percolating acidic groundwater, which would have dissolved the aragonite. There are also fish teeth and scales, and pieces of fossilized whale bone.

One of the difficulties in determining the age of Coastal Plain fossils and sediments is finding sedimentary materials that can be radiometrically dated. For example, typically the grains of sand that make up sandstone were derived from the weathering of preexisting rock, like granite, somewhere upstream. If a geologist age dated the sand grains, the result would represent how long ago the granite cooled and crystallized, not when the sandstone was deposited. Sedimentary rocks that can be dated are rare.

However, the Twiggs Clay, exposed in the quarry, contains sand-sized grains of an unusual green mineral called glauconite, which forms on the seafloor when clay mixes with organic matter in the fecal pellets of seafloor dwellers, such as clams and worms. Besides iron, the clay absorbs potassium, some of which is radioactive. Glauconite is one of the few minerals in sedimentary rock that can be radiometrically dated. Glauconite in the Twiggs Clay here ranges between 34 and 23 million years old.

The Twiggs Clay is covered by a layer of reddish brown residuum with pieces of fossiliferous chert and phosphate nodules, indicating that it was derived from the chemical weathering of limestone. This unit may be part of the Altamaha Formation, exposed to the south in the Vidalia Upland.

At Clinchfield, US 341 rises up over the Perry Escarpment and passes southward into the Fall Line Hills, with its higher elevations and rolling hills. Near Hawkinsville, US 341 leaves the Fall Line Hills and passes southeastward into the Vidalia Upland, which is dominated by reddish orange, gravelly, clayey sand of the Altamaha Formation. The Altamaha Formation blankets much of the upper Coastal Plain. Along US 341 it extends from just east of Hawkinsville to Hazlehurst, and orangish red, pink, and white soils weathered from it are exposed along the road.

Fossiliferous Tivola Limestone in the CEMEX quarry.

A riverine dune at the entrance to Little Ocmulgee State Park.

Little Ocmulgee State Park

Riverine dunes are exposed along the east side of the Little Ocmulgee River in and near Little Ocmulgee State Park, about 2.1 miles north of McRae on US 441/ US 319. In places the sand is as much as 75 feet thick. Drought-tolerant plants inhabit the well-drained, sandy, nutrient-poor soils of the dunes. Stunted turkey oaks are widely spaced so their root systems have a large area from which to gather water. The gopher tortoise (*Gopherus polyphemus*), Georgia's state reptile, burrows in the sand here and is the only tortoise living east of the Mississippi River. The nonnative nine-banded armadillo (*Dasypus novemcinctus*) is another animal that digs in the dunes.

McRae to Hazlehurst

Oil and gas seeps have been reported in Telfair and Pulaski counties near springs issuing from Miocene-age sands and clays, and in the 1930s a petroleum company drilled an exploration well near Scotland, east of McRae. The well ended at a depth of 3,384 feet in hard sandstone and was the deepest well in Georgia at the time. However, it wasn't commercially developed because it was a dry hole.

East of McRae, as you approach Lumber City, you can see the red soil and sediment typical of the Vidalia Upland exposed in excavations and along the side of the road. Along this stretch the Little Ocmulgee River, about 1 mile to the north, parallels US 341. At Lumber City the Little Ocmulgee flows into the Ocmulgee River in nearly a T-junction, which is unusual for rivers. Lumber City lies along a 30-mile section of the Ocmulgee River that anomalously flows northeast, nearly at right angles to the direction of other Coastal Plain rivers. The Ocmulgee River intersects the Oconee River about 7 miles east of Lumber

City at another T-junction, forming the Altamaha River, which follows a typical southeasterly trend.

What is the reason for this peculiar river pattern? It may be due to a subsurface structural feature, such as a fault. Unfortunately, the featureless sediment in the Vidalia Upland does not provide any clues. Another possibility is that the pattern is the result of stream capture. The Ocmulgee and Little Ocmulgee rivers appear to have once flowed southeastward into what is now the Alapaha River, toward the Gulf of Mexico. Headward erosion by the Altamaha River may have captured the headwaters of the Alapaha River, including the flows of the Ocmulgee and Little Ocmulgee rivers, redirecting them into their odd configurations.

Once across the Ocmulgee River the road climbs from the Vidalia Upland to the Hazlehurst terrace of the Bacon Terraces District. The town of Hazlehurst sits on the drainage divide between the Ocmulgee and Altamaha rivers. The soil is red, the terrain is hilly, and you can see layers of sediment in some of the road cuts and excavations.

Broxton Rocks Preserve

The Broxton Rocks Preserve lies southwest of Hazlehurst, south of the Ocmulgee River, in northeastern Coffee County near GA 107 and Rocky Creek Road. Rock outcrops are rare in the Coastal Plain, so the preserve is something of a natural wonder. The rocks that make this area so unique are part of the Altamaha Grit, a well-cemented sandstone unit of the Altamaha Formation, which

Altamaha Grit at the Broxton Rocks Preserve. —Courtesy of Alan Cressler

may be hard and flinty in places and crumbly in others. Rocky Creek flows over sandstone ledges in the preserve, creating waterfalls. Weathering and erosion have formed cavelike recesses. The tourism office in the city of Douglas, 31 miles south of Hazlehurst on US 221, offers tours several times a year.

Near Broxton Rocks you can see features of the Altamaha Grit without a guide at the Flat Tub Wildlife Management Area. This 3,597-acre tract lies between the Ocmulgee River and GA 107 near Flat Tub Road and straddles the Coffee–Jeff Davis county line. To get there, from Hazlehurst head southwest on US 221 for 9.5 miles. Turn right onto GA 107 and continue 9.7 miles. Turn right onto CR 177 and continue along Flat Tub Road into the wildlife management area. Be careful, because this region is home to the nation's most venomous snake, the coral snake, along with other venomous and nonvenomous snakes.

Hazlehurst to Jesup

Between Hazlehurst and Jesup, the Altamaha Grit is exposed in bluffs along the south side of the Altamaha River. East of Hazlehurst, you know that you are in the Bacon Terraces District when you see pine flatwoods—flat land inhabited by a monotonous pine forest with saw palmettos and other shrublike groundcover. Locals call this area "slash pine country." Slash pines grow well and quickly in wet areas. Organic hardpan commonly underlies the sandy soil a foot or two below the surface. This weakly cemented, clayey layer prevents surface water from percolating down, causing the soils of the flatwoods to be saturated during the wet season. Tree farms with trees planted in rows are present along both sides of the road for miles.

The Bacon Terraces are flat or gently inclined areas separated by east-facing slopes with about a 25-foot drop in elevation. Each flat terrace (or step) marks a former seafloor, and the east-facing slopes former shorelines cut by waves. The terraces are younger to the east and date to Pliocene or early Pleistocene time. Orangish red sediment of the Altamaha Formation lies beneath them.

US 341 crosses the Hazlehurst terrace between Hazlehurst and Graham, a distance of about 5.5 miles. Next is the Pearson terrace, which extends from Graham about 10 miles to the Baxley area. Carolina bays are present on this terrace. Big Pond, one of the larger Carolina bays in the state, is about 2 miles north of US 341 and about halfway between Hazlehurst and Baxley. Big Pond is a densely vegetated wetland about 1.5 miles across, with tupelos, cypress, and shrubs. US 341 crosses the Claxton terrace between Baxley and Surrency. The Argyle terrace extends from Surrency to just west of Jesup. Much of this area is low and swampy. There are several Carolina bays south of US 341 along this stretch.

About 18 miles east of Surrency, and 2 to 3 miles west of Jesup, US 341 enters the Barrier Island Sequence District. The road rises about 20 feet and then descends about 30 feet over about 1.5 miles. This is Trail Ridge, the largest sand ridge in Georgia. Formed during Pleistocene time, this former barrier island extends southward, forming the eastern boundary of the Okefenokee Swamp. Trail Ridge ends just north of US 341 and merges with the Orangeburg Escarpment, a topographic feature that waves cut into the landscape about 3.5 to 3 million years ago. Though the escarpment is a subdued feature today (see the I-16: Macon—Savannah road guide), visualize it about 3 million years ago as a wave-cut cliff, with Trail Ridge extending to the south as a sandy spit.

The town of Jesup sits on a sandy, Pleistocene-age barrier island of the Wicomico shoreline complex. The swampy area between Jesup and Trail Ridge was the muddy back-barrier marsh of this shoreline complex.

Griffin Ridge Wildlife Management Area

You can explore a Pleistocene-age riverine dune field at the Griffin Ridge Wildlife Management Area on US 301 north of Jesup, just across the Altamaha River. A gated dirt road leads into the woods, toward the dunes, on the west side of US 301. Several companies mine the sand for construction and industrial uses. These excavations into the yellowish quartz sand are visible along US 301 just north of the Altamaha River.

There are several distinct bands of parabolic dunes, up to 50 feet high, paralleling the northeast side of the river. The dunes nearest the river are youngest, and they become older to the northeast. The Altamaha was a braided river when the dunes were active. Sandbars in the river channel were the source of the windblown sand.

The dunes are between 15,000 and 45,000 years old, dating back to the last glaciation. The ancient sand dunes are sparsely vegetated with drought-resistant plants, including cactus. Several archaeological sites in the dunes indicate they were popular with Native Americans.

Jesup to Brunswick

From Jesup to the coast, US 341 continues over the former seafloor, but instead of stepping down from one terrace to another, the road crosses five distinct shoreline complexes. Each complex is composed of a former barrier island, now a sand ridge, separated by the former marshes that lay behind each island. I-95 runs along the sandy Pamlico barrier island.

US 1
South Carolina State Line—
Florida State Line
225 miles

US 1 enters Georgia at the Savannah River in Augusta, near the Fall Line, and passes through the heart of the city. About 6 miles southwest of I-520, US 1 passes the entrance to a county landfill on the southeast side of the road. The landfill is one of the busiest in the state, processing more than 1,000 tons of waste each day. In order to protect the groundwater, the landfill was sited on the relatively impermeable Twiggs Clay.

Southwest of the landfill, US 1 rises to a higher elevation, providing a panoramic view to the northwest, across the valley of the Savannah River to South Carolina in the distance. US 1 heads southwest toward Wrens and enters the Kaolin Belt. There are a number of kaolin mines in this region, including one

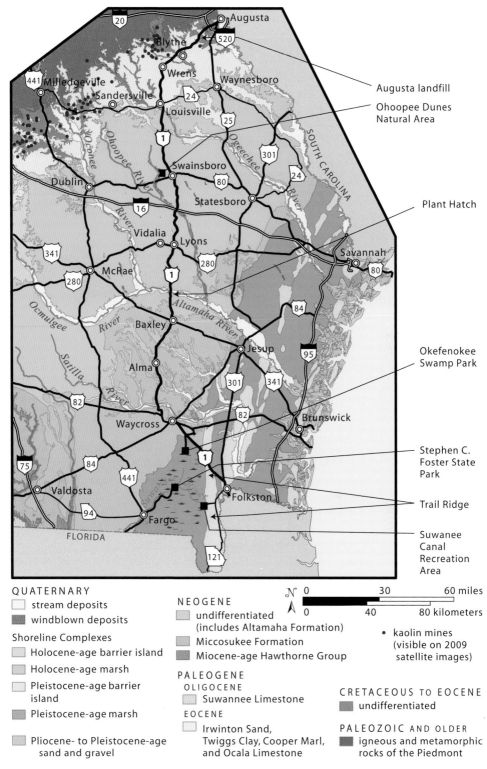

QUATERNARY
☐ stream deposits
▨ windblown deposits

Shoreline Complexes
☐ Holocene-age barrier island
☐ Holocene-age marsh
☐ Pleistocene-age barrier island
▨ Pleistocene-age marsh

☐ Pliocene- to Pleistocene-age sand and gravel

NEOGENE
☐ undifferentiated (includes Altamaha Formation)
☐ Miccosukee Formation
▨ Miocene-age Hawthorne Group

PALEOGENE
OLIGOCENE
☐ Suwannee Limestone
EOCENE
☐ Irwinton Sand, Twiggs Clay, Cooper Marl, and Ocala Limestone

0   30   60 miles
0   40   80 kilometers

• kaolin mines (visible on 2009 satellite images)

CRETACEOUS TO EOCENE
▨ undifferentiated

PALEOZOIC AND OLDER
▨ igneous and metamorphic rocks of the Piedmont

Geologic map of the central and eastern Georgia Coastal Plain.

Stunted turkey oaks growing in white, sandy soil in Ohoopee Dunes Natural Area. —Courtesy of Alan Cressler

on the west side of US 1, off Old Warrenton Road between Fort Gordon and Wrens. South of Louisville the road crosses the Ogeechee River and rises onto the Vidalia Upland, which is underlain by red clay of the Miocene-age Altamaha Formation.

Ohoopee Dunes Natural Area

Ohoopee Dunes Natural Area is off US 80 about 4 miles west of Swainsboro. This is part of the largest Pleistocene-age riverine dune field in Georgia, extending more than 65 miles along the eastern side of the Little Ohoopee and Ohoopee rivers, from northwest of Swainsboro south to the Altamaha River. The footprints of some of the individual dunes are as large as 4.5 by 1.5 miles. Drought-tolerant plants are adapted to live in the porous, nutrient-poor, sandy soils of the dune field. The crescent-shaped dunes formed between about 1 million and 10,000 years ago, when the climate was cool and dry and there was less vegetation. River discharge fluctuated, and at times the rivers of the Coastal Plain nearly dried up. Prevailing winds from the west blew the riverbed sand into vast dune fields, such as the one seen here.

Swainsboro to Waycross

South of Swainsboro, near mile 89 just north of I-16, the white, sandy soil along the sides of US 1 and along the Ohoopee River is part of the same dune field at Ohoopee Dunes Natural Area. South of I-16, US 1 passes near the town of Vidalia, which is known for its sweet onions. What makes a Vidalia onion so sweet? The short answer is the soil. The low sulfur content of the soil has allowed farmers to cultivate what is called a "sweet" onion. Much as regions of France with particular combinations of climate and soil are known for the types of wine they produce, Vidalia is also known as a unique agricultural area. By law, only onions grown in a specific twenty-county area around Vidalia—part of the Vidalia Upland—can be labeled Vidalia onions. Some people actually eat the onions like apples. The same type of onion grown in a different climate and soil type is said to lack the special characteristics of the sweet Vidalia onion. You can learn more at the Vidalia Onion Museum in Vidalia.

About 19 miles south of Lyons, US 1 crosses the Altamaha River, which has the largest drainage basin in the state. Of the rivers on the east coast of the United States, the Altamaha is the third-largest contributor of freshwater to the Atlantic Ocean. From the Altamaha River bridge, you can see a tall tower and power lines just downstream on the south bank. These are the first signs that you are approaching Plant Hatch, also known as the Edwin I. Hatch Nuclear Electric Generating Plant. There is a reason why Plant Hatch is located on the Altamaha River. Nuclear power plants require a large supply of water for steam generation and for cooling the steam. The Altamaha, Georgia's largest river, discharges about 100,000 gallons per second.

South of the Altamaha River, US 1 heads toward Baxley through hilly terrain. You can see several feet of orange and white clayey sediment of the Altamaha Formation in road cuts on both sides of the road. In places the clay contains gravel or chert. Between the river and Baxley, US 1 leaves the Vidalia Upland, passing into the Bacon Terraces District just south of Tenmile Creek on the northern outskirts of Baxley. This series of flat areas separated by slopes that step down toward the east are old marine terraces that waves cut during sea level high-stands.

Near Baxley and southward, the soil is sandy because sand and gravel deposits of Pliocene and Pleistocene age overlie the red clay (Altamaha Formation) of the Vidalia Upland. The Alma area, south of Hurricane Creek, sits on the flat Claxton terrace. There are about 4,500 acres of blueberries in the Alma area because they grow best in slightly acidic soil. US 1 crosses the Satilla River about 17 miles south of Alma. Pleistocene-age riverine dunes are being excavated for sand on both sides of the road on the north side of the river. Waycross sits on the south bank of the Satilla River on the Waycross terrace. US 1 crosses US 82 at Waycross and leaves the Bacon Terraces, entering the Okefenokee Basin.

Okefenokee Swamp Park

About 8 miles south of Waycross turn onto GA 177 from US 1. Head south for about 5 miles. The road passes through the Dixon Memorial State Forest with its slash pines and swampy areas with cypress, black gum, and bay trees. The road ends at Okefenokee Swamp Park, which has a nature center, live animal displays, a boardwalk, boat tours, a gift shop, and more. You will definitely see live alligators, both enclosed and loose. This park is well worth a visit, but insect repellant is recommended.

The Okefenokee Swamp is one of the largest freshwater wetlands in the United States, covering an area of roughly 700 square miles. The swamp is habitat for birds, mammals, reptiles, amphibians, and carnivorous plants, many of

The highly reflective black water of the Okefenokee Swamp.

which are endangered or threatened. Most of the swamp is covered by water less than 2 feet deep. Cypress trees and hardwoods dominate the western part of the swamp, and prairies with ponds and grasses dominate the eastern portion. There are numerous lakes within the swamp, in addition to channels with flowing water and sandy islands covered in pine trees.

The Okefenokee is what's known as a blackwater swamp. Its water is dark, resembling tea due to the tannic acid that comes from decaying vegetation. Masses of partially decayed plant matter that accumulate on the bottom occasionally rise to the surface, floated by gases that form during decomposition. This produces floating "islands" that may be firm enough to walk on. These floating islands are responsible for the American Indian name Okefenokee, meaning "land of trembling earth." A thick layer of peat has accumulated in the swamp. It ranges from 5 to 15 feet thick. The Okefenokee Swamp is a modern analogue for ancient swamps that are today mined for coal.

Trail Ridge

About 12 miles southeast of GA 177, near the intersection with GA 121/GA 15, US 1 climbs Trail Ridge. This former barrier island was a shoreline 2.1 million years ago. US 1 follows Trail Ridge southward for several miles, leaving it as the road descends to Spanish Creek (near Mattox), which flows through the former marsh area behind the Penholoway shoreline complex. This complex is younger than that of Trail Ridge. US 1 continues south through flat pinewoods to Folkston, which sits on the Penholoway barrier island, and on to the Florida state line.

The eastern entrance to Okefenokee National Wildlife Refuge is about 8 miles south of Folkston, off GA 121, at the Suwannee Canal Recreation Area, where a 5-mile-long canal was dug through Trail Ridge in the 1890s in an attempt to drain the swamp. The project was abandoned because the sides of the canal kept collapsing due to underground springs, and the company ran out of money. The swamp was subjected to logging and dredging after that. The recreation area has a visitor center, nature trails, and a boardwalk through the upland and swamp areas, and guided tours can be arranged.

US 82
Alabama State Line—Brunswick
239 miles

US 82 cuts a west to east swath across the southern part of the state, starting at Walter F. George Reservoir on the Chattahoochee River. The route begins in Cretaceous-age sediments of the Fall Line Hills and proceeds eastward through ever-younger deposits, culminating in Quaternary-age sediments near the coast. At the west end of the route, streams have created the hilly topography by cutting into the easily eroded Providence Sand and overlying sediments, which top the hills between Georgetown and Cuthbert. Near Cuthbert, the road ascends onto the Dougherty Plain.

The level land and fertile soils of the Dougherty Plain make it some of the best agricultural land in the state. The area is underlain by Eocene-to-Oligocene-age limestone, which has weathered to form a clayey residuum, or soil. Characteristic landforms, such as bowl-shaped lime sinks, small lakes, springs, and underground caverns, collectively known as karst topography, tend to develop in humid areas underlain by limestone.

Albany and the Flint River

US 82 Business Route (Oglethorpe Boulevard) crosses the Flint River in the heart of Albany. Its headwaters are about 150 miles upstream near Atlanta. The Flint River derives its name from the "flint" (technically chert) found in and along its banks, weathering out of the local limestone. Indian artifacts made of the chert have reportedly been found in sandbars along the river.

At the Albany Civic Center, on the south side of US 82 just west of the Flint River, there is a granite monument commemorating the worst flood in the history of the state. Between July 1 and July 7, 1994, Tropical Storm Alberto dropped more than 27 inches of rain north of Albany. The floodwaters crested at 23 feet above flood stage in Albany on July 7, exceeding 500-year-flood levels. The campus of Albany State University, just downstream on the east side of the river, was under 10 feet of water.

Numerous lime sinks are present in and around Albany (some are visible from US 82 west of Albany). Most are prehistoric, but more than three hundred new lime sinks opened following the 1994 flood.

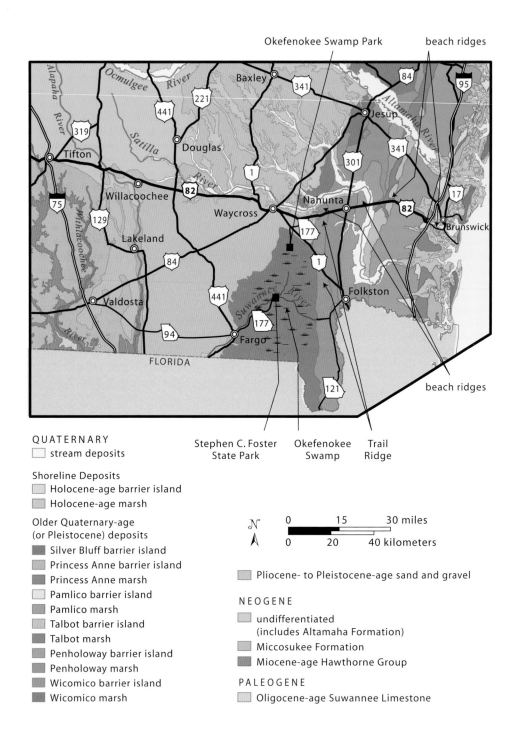

Okefenokee Swamp Park

beach ridges

Alapaha River

Ocmulgee River

Baxley

341

84

95

221

441

Jesup

319

Satilla River

Douglas

341

Tifton

1

301

82

Willacoochee

Waycross

Nahunta

82

75

129

177

Brunswick

17

Lakeland

Suwannee River

1

Folkston

84

Withlacoochee River

441

177

Valdosta

94

177

Fargo

121

FLORIDA

beach ridges

QUATERNARY

□ stream deposits

Stephen C. Foster
State Park

Okefenokee
Swamp

Trail
Ridge

Shoreline Deposits
☐ Holocene-age barrier island
☐ Holocene-age marsh

Older Quaternary-age
(or Pleistocene) deposits
■ Silver Bluff barrier island
▦ Princess Anne barrier island
■ Princess Anne marsh
☐ Pamlico barrier island
■ Pamlico marsh
▦ Talbot barrier island
■ Talbot marsh
▦ Penholoway barrier island
▦ Penholoway marsh
▦ Wicomico barrier island
■ Wicomico marsh

N

0   15   30 miles

0   20   40 kilometers

▦ Pliocene- to Pleistocene-age sand and gravel

NEOGENE
☐ undifferentiated
(includes Altamaha Formation)
☐ Miccosukee Formation
■ Miocene-age Hawthorne Group

PALEOGENE
☐ Oligocene-age Suwannee Limestone

*Geology along US 82 east of I-75. See the map on page 75 for the
geology along the western end of US 82.*

The plaque above the entranceway of a building at Albany State University marks the water level on July 7, 1994.

Radium Springs

Radium Springs is the largest natural spring in Georgia and one of the Seven Natural Wonders of the state. It is 3.5 miles south of US 82 Business Route on Radium Springs Road at Holly Drive. Radium Springs is a natural blue hole spring, a name given because blue light penetrates deep into the water and is reflected off white sand on the bottom, whereas light at the red end of the spectrum is absorbed by the water. The springwater comes from caverns in the underlying limestone that is part of the Floridan aquifer. The springs discharge about 70,000 gallons per minute, with a constant temperature of 68°F.

Radium Springs draws its name from the presence of radium-226, a natural radioactive isotope. The radium is the product of the radioactive decay of uranium-bearing minerals in the aquifer. The ultimate source of the uranium is granite and other crystalline rocks in Georgia's Piedmont. Uranium-238, derived from the weathering of these rocks, was incorporated into the crystal structure of phosphate pellets in Miocene-age Hawthorne Group sediments overlying the limestone. Groundwater percolating through the Hawthorne Group dissolves the uranium and carries it downward into the Floridan aquifer.

Radium Springs. —Courtesy of the Albany Convention and Visitors Bureau

The amount of radioactive elements in the local water supply exceeds drinking water standards, and some wells east and northeast of Albany have had to be plugged. At one time, some people thought the radium in the water had healing properties, and a resort was built in the 1920s to cater to these individuals, but it was heavily damaged by flooding in 1994 and later demolished.

Albany to Brunswick

East of the Flint River, on the south side of US 82, you can see 20- to 30-foot-tall Pleistocene-age riverine dunes covering approximately 1 square mile. The sand is mined here for construction and other uses.

About 12 miles east of Albany, just west of Sylvester, US 82 ascends the Pelham Escarpment, passing from the limestone of the Dougherty Plain up onto higher ground of the Tifton Upland, which is underlain by Neogene-age sand and clay.

Near the town of Willacoochee, US 82 leaves the Tifton Upland and enters the Bacon Terraces District. The terraces are hard to see on the ground, but US 82 passes down a series of about seven steps as it heads toward the coast. (The Okefenokee Swamp is about 10 miles southeast of Waycross off US 1. See the US 1: South Carolina State Line—Florida State Line road guide.)

Just east of Hoboken, and about 4 miles west of Nahunta, US 82 rises over Trail Ridge. Look for the Trail Ridge historical marker near the fire tower on the south side of the road. The road drops as it descends the east side of the ridge to the swampy area known as Caney Bay. Trail Ridge and the Okefenokee Swamp to the west compose the Wicomico shoreline complex, the oldest and westernmost complex of the Barrier Island Sequence District, which preserves six Pleistocene-age shoreline complexes that record higher sea levels. Between Trail Ridge and the coast, the subtle rise and fall of the road marks the positions of the barrier islands and back-barrier marshes of the former shorelines. The hills were the barrier island beaches, and the river valleys or low swampy areas were the back-barrier marshes and lagoons.

Caney Bay is the back-barrier marsh of the Penholoway shoreline complex, and the Penholoway barrier island lies between Nahunta and the Satilla River. This barrier island is mined for heavy mineral sand on the north side of the road near the Brantley County Airport. US 82 merges with US 17 and continues eastward to Jekyll Island (see the Jekyll Island road guide in the Sea Islands chapter).

I-95
South Carolina State Line—
Florida State Line
112 miles

Between the South Carolina and Florida state lines, I-95 passes north to south through the eastern edge of the Barrier Island Sequence District and is a convenient gateway to the Sea Islands. See the Sea Islands chapter for a more detailed

The tidal salt marsh at the North Newport River, where you can see the S-shaped curve of a tidal channel. The smokestacks of a paper mill are visible upstream.

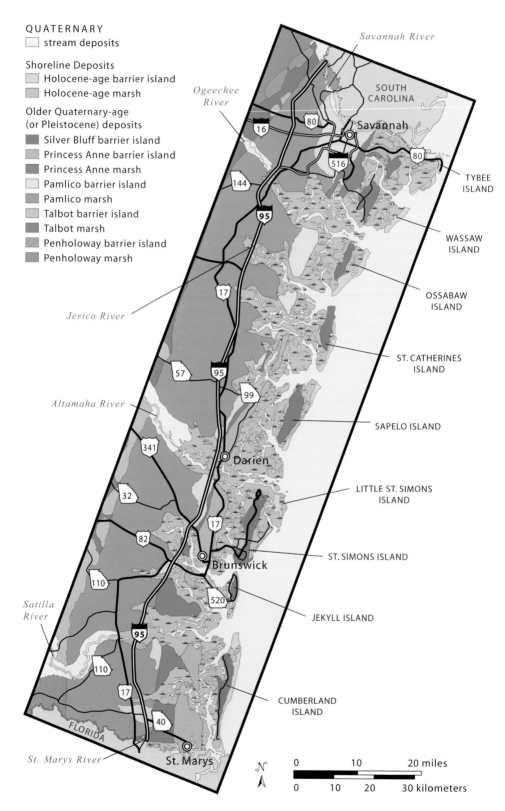

QUATERNARY
☐ stream deposits

Shoreline Deposits
☐ Holocene-age barrier island
☐ Holocene-age marsh

Older Quaternary-age
(or Pleistocene) deposits
☐ Silver Bluff barrier island
☐ Princess Anne barrier island
☐ Princess Anne marsh
☐ Pamlico barrier island
☐ Pamlico marsh
☐ Talbot barrier island
☐ Talbot marsh
☐ Penholoway barrier island
☐ Penholoway marsh

Savannah River

SOUTH CAROLINA

Ogeechee River

Savannah

TYBEE ISLAND

WASSAW ISLAND

Jerico River

OSSABAW ISLAND

ST. CATHERINES ISLAND

Altamaha River

SAPELO ISLAND

Darien

LITTLE ST. SIMONS ISLAND

ST. SIMONS ISLAND

Brunswick

JEKYLL ISLAND

Satilla River

CUMBERLAND ISLAND

FLORIDA

St. Marys River

St. Marys

N

0   10   20 miles

0   10   20   30 kilometers

Geology along I-95.

description of them. I-95 passes over Pleistocene-age shoreline deposits, mostly those of the Pamlico and Princess Anne shoreline complexes. The former barrier islands are higher and sandier, and the former marshes are lower and muddier. I-95 crosses salt marsh, a sea of *Spartina* grass flooded and drained by tides twice each day, at the Jerico, North Newport, and Altamaha rivers.

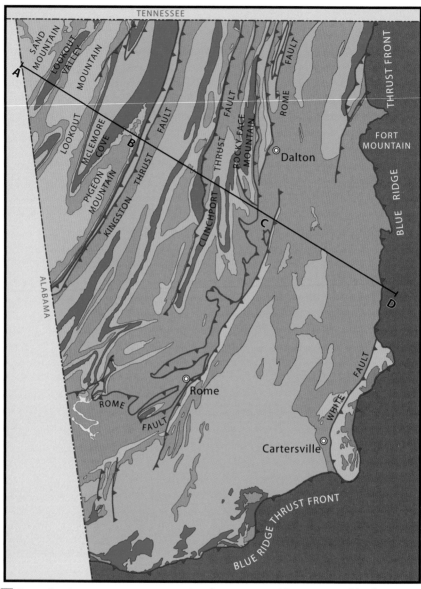

TENNESSEE

SAND MOUNTAIN
LOOKOUT VALLEY
LOOKOUT MOUNTAIN

A

B

McLEMORE COVE

PIGEON MOUNTAIN

KINGSTON THRUST FAULT

CLINCHPORT THRUST FAULT

ROCKY FACE MOUNTAIN

ROME FAULT

THRUST FRONT

FORT MOUNTAIN

BLUE RIDGE

⊙ Dalton

C

ALABAMA

D

⊙ Rome

ROME FAULT

WHITE FAULT

Cartersville ⊙

BLUE RIDGE THRUST FRONT

Pennsylvanian-age
sandstone and shale

Mississippian-age limestone,
shale, and chert

Silurian- to Devonian-age
sandstone, shale, and chert

Ordovician-age limestone,
shale, and sandstone

Cambrian- to Ordovician-age
dolostone with chert (Knox Group)

thrust fault; barbs occur on
the overriding fault block

𝒩
∧

0   10   20 miles

0   10   20 kilometers

Cambrian-age limestone,
shale, and sandstone (Conasauga Group)

Cambrian-age shale, dolostone, and sandstone
(Shady Dolomite and Rome Formation)

Cambrian-age sandstone
and shale (Chilhowee Group)

Precambrian- and Paleozoic-age
metamorphic and igneous rocks

location of cross section on page 125

The pattern of alternating ridges and valleys in the Valley and Ridge and Appalachian Plateau reflects the folding and faulting of rock layers, which occurred as Africa finished colliding with North America about 265 million years ago, and the differing resistance to erosion of various rock types.

VALLEY AND RIDGE AND APPALACHIAN PLATEAU

Northwest of the mountainous Blue Ridge, Georgia's physiographic provinces are the Valley and Ridge, consisting of long, parallel ridges separated by flatlands, and the Appalachian Plateau, a region of flat-topped mountains interrupted by widely separated straight valleys. Both are underlain by the same sequence of sedimentary rocks, ranging in age from Cambrian to Pennsylvanian.

A bird's eye view of northwest Georgia reveals four distinctive landscapes, or districts. The part of the Appalachian Plateau that lies in Georgia and adjacent states is also called the Cumberland Plateau. The remaining three districts comprise the Valley and Ridge. The Chickamauga Valley is a low, flat area interrupted by low, northeast-trending ridges. The Armuchee Ridges (pronounced Ar-MUR-chee) have odd zigzag patterns. And to the southeast is the Great Valley, a broad, low area of flatlands or gently rolling hills with a few straight ridges.

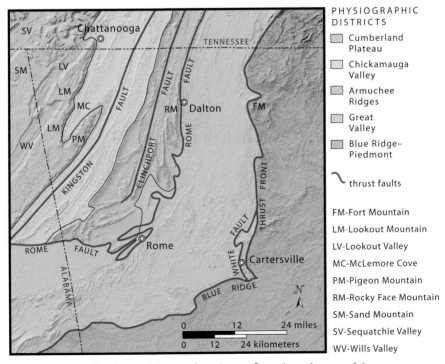

The landscape districts of northwest Georgia and some of the mountains, valleys, and faults mentioned in the text.

121

Comparing the landscape with a geologic map reveals the relationship of the topography to underlying bedrock. Plateaus and mountains are capped by erosion-resistant sandstone and conglomerate. Valleys are underlain by more easily eroded limestone and shale. Low ridges and rolling hills are underlain by limestone and dolostone residuum with erosion-resistant chert.

THE LANDSCAPE IN RELATION TO ANTICLINES AND SYNCLINES

The sedimentary formations of northwest Georgia, originally deposited in relatively flat layers, look like they have been through the tectonic mill. They have been sliced and shifted by faults and folded into arched layers called anticlines and trough-like folds—the opposite of anticlines—called synclines.

Contrary to what you might expect, many of today's valleys developed along the axes, or hinge lines, of anticlines. The Sequatchie Valley is the northwesternmost and longest valley of the Cumberland Plateau. It lies just northwest of Georgia and is practically arrow-straight for its 140-mile length. Before it became a valley, its layers of sedimentary rock were arched upward into an anticline. The crest of the fold began eroding because of its higher elevation (streams erode highlands more quickly because their gradients are steeper) and the presence of fractures that had formed along the hinge line during folding (weathering and erosion exploit such weaknesses in rock). Once erosion had cut through the relatively tough Pennsylvanian-age sandstone that capped the

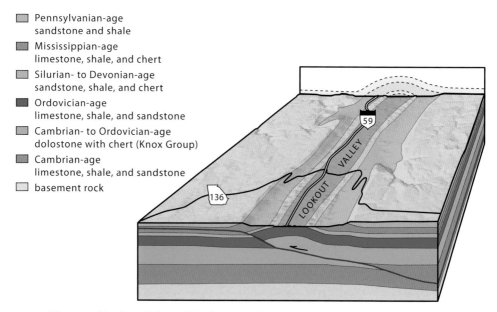

Pennsylvanian-age
sandstone and shale

Mississippian-age
limestone, shale, and chert

Silurian- to Devonian-age
sandstone, shale, and chert

Ordovician-age
limestone, shale, and sandstone

Cambrian- to Ordovician-age
dolostone with chert (Knox Group)

Cambrian-age
limestone, shale, and sandstone

basement rock

Diagram of Lookout Valley, which developed in the fold called the Lookout Valley anticline. The portion of the anticline that has eroded is shown at the rear of the diagram (dashed lines). Movement along the thrust fault (red line) is responsible for the anticline.

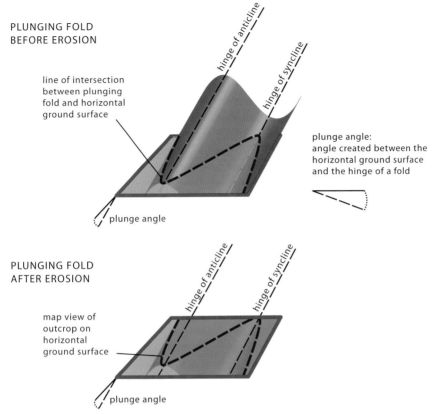

PLUNGING FOLD
BEFORE EROSION

hinge of anticline

hinge of syncline

line of intersection
between plunging
fold and horizontal
ground surface

plunge angle:
angle created between the
horizontal ground surface
and the hinge of a fold

plunge angle

PLUNGING FOLD
AFTER EROSION

hinge of anticline

hinge of syncline

map view of
outcrop on
horizontal
ground surface

plunge angle

*When erosion planes off the tops of adjacent plunging folds, the V patterns
for the anticline and syncline point in opposite directions. This is how zigzag
map patterns develop.*

anticline, it proceeded more rapidly in the softer shale and limestone underneath, resulting in the long, straight valley. Lookout Valley and Wills Valley,
which are the major valleys of Georgia's portion of the Cumberland Plateau,
developed in the hinge lines of similar anticlines.

The zigzag map patterns of some of Georgia's ridges and valleys, such as
McLemore Cove and Pigeon Mountain, or the Armuchee Ridges, developed
where differential erosion has operated on a pattern of folding known as plunging folds. A fold is said to be "plunging" if its hinge line is itself tilted. When the
hinge line of a fold is horizontal, the map pattern will consist of parallel stripes
of valleys and ridges. For example, Sequatchie Valley formed in the untilted
hinge line of an anticline, which explains its nearly straight, 140-mile length.
However, if the hinge line is inclined, a V-shaped geologic-map pattern will
develop as erosion has its way with the rock layers. This is purely a 3D geometric effect that develops as plunging, folded layers are planed down to a more or
less horizontal land surface. The wavy lines in a piece of lumber create roughly
the same kind of geometric effect.

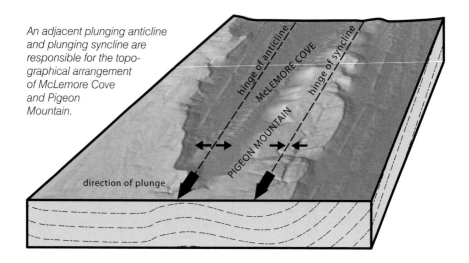

An adjacent plunging anticline and plunging syncline are responsible for the topographical arrangement of McLemore Cove and Pigeon Mountain.

If you study the accompanying diagram, you can see that for a plunging anticline, the point of the V lies in the direction that the hinge line plunges beneath other strata. For the adjacent plunging syncline in the same diagram, you can see that the V points opposite the direction the syncline plunges.

The major faults in the Valley and Ridge and Appalachian Plateau are thrust faults that formed in response to tectonic compression. Movement along thrust faults pushes rock layers up and over the adjacent rock layers. After the tectonic event had ended, packages of layered rocks, called thrust sheets, had been moved several miles northwest along the faults—picture an area in which thrust sheets resemble overlapping shingles on a roof.

This happened in Permian time, probably around 265 million years ago during the final stages of the collision between North America and Africa. Rather than occurring all at once, fault movement occurred bit by bit over hundreds of thousands of years. Most episodes of movement would have been at most a few feet at a time, but larger movements would have caused major earthquakes.

Opposite: Cross section across northwest Georgia; see map on page 120 for location of cross section. (Modified after Mittenthal and Harry, 2004, and Woodward, 1985.)

A-B: Sand Mountain and Lookout Mountain have nearly flat layers, little disturbed except they slid a few miles to the northwest. This contrasts with the Lookout Valley anticline between the mountains, where a thrust fault caused layers to overlap.

B-C: Movement on the Kingston thrust fault and other faults to the west produced the eastward dip of the layers beneath Chickamauga Valley. The parallel ridges forming the topographical relief east of this valley are composed of cherty dolostone of the Knox Group, which resisted erosion compared to neighboring layers. Other thrust faults are responsible for the plunging folds (zigzag map patterns) of the Armuchee Ridges.

C-D: Beneath the Great Valley the Rome fault carried a wide swath of rocks westward, and later fault movements gently rumpled both the Rome fault and the swath of rocks. The gentle folds have left either resistant Knox Group rocks or neighboring less-resistant shale and limestone at the surface, causing broad areas of rolling hills or flatlands, respectively, which characterize the Great Valley.

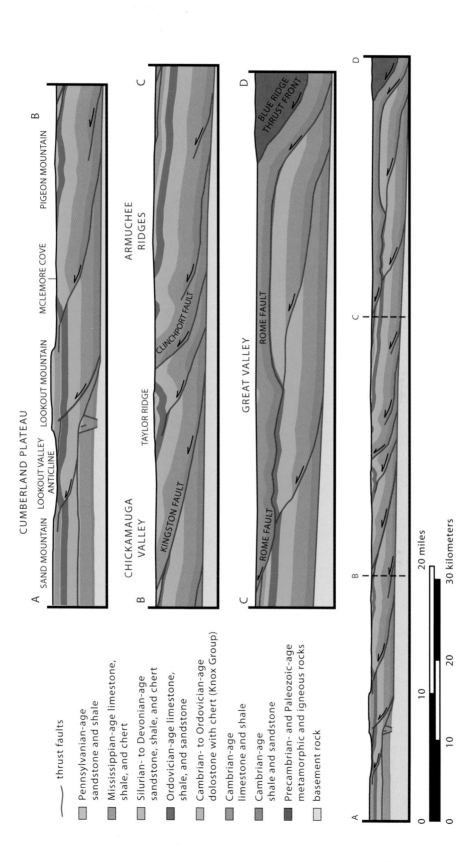

CUMBERLAND PLATEAU

SAND MOUNTAIN LOOKOUT VALLEY LOOKOUT MOUNTAIN MCLEMORE COVE PIGEON MOUNTAIN
ANTICLINE

A — B

CHICKAMAUGA
VALLEY

ARMUCHEE
RIDGES

TAYLOR RIDGE

KINGSTON FAULT

CLINCHPORT FAULT

B — C

GREAT VALLEY

ROME FAULT

ROME FAULT

BLUE RIDGE
THRUST FRONT

C — D

A — D

— thrust faults

Pennsylvanian-age
sandstone and shale

Mississippian-age limestone,
shale, and chert

Silurian- to Devonian-age
sandstone, shale, and chert

Ordovician-age limestone,
shale, and sandstone

Cambrian- to Ordovician-age
dolostone with chert (Knox Group)

Cambrian-age
limestone and shale

Cambrian-age
shale and sandstone

Precambrian- and Paleozoic-age
metamorphic and igneous rocks

basement rock

0   10   20

0   10   20   30 kilometers

20 miles

Movement on these thrust faults is what caused the folding and tilting of the sedimentary strata in northwest Georgia. Major folds, such as the Lookout Valley anticline, developed from thrust faults directly beneath them.

A complex of major thrust faults, referred to as the Blue Ridge thrust front, separates the sedimentary rocks of the Valley and Ridge and Appalachian Plateau from the metamorphic and igneous rocks of Blue Ridge and Piedmont provinces to the southeast. The sedimentary rock layers of northwest Georgia extend for a more than a hundred miles to the southeast below the Blue Ridge–Piedmont rocks.

SEDIMENTARY ROCKS AND ANCIENT GEOGRAPHY

Thanks to faulting, folding, and erosion, you can see rocks from most of the geologic periods of the Paleozoic era in northwest Georgia. By looking at rocks of the same age in different areas, it is possible to see differences caused by the varied geographic features of that particular time. For example, where rocks in one area contain more sand than rocks of the same age in another area, you can infer that the sandy rock was deposited relatively close to the eroding landscape providing the sand. At sites that were more distant from the ancient eroding landscape, there is a higher proportion of clay because clay remains in suspension in water longer and can travel much farther.

Cambrian-age sediment in northwest Georgia becomes coarser grained toward the west, implying that eroding land lay west of northwest Georgia during Cambrian time. Crossbeds in these strata, deposited by flowing water, indicate that streams and rivers were draining across Georgia from the west, corroborating the other evidence. Geologists infer that this source of sediment disappeared over time as its hills eroded away, because first sand and then clay diminish upward in Cambrian strata.

From Late Cambrian to Early Ordovician time, carbonate rock was deposited in a warm, shallow sea in northwest Georgia. Sand and clay become increasingly abundant in the rock record from Middle Ordovician through Silurian time, particularly in rocks toward the southeast, implying that land had appeared in that direction and had begun eroding. This landmass was probably the volcanic island chain that collided with North America during the Ordovician, deforming the edge of the continent and pushing thrust sheets to the west. The sea that covered northwest Georgia began to fill with material (sand and clay) eroded from both the rising thrust sheets and the volcanic islands.

First sand and then clay diminish in Devonian-age strata and much of the strata from Mississippian time, indicating that erosion had lowered the highlands to the east. The fossils in Mississippian-age limestone show that a clear, warm, shallow sea once again covered most of northwest Georgia. In rocks of Late Mississippian to Pennsylvanian age, first clay, and then sand and gravel appear. The record shows that the deposition of coarser sediment started in the southeast and progressed through northwest Georgia. The change in sediment size indicates that Africa was approaching and eventually colliding with North America, producing highlands.

STRATIGRAPHIC COLUMN FOR THE VALLEY AND RIDGE OF GEORGIA			
ERA	PERIOD	FORMATION OR GROUP	DESCRIPTION AND DEPOSITIONAL ENVIRONMENT
PALEOZOIC	PENNSYLVANIAN	Crab Orchard Mountain Formation	Shale, sandstone, and coal. Delta.
		Gizzard Group	Crossbedded sandstone, conglomerate, gray shale, and coal. Delta, barrier island, and lagoon.
	MISSISSIPPIAN	Pennington Formation	Red and green shale with interbedded limestone and mudstone. Prodelta and delta deposits.
		Bangor Limestone	Limestone, with chert nodules and dolostone in the lower part. Grades into shale. Marine shelf deposition.
		Monteagle Limestone	Crossbedded limestone with ooids.
		Tuscumbia Limestone	Limestone with black chert nodules.
		Fort Payne Chert (Maury Shale Member)	Bedded chert and cherty dolostone. Shallow marine shelf and tidal flat deposit. (Green shale with phosphate nodules at the base.)
	DEVONIAN	Chattanooga Shale	Black shale and mudstone. Deposited in shallow sea with oxygen-poor water.
		Armuchee Chert	Gray chert with scattered thin sandstone layers.
	SILURIAN	Red Mountain Formation	Red sandstone, siltstone, shale, and limestone. Some ironstone (hematite). Barrier islands and lagoons.
	ORDOVICIAN	Sequatchie Formation	Red, green, and gray calcareous siltstone, sandstone, shale, and limestone. Shallow marine passing eastward into tidal flat and alluvial deposits. The result of erosion of a landmass to the east.
		Chickamauga Group / Rockmart Slate	Limestone. Shallow marine. Bentonite is weathered volcanic ash indicating collision of volcanic arc with North America. Slate (metamorphosed shale) is deeper marine beds thrust faulted into contact with Chickamauga.
		Knox Group	Cherty limestone and dolostone. Shallow marine and tidal flats.
	CAMBRIAN	Conasauga Group	Interbedded greenish gray shale, limestone, and dolostone with black chert. Tan-weathering shale with trilobites.
		Rome Formation	Maroon to yellow sandstone, siltstone, and shale.
		Shady Dolomite	Dolostone and dolomitic limestone with chert, barite, iron ore, ochre, umber, and manganese. Weathers to a red clay residuum. Archaeocyathids, trilobites, and stromatolites.
		Chilhowee Group (Weisner Quartzite and other units)	Quartzite and gray shale.

Generalized geologic column for the Valley and Ridge and Appalachian Plateau of Georgia.
Wavy lines represent unconformities.

Fossils

You can find a wide variety of Paleozoic-age fossils in northwest Georgia. Certain beds are more fossiliferous than others. Limestone and shale are more likely to contain fossils than other rock types.

Fossils of Cambrian Age

Trilobites were sea-dwelling animals that only lived during Paleozoic time. They were arthropods with an exoskeleton and jointed limbs. Insects, spiders, and crabs are related to them. Like many arthropods, trilobites shed their exoskeleton multiple times as they grew, a process called molting. The molted exoskeletons are preserved as impressions in weathered tan shale of the Conasauga Group.

Trilobite fossils from the Conasauga Group.

Fossils of Ordovician Age

A great variety of fossils are present in limestone of the Chickamauga Group and overlying Sequatchie Formation. Slight weathering makes the fossils stand out in relief on bedding planes. We describe the most common fossils of this period below.

Most brachiopods became extinct at the end of Paleozoic time, although a few forms thrive today. Though their ribbed shells superficially resemble those of clams, brachiopods are not related to mollusks. Brachiopods use a fleshy stalk to remain attached to the sea bottom throughout their adult lives, and they feed upon organisms they filter from seawater.

Brachiopods on a slab of Ordovician-age limestone from the Sequatchie Formation.

Crinoid columnals. The largest is about 0.75 inch in diameter.

Bryozoans remain abundant in the oceans to this day. Like modern corals, bryozoans are actually colonies of many tiny individuals. Some bryozoan fossils are mosslike patches that may encrust other fossils (hence the name *bryozoa*, meaning "moss animals"). Others resemble long twigs with or without branches. The tiny wafflelike texture of these fossils, easily seen with a magnifying glass, is an identifying feature.

Related to sea stars, sea urchins, and sand dollars, crinoids are animals that resemble flowers. Their many arms gather food and feed it to the cuplike main part of their body, which is called a calyx. Both the calyx and the cylindrical stem (or columnal) that supports it are made of the mineral calcite. Crinoids exist today but were far more abundant in Paleozoic time, reaching their peak in the Mississippian period when reefs were built mainly of crinoid skeletal material. Crinoid columnals are common fossils in the red clay residuum of weathered limestone. Often the calcite has been replaced by silica.

Clams are the most familiar modern bivalve mollusk, and their ancient relatives lived in much the same way, moving about the seafloor or burrowing into the sediment. The fossils are often curved traces on weathered rock surfaces.

Horn (or rugose) corals were solitary, jellyfish-like animals that lived in the opening of a cone that contained radiating dividers (septa), which strengthened the structure. In limestone, cross sections of the fossil can appear in two forms: one cuts across the width of the cone and resembles the cut surface of a halved grapefruit, and the other cuts the length of the tapering cone, with septae converging toward the point. Whole cones can be found where the surrounding rock has weathered to soil.

Ostracods are crustaceans that have a smooth shell similar in shape to, but less than half the size of, a jellybean.

Burrows are roughly cylindrical areas of lighter-colored material that gives some rocks a mottled appearance. They typically represent the pathway of an organism as it passed through sediment, eating and digesting the organic matter it contained while excreting lighter-colored material. Burrows are an example of a trace fossil, meaning that the organism has left a trace without necessarily leaving any body parts behind.

Gastropod shells are coiled in a helix. As with modern gastropods, such as land and marine snails, ancient gastropods had a fleshy foot that extended from the shell's opening.

A straight-coned cephalopod resembled a modern squid but, like its modern relative the nautilus, it had a chambered shell. The fossil is shaped like a long, thin cone with a rounded tip. It is partitioned by nested, curved surfaces that are cupped toward the large end of the cone. Like the squid, the cephalopod was a predator of the open ocean. The animal added chambers as it grew, living in the outermost chamber and pumping gas into or out of the unoccupied chambers to control its buoyancy.

Fossils of Mississippian Age

Some dirt banks on ridges underlain by the Fort Payne Chert will yield dozens of fossils in half an hour of inspection. Following a hard rain, the fossils may be perched on little spikes of dirt. These include crinoids, brachiopods, and horn corals.

Fossils are frequently seen, but not easily collected, from outcrops of limestone. They stand out in low relief on faces of rock that have been etched and rounded by groundwater. Crinoid, brachiopod, tabulate coral, and horn coral fossils are the most abundant. Tabulate corals, like today's corals, were colonies of many individual jellyfish-like animals. Their fossils resemble honeycombs.

You may find a screwlike fossil of the bryozoan *Archimedes*. The shape provided support for a radiating, spiral-shaped, lacy colony of tiny bryozoans.

Fossils of Pennsylvanian Age

Pennsylvanian-age shale commonly contains plant fossils, especially in areas containing coal. Many of the plants in coal swamps reproduced by spores and were relatives of today's horsetails, club mosses, and ferns. Others were part of an extinct group called seed ferns, which superficially resembled ferns but reproduced by pecan-sized seeds. A few plants were gymnosperms, relatives of modern conifers. All of these groups included tree-sized specimens, with some types growing up to 300 feet tall.

Plant fossils are preserved as thin films of coal, which show up well on gray shale. One of the most common plant fossils has tiny parallel lines spaced less than 1 millimeter apart. These could be leaf prints from a giant club moss called *Lepidodendron*, or impressions from the bamboolike stem of a giant horsetail called *Calamites*. The bark impression of *Lepidodendron* (also called scale-bark tree) is a grid of elongated diamond shapes.

Also common are oval leaves from a seed fern called *Neuropteris*, found singly or as part of a fernlike array of leaves. Under a magnifying glass the leaves, less than 0.5 inch long, have a symmetrical pattern of curved veins. A radiating array of small leaves is most commonly from the horsetail *Sphenophyllum*.

Lepidodendron on display at the Tellus Science Museum.

Pennsylvanian plant fossils on display at the Tellus Science Museum.

Sometimes the tracks of amphibians and reptiles, which resemble the three-toed tracks of small birds, are preserved in Pennsylvanian-age rocks. Amphibian tracks were found in sandy siltstone on Lookout Mountain and are on display at the Tellus Science Museum. One of the world's best track sites, the Steven C. Minkin Paleozoic Footprint Site, with thousands of tracks, was discovered in 1999 in Walker County, Alabama, in rocks similar to those atop Lookout and Sand mountains.

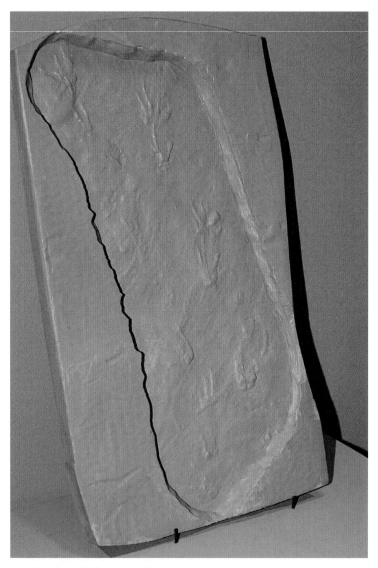

Replica of amphibian tracks found in Pennsylvanian-age rocks in Chattooga County, on display at the Tellus Science Museum.

I-24 and I-59
Tennessee State Line—
Alabama State Line
23 miles

The route along I-24 and I-59 passes more than twenty superb road cuts in Paleozoic strata, more than nearly any other route in Georgia. The route lies mostly within Lookout Valley, bordered by Sand Mountain on the west and Lookout Mountain on the east. Lookout Valley formed due to erosion along the crest of the Lookout Valley anticline. The oldest rocks in the area, of Ordovician age, are along the center of the valley, flanked by rocks of Silurian through Mississippian age. Pennsylvanian-age sandstone and conglomerate cap the mountains on either side. You can see cliffs of Pennsylvanian-age strata along the entire route. These are popular launching spots for hang gliders.

Limestone, shale, and siltstone of the Sequatchie Formation are exposed in road cuts that stretch for about 1.3 miles on the northwest side of I-24 at the Tennessee border. These rocks were deposited in a shallow Ordovician sea that covered much of the continental interior. The Red Mountain Formation, which overlies the Sequatchie Formation, is exposed 1.7 miles south of the state line on the west side of the road. The rocks consist of shale, siltstone, and limestone with a few sandstone beds. The base of a prominent orange sandstone bed near the bottom of this road cut coincides with the boundary between the Ordovician and Silurian periods. This boundary marks Earth's third-largest extinction event. About half of known genera became extinct, with huge losses of brachiopods, bryozoans, and trilobites. The causes of the extinction are not well understood but may be related to glaciations. The beds dip about 15 degrees to the northwest because the outcrop lies on the northwest side of the northeast-southwest-trending Lookout Valley anticline.

These rocks have less sand and more clay than rocks of the same age along I-75 to the southeast at Ringgold and near Dalton. Geologists infer that they were deposited farther from an eroding land area located to the southeast, meaning rivers and sea currents had dropped the larger particles before reaching this area.

The Red Mountain Formation is named after a mountain in Birmingham, Alabama, where the reddish brown mineral hematite (also known as iron oxide or rust) was mined from layers several feet thick. When combined with abundant Pennsylvanian-age coal nearby, the hematite was the basis for Birmingham's iron and steel industry. From the 1840s until about 1910, thinner layers of hematite were mined from the Red Mountain Formation from the valleys on both sides of Lookout Mountain. The town of New England, 3 miles north of Trenton, was named after the company that bought most of the mineral rights around 1890, and a blast furnace operated at Rising Fawn. Other blast furnaces operated in McLemore Cove (see the GA 136: Alabama State Line—Carters Dam road guide) and Chattanooga. Hematite ore is not exposed in this road cut.

Just west of this outcrop, I-24 passes through a gap in a low ridge composed of the Fort Payne Chert (Mississippian age) and underlying Red Mountain Formation. Follow the signs for I-59. Near the south end of the I-24/I-59 interchange you can see road cuts in Mississippian-age limestone. Four miles south

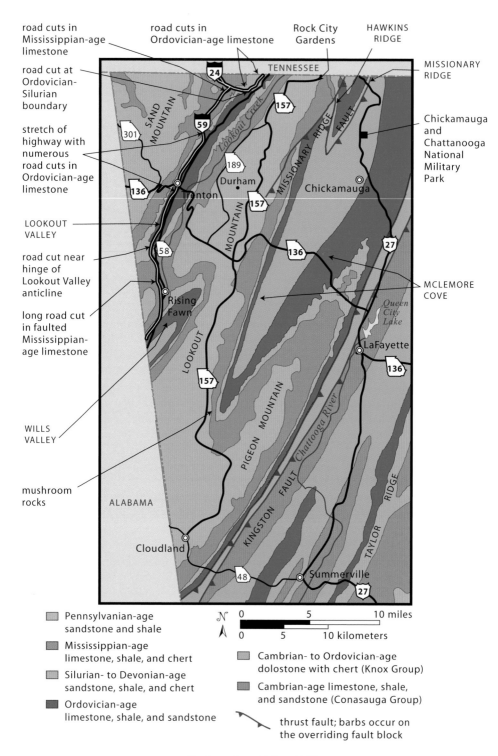

road cuts in Mississippian-age limestone

road cuts in Ordovician-age limestone

Rock City Gardens

HAWKINS RIDGE

MISSIONARY RIDGE

road cut at Ordovician-Silurian boundary

stretch of highway with numerous road cuts in Ordovician-age limestone

Chickamauga and Chattanooga National Military Park

LOOKOUT VALLEY

road cut near hinge of Lookout Valley anticline

long road cut in faulted Mississippian-age limestone

WILLS VALLEY

mushroom rocks

TENNESSEE

SAND MOUNTAIN

Lookout Creek

Durham

Trenton

LOOKOUT MOUNTAIN

Rising Fawn

MISSIONARY RIDGE FAULT

Chickamauga

MCLEMORE COVE

Queen City Lake

LaFayette

PIGEON MOUNTAIN

Chattooga River

KINGSTON FAULT

TAYLOR RIDGE

ALABAMA

Cloudland

Summerville

Geology along I-24 and I-59.

Pennsylvanian-age sandstone and shale

Mississippian-age limestone, shale, and chert

Silurian- to Devonian-age sandstone, shale, and chert

Ordovician-age limestone, shale, and sandstone

Cambrian- to Ordovician-age dolostone with chert (Knox Group)

Cambrian-age limestone, shale, and sandstone (Conasauga Group)

thrust fault; barbs occur on the overriding fault block

N

0   5   10 miles

0   5   10 kilometers

of the I-24 and I-59 interchange, just south of exit 17, I-59 crosses the same ridge of Fort Payne Chert and crosses back into a region underlain by limestone of Ordovician age. Limestone of the Chickamauga Group, which underlies the Sequatchie Formation, is exposed in a series of road cuts on both sides of the road, from 4.4 miles south of I-24 all the way to Trenton.

The turnoff at Trenton (GA 136, exit 11) is geologically and scenically reward-ing in both directions; Cloudland Canyon State Park (see the GA 136: Alabama State Line—Carters Dam road guide) is a fifteen-minute drive east. The south-bound on-ramp at exit 11 was a sampling site for a study of the Deicke (rhymes with Nike) bentonite, which is a type of clay formed from volcanic ash. The ash came from one of the two largest explosive volcanic eruptions known to have occurred during the last 542 million years. Bentonite is easily weathered, and only a few boulders of Ordovician-age Chickamauga Group limestone, within which the clay layer occurred, are still visible. Based on mineralogy and trace elements, bentonite from these eruptions has been correlated over much of eastern North America and even Scandinavia, which was joined to Greenland at the time of the eruptions. The bentonite is thickest in southwest Virginia, suggesting that the region was close to the volcanic islands that erupted. The islands must have been located in the Piedmont area, which was covered by an ocean at the time.

About 5 miles south of GA 136, on the east side of I-59, is a road cut in Sequat-chie Formation rocks. Lookout Valley ends about 5 miles south-southwest of here, where the hinge of the Lookout Valley anticline plunges beneath the Pennsylvanian-age rocks of Sand Mountain. From the road cut I-59 turns southeast, entering Wills Valley near the north end of the Wills Valley anticline. The relationship between the two anticlines is not hard to understand if you visualize a wrinkled throw rug on a floor. The strata of Lookout Mountain were transported a few miles northwestward on a buried thrust fault and wrinkled in the process. The anticlines along the two valleys are analogous to the tops of the wrinkles in such a rug. This is counterintuitive because the valleys are the low spots. However, before erosion had its way with the wrinkled rocks, the strata crested far above the modern valleys.

Mississippian-age limestone that lies between the ends of the two anticlines is exposed in a 0.25-mile-long road cut along the west side of I-59 just south of exit 4 near Rising Fawn. Perhaps reflecting the tectonic stresses that occurred as the two anticlines formed, at least three thrust faults are exposed in the road cut. The faults are interconnected, each ramping up from the same underlying bedding plane and flattening out into a common bedding plane above.

Fox Mountain is on the west side of I-59 between Rising Fawn and the Ala-bama border. Several caves, including some of Georgia's longest and deepest, underlie the mountain on property owned by the Southeastern Cave Con-servancy. A trail passes under the interstate leading to the caves. Caves can be extremely dangerous and should only be visited with well-equipped, experi-enced cavers.

At the Alabama border, I-59 exits the state much as I-24 entered it, at a road cut in Sequatchie Formation limestone. From the road near here, you can see low hills to the south, in the center of Wills Valley, that are underlain by

Inside Hurricane Cave. —Courtesy of Alan Cressler

One of three thrust faults seen in a cut along I-59 near Rising Fawn. The thrust fault (red line) flattened out upward into a bedding plane (green line) above. It folded slightly that bedding plane.

chert-bearing carbonate rocks of the Knox Group of Cambrian to Ordovician age. These are the oldest strata exposed along this route. They are at the surface because the Wills Valley anticline was bunched up more tightly than the Lookout Valley anticline when it formed, bringing slightly lower, older rocks to the surface along its hinge.

I-75
Tennessee State Line—
Cartersville (US 411)
61 miles

I-75 parallels the route of the historic Western and Atlantic Railroad (now CSX), crossing the Armuchee Ridges through low gaps from the Tennessee state line to Dalton. From there, it follows the Great Valley to the southeastern corner of the Valley and Ridge near Cartersville. This route was important both in America's westward expansion and the Civil War. From the need to build the South's first railroad tunnel in a particular location to the use of boulders as weapons in battle, geology has played a key role in this region's history.

I-75 crosses the Kingston fault about 2 miles south of the Tennessee line, on the east edge of the ridge topped by exit 353 (GA 146). The fault is more than 150 miles long and has carried the rocks on its southeastern side between 10 and 20 miles to the northwest. Here, the less erosion-resistant Cambrian-age shale underlying the valley moved westward over erosion-resistant, chert-bearing carbonate rock of the Knox Group, which forms the ridge to the west.

As you drive southeast you cross progressively younger rocks, which dip southeastward. The lanes split as the interstate ascends into hilly country that developed in the Knox Group, which surrounds exit 350 (GA 2). The highway descends eastward near exit 348 (the western of two Ringgold exits, GA 151) into a fertile valley underlain by Chickamauga Group limestone. Ringgold Gap is visible to the east.

Ringgold Gap

Ringgold Gap is a natural gap through a ridge, with White Oak Mountain to the northeast and Taylor Ridge to the southwest. East Chickamauga Creek runs through the gap, so it is the type of feature called a water gap. The ridge is underlain by Red Mountain sandstone. When I-75 was built, Ringgold Gap proved too narrow to accommodate the stream, railroad, US 41, and the interstate. The result is the largest road cut on a Georgia interstate. It towers more than 200 feet above the road and is more than 2,600 feet long. Unfortunately, a chain-link fence blocks access from the road cut. Because the layers dip southeastward, more than 1,300 feet of the Chickamauga Group, Sequatchie Formation, and Red Mountain Formation are exposed. A smaller road cut less than 1,000 feet to the southeast exposes the younger Chattanooga Shale and Fort Payne Chert.

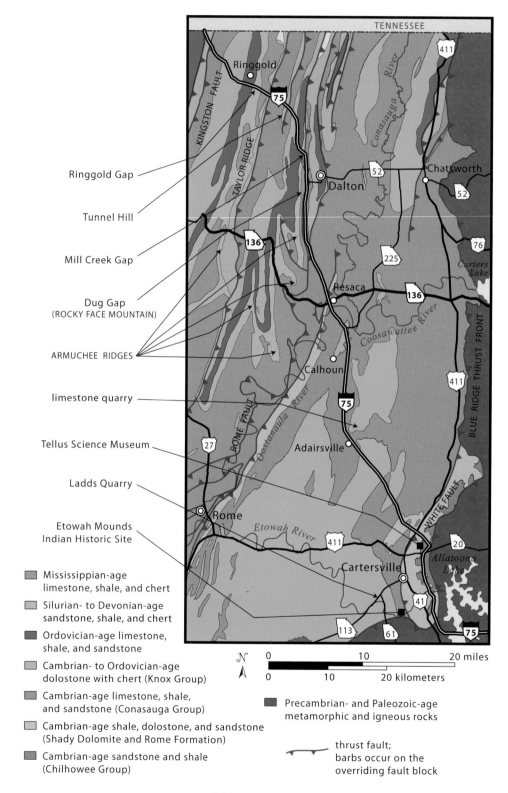

Geology along I-75.

From bottom to top, the dominant rock types include gray limestone, multi-colored shale and siltstone, and red sandstone. Fossils in the limestone record the presence of a warm, clear, shallow sea, rich with life, similar to tropical areas today such as the Bahamas. Changes in fossils and the rock layers document fluctuations in water depth, which were tied to worldwide sea level changes. The change from limestone through shale, siltstone, and sandstone tells the story of the rise of eroding highlands to the east. These highlands formed through the collision of a chain of islands with the North American continent. As they rose, the sea retreated, and ever-larger particles of sediment were delivered to the Ringgold region. The influx of clay, silt, and sand wiped out the rich marine ecosystem, so fewer fossils are present near the top of the road cut. The rocks of the islands are now part of the eastern Blue Ridge.

Ringgold Gap is the only location in Georgia where a stream cuts across the White Oak Mountain–Taylor Ridge trend. It is fair to ask how a stream can cut a path through a ridge without, at some stage, water flowing uphill. There are a couple of possible scenarios. One possibility is that the headwaters of East Chickamauga Creek formerly lay on the west slope of this ridge, and that erosion at the headwaters eventually broke through and caused the creek to capture a stream flowing on the other side of the ridge.

The other possibility is that East Chickamauga Creek initially set its westward course when the area was relatively flat, before erosion brought the erosion-resistant sandstone to light. As softer rocks eroded away from around the tough sandstone, Taylor Ridge and White Oak Mountain slowly emerged. Meanwhile, the stream had enough erosional power to continue to cut a channel across the sandstone. The meandering character of East Chickamauga Creek in the area supports this scenario. Streams tend to meander when they are flowing across flat areas underlain by uniform, easily eroded material. The meanders remained in place after the stream began to cut through resistant layers of sandstone, locking the creek in its meandering pattern.

Ringgold Gap is a significant gateway for both the railroad and the highway, which connect the Atlantic Coast to Chattanooga and beyond. It also served as a natural gateway for travel and trade long before European settlement. Excavations during construction of I-75 just west of the gap destroyed an Indian mound that was about 50 to 60 feet across and 20 feet high. Construction also uncovered an Indian village between East Chickamauga Creek and I-75, near Anderson Memorial Gardens. Some artifacts were 3,000 years old and are on display in the Old Stone Church Museum east of Ringgold Gap, at the intersection of US 41 and GA 2.

At least two pre–Civil War structures built of local Red Mountain sandstone remain standing. The Old Stone Church Museum, built as a Presbyterian church circa 1849, served as a hospital during the Civil War; bloodstains from the injured in the 1863 Battle of Ringgold remain on the floor. The 14-inch-thick brown sandstone walls of the Ringgold Railroad Depot were damaged by Union cannons and subsequently repaired with light gray Ordovician-age limestone. The irregular white marks in the limestone are burrows made by ancient worms. The worms ate the darker organic matter as they moved through the soft sediment shortly after it was deposited.

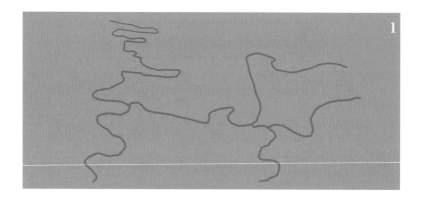

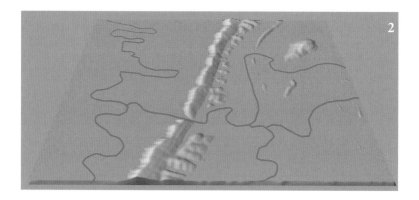

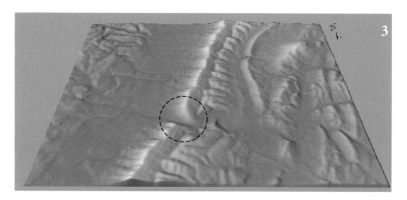

A hypothetical history of the development of Ringgold Gap: (1) Meandering streams establish courses through flat country. (2) As erosion exposes a line of more resistant rocks and a ridge appears, the stream continues cutting. (3) Ringgold Gap today (circled).

Close-up of the original dark sandstone beside the light limestone used to repair Civil War damage to the Ringgold Railroad Depot. The irregular white patches in the limestone are ancient worm burrows.

Tunnel Hill to Dalton

Exit 341 (GA 201) is located astride Tunnel Hill Ridge, which is underlain by cherty carbonate rock of the Knox Group. Rising less than 200 feet above the adjacent valleys, it seems insignificant in comparison to neighboring Taylor Ridge (west) and Rocky Face Mountain (east), both underlain by more-erosion-resistant Ordovician- and Silurian-age sandstone. Nevertheless, the ridge is a major drainage divide. Water on the west side flows through Ringgold Gap and continues into the Tennessee River, eventually reaching the Gulf of Mexico at the mouth of the Mississippi River. Water on the east side flows through Mill Creek Gap in Rocky Face Mountain and joins the Coosa River, entering the Gulf of Mexico at Mobile Bay.

With no water gap, Tunnel Hill Ridge was an obstacle to the first railroad linking the Atlantic Coast to the West. As a result, the first railroad tunnel in the South was constructed through the ridge in 1848–49. An engineering marvel of its day, the 1,477-foot-long tunnel still exists, although trains now pass through a larger tunnel built in the 1920s.

You can see the tunnel at the Tunnel Hill Heritage Center near the end of South Varnell Road (GA 201), 2 miles south of I-75. The stone-block facing

The 1848–49 tunnel at Tunnel Hill (right) faced with limestone of Cambrian age. A train is passing through the 1920s tunnel on the left.

around the tunnel opening consists mostly of limestone from the Conasauga Group. It was probably quarried northeast of Dalton, where limestone in the old quarries has calcium carbonate structures similar to those seen in the stone blocks. Called oncolites, these pebble-sized spherical structures with concentric internal layering were made by a type of algae called *Girvanella.*

About 1 mile north of exit 336 (US 41), I-75 passes through a water gap that Mill Creek cut across Rocky Face Mountain. Red sandstone exposed on the east side of I-75, like some of the red rocks at Ringgold Gap, is the Sequatchie Formation.

Rocky Face Mountain at Dug Gap

You can drive to the ridge crest of Rocky Face Mountain, which offers views west to Lookout Mountain and east to the Blue Ridge–Piedmont region. Turn west from I-75 at exit 333 (GA 52) and continue 1.8 miles to the entrance and parking area for Dug Gap Battlefield Park on your right, a few hundred yards before the gap at the crest of the ridge. The road passes through Dug Gap, so named because the ridge was dug out in the mid-nineteenth century to improve the road from Dalton to LaFayette.

Sandstone, siltstone, and shale of the Sequatchie Formation are well exposed along Dug Gap Battle Road on the west side of the mountain. On the west side

A palisade-like rock outcrop in Dug Gap Battlefield Park.

of the ridge within the park, sandstone forms a palisade of 10- to 15-foot-high cliffs separated by narrow notches. Overall, beds dip to the east, tilted in that direction as a result of movement on thrust faults in late Paleozoic time. The east side of the mountain is gentler because the topography mimics the eastward dip of the strata. The slopes are also less stable because some of the rock layers are prone to sliding along bedding planes.

The sandstone contains crossbeds that generally slope westward relative to the beds that contain them. This suggests that the streams that deposited the beds flowed west, away from the eroding highlands—the source of sediments—that geologists infer existed to the east.

Dug Gap Battlefield Park commemorates an event of the Civil War that was especially influenced by the region's geology. On May 8, 1864, the outgunned and outnumbered Confederate Army used sandstone boulders as weapons, rolling them down the mountain and onto Union soldiers passing through the gap.

Resaca to Cartersville

Around mile 322, I-75 crests the siltstone and sandstone ridge of the Rome Formation that the Confederate Army defended in the Battle of Resaca. Some of the formation is exposed on the east side of the road.

GEOLOGY AND THE CIVIL WAR IN GEORGIA

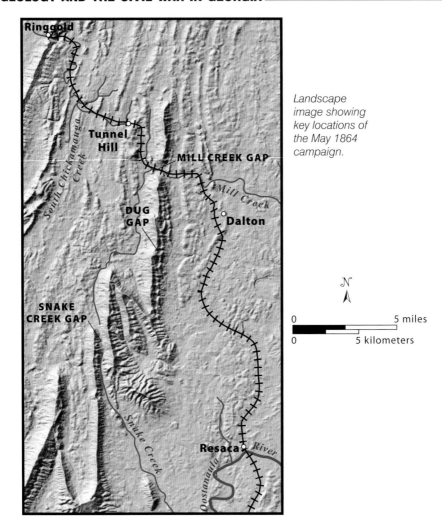

Landscape image showing key locations of the May 1864 campaign.

Over ten thousand soldiers are estimated to have died in Georgia in May 1864. The geology helped decide where men fought and died, and in some cases, the outcome of battles. The stage was set for the May 1864 battles in November 1863, after Union victories in the Chattanooga area forced the Confederate Army of Tennessee to retreat through Ringgold Gap. At Ringgold Gap, General Patrick Cleburne's four thousand troops held off poorly coordinated attacks by twelve thousand Union troops for five hours. As they did later at Dug Gap, Confederate troops rolled huge sandstone boulders down the ridges.

The five-hour delay saved the slow-moving Confederate supply trains and artillery from capture and gave the remainder of the Army of Tennessee the opportunity to secure a defensive position behind Rocky Face Mountain at Dalton, where they remained over the winter. On May 7, the Union Army began advancing from Ringgold. With its numerical advantage, the Union Army was able to feint an advance around the north end of Rocky Face Mountain at the same time it attacked at Mill Creek Gap and Dug Gap, while yet another force continued through Snake Creek Gap around the south end of the mountain.

Near the south end of Rocky Face Mountain, Snake Creek threads between two small folds that are responsible for the outcrop pattern of Ordovician and Silurian sandstone. This pattern creates a zigzag in the Armuchee Ridges (see the GA 136: Alabama State Line—Carters Dam road guide). Through this sparsely inhabited gap, along a simple wagon track, twenty thousand troops under General McPherson moved on May 9 without encountering a Confederate patrol.

McPherson's instructions were to destroy the railroad at Resaca, cutting off the supply line to the bulk of the Confederate force, which was encamped at Dalton. After one failed attempt, the Union troops attacked from a low area underlain by Cambrian-age Conasauga limestone. To the east, the Confederates defended a ridge underlain by sandstone and siltstone of the older Rome Formation. Thousands of lives were lost in the Battle of Resaca. After the Confederate army succumbed, they retreated along the railroad line. The thirty-mile stretch from Resaca lacked positions that could be defended. Even though the same cherty Knox Group limestone composed the terrain, it lacked the narrow, defensible ridges the troops had known to the west, such as Missionary Ridge and Tunnel Hill. Instead, wide belts of high ground characterized the retreat. This is due to the broad, gentle folds that are typical in this part of the Great Valley.

The retreating army considered defending a ridge south of Cassville underlain, like much of the high ground in Resaca, by sandstone and siltstone of the Rome Formation. However, Union artillery positions on Knox Group high ground to the north forced the Confederates to retreat into the more mountainous areas in the Blue Ridge–Piedmont south of Cartersville.

Between Calhoun and Adairsville, I-75 follows a valley along the crest of an anticline. The valley is underlain by Conasauga Group shale and limestone flanked on either side by hilly country underlain by the younger chert-bearing carbonate rock of the Knox Group. A large quarry in the Knox Group is visible at Adairsville, just southeast of exit 306 (GA 140). Two miles south of the exit, I-75 is on higher ground in the Knox Group, affording a fine view to the northwest.

The Tellus Science Museum is just off I-75 at exit 293 (US 411). It has exhibits of fossils and minerals of Georgia (see the appendix for more information). I-75 crosses the White thrust fault at this exit. The White thrust sheet to the southeast is a transition between the Valley and Ridge and the Blue Ridge–Piedmont provinces, both topographically and geologically. For more information, see the I-75: Cartersville (US 411)—Kennesaw road guide in the Blue Ridge–Piedmont chapter.

Cartersville Mining District

Red clay pits are common in the Cartersville area, especially near exit 288 (GA 113/East Main Street). The red clay is residuum of the Shady Dolomite, what was left after chemical weathering dissolved the carbonate minerals calcite and dolomite. The residuum contains economic minerals that groundwater had deposited in the Shady earlier, just after the rock was deposited or during thrust faulting. These minerals include limonite (brown iron ore), manganese oxides, barite, umber, and ochre. These are the basis of the Cartersville Mining District.

Limonite was the main ingredient for the Cartersville iron industry from 1836 until the twentieth century. Manganese, an essential ingredient in steel, was mined here (among other places) when the two world wars interrupted richer overseas supplies. Barite, or barium sulfate, occurs as nodules, veins, and geodes in the Shady Dolomite. Barite has many uses because it is one of the heaviest nonmetallic minerals. Its density and nontoxic nature make it an essential ingredient in the barium milkshake patients ingest in order to have their digestive tract X-rayed. Its primary use is to make drilling mud heavy enough to prevent oil well blowouts, but it is also used to add weight to tennis balls, bowling balls, and carpet. The barite deposits were mined here as recently as the turn of the twenty-first century.

Ochre is a form of iron oxide. It is a brown pigment used in construction materials, paint, paper, and cosmetics. As of 2010, the only year-round ochre mining and processing operation in the United States was the New Riverside Ochre Mine, east of US 41 and just north of the Etowah River. The ochre occurs within red clay residuum near the base of the Shady Dolomite.

You can see former mining pits, as well as some of the geology related to the mining district, along East Main Street, just east of exit 288 and at the base of Bee Mountain. The road cut on the north side of the road is in Chilhowee Group phyllite and sandstone. These Cambrian-age rocks belong to the White thrust sheet. Unlike most rocks of the Valley and Ridge, they have been affected

Ochre being mined at the New Riverside Ochre Mine on the north side of the Etowah River.

by metamorphism. The phyllite started as shale. The metamorphism has contributed to the difficulty in differentiating them from rocks of the Blue Ridge, which lie just past the crest of Bee Mountain to the east.

In the soil just east of the outcrop, you can see the abrupt transition to the brick red color that marks the base of the Shady Dolomite. The flat ground north and south of the road, now covered with grass, was excavated to mine ochre and barite from the dolomite's residuum.

About 0.2 mile east of the exit there is a 15-foot-high boulder of Shady Dolomite next to another former pit that is now being reforested. The boulder contains fragments and veins of barite and limonite. To get there, follow East Main Street to the dead end of the divided road and turn right. Continue to a gravel parking lot on your left, which is at a trailhead. The boulder sits on the east side of the piney area a few steps past the north end of the lot.

The dolostone boulder (foreground) with white barite patches, east of exit 288. The red soil in background is residuum of the Shady Dolomite.

Ladds Quarry and Etowah Mounds

Ladds Quarry, about 2.5 miles southwest of downtown Cartersville on the side of Quarry Mountain, is highly visible on the north side of GA 61. The quarry, active until the 1950s, is in cherty dolostone of the Knox Group. The dolostone was processed into magnesium-rich lime used to make baking powder and, reportedly, the gas that gave Coca-Cola its fizz. The quarry cut into several caves that have yielded a wide range of Pleistocene-age vertebrate fossils, including those of at least forty-eight species of mammals, five species of birds, and twenty-three species of reptiles and amphibians. Twenty-five species of mollusks have been found as well. Excluding the mollusks, the animals either lived in the caves or became trapped in sinkholes that opened into the caves. The site is also popular with mineral collectors because cave formations (stalactites, stalagmites, etc.) unearthed by quarrying may be collected legally, unlike those found in caves. Arrangements must be made in advance with Bartow County to enter the quarry area.

Etowah Indian Mounds Historic Site, about 3.4 miles south of Cartersville off GA 61, is the most intact archaeological site from the Mississippian Indian culture (900 to 1650 AD) in the southeastern United States. The site was home to thousands of people at its peak in the fifteenth century and was one of the places the Hernando de Soto expedition visited in 1540.

Geology has favored this site for settlement. It is close to the point where the Blue Ridge thrust front abruptly changes from a north-south to an east-west orientation, making it the Valley and Ridge location closest to the Atlantic coast. This made it a natural location for peoples in the Valley and Ridge to trade

Ladds Quarry at Quarry Mountain near Cartersville.

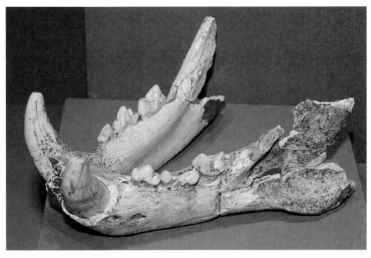

A jaguar jaw found at Ladds Quarry, on display at the Tellus Science Museum.

with those traveling from the Coastal Plain. It also lies where the Etowah River crosses a 1-mile-wide belt of limestone and shale of the Conasauga Group. The terrain that develops in this group is flatter, and the soil less rocky, than that which develops in either the cherty Knox Group carbonates on the west or the siltstone and sandstone-bearing Rome Formation on the east. This made for good farmland that could support a large population center.

GA 136
Alabama State Line—Carters Dam
76 miles

See the maps on pages 134 and 138.

This route provides a fine cross section of northwest Georgia's geology, with spectacular views as well as outcrops revealing sedimentary rocks from nearly every time period of the Paleozoic era.

At the Alabama border, GA 136 begins on the tabletop of Sand Mountain, around 1,400 feet above sea level. About 2.2 miles into Georgia, from the eastern edge of Sand Mountain, you can see across Lookout Valley (about 700 feet below) to Lookout Mountain. GA 136 skirts an outcrop of the Pennsylvanian-age sandstone that caps Sand Mountain (on the north side of the road) and then descends into the valley. Beneath the sandstone outcrops, the poorly exposed bedrock is Mississippian- to Pennsylvanian-age shale, but the more visible rocks are great blocks of sandstone that have fallen from the cliffs above and then ridden the loose shale downslope, creeping along at an imperceptibly slow pace.

Mississippian limestone is well exposed on the north side of the road about 1.5 miles farther east, before the second of two switchbacks. If you stop and examine the white patches on the rock, they may remind you of travertine deposits that form inside caves (see Cave Spring in the US 27: Tennessee State Line—Cedartown road guide), and for good reason: before erosion and blasting for the road, the patches probably were inside a cave. You may also see fossils of horn corals and crinoids embedded in the rock. Thin black ridges of chert (also called "flint") protrude from the rock.

On the east side of Lookout Valley, GA 136 ascends Lookout Mountain, which offers views of the valley and outcrops of Mississippian limestone and Pennsylvanian sandstone.

Cloudland Canyon State Park

Cloudland Canyon State Park, one of the Seven Natural Wonders of Georgia, is 6.4 miles east of I-59. Turn left (north) onto Cloudland Canyon Park Road and continue 1.5 miles to the overlook parking area.

From the parking lot, walk to the railing along the cliff edge and follow the path to the right. This leads to the Point, the overlook near the intersection of the canyons cut by Daniel Creek, on your left, and Bear Creek, straight ahead. You can see Lookout Valley beyond the canyons, with its lush pastures underlain

The canyon cut by Daniel Creek. Sand Mountain is visible in the distance.

by Ordovician-age carbonate rocks. The ridge in the distance beyond the valley is Sand Mountain, topped by Pennsylvanian-age sandstone and conglomerate.

The layers of sandstone in the bluffs across the canyon are not fully continuous. That's because there are many independent bodies of sandstone, each the product of a particular event, such as the filling of an abandoned river channel. You can see such a sandstone body, with its convex bottom projecting below nearby sandstone layers, directly across the canyon from the Point.

At the Point you are standing on crossbedded sandstone. What resembles wood grain beneath your feet is actually the eroded top edges of numerous crossbeds. The edges of these former ripples are curved in a U shape because they formed in a stream and were sculpted as they migrated; the concave portion points downstream. On the way back to the parking lot, at the first steps you encounter, look to your left. Here a cross section of the crossbeds is exposed. The orientation of the crossbeds indicates the stream that deposited them flowed south, toward the parking lot.

There is a path to the bottom of the canyon where you can see two waterfalls on Daniel Creek. The complete hike will take about ninety minutes round-trip, but about 300 yards down the trail is an interesting river channel deposit. From the parking lot, head left along the canyon rim. Continue in the same direction (south) along the Waterfalls Trail and follow the switchback that heads

Crossbedding seen from above at the Point.

*Crossbedded strata (lower right) in Cloudland Canyon State Park are cut by an unlay-
ered sandstone. The crossbeds were deposited by a current that flowed to the right.*

north. Just below the switchback is a long outcrop about 15 feet high, which,
where you first see it, is made entirely of crossbedded sandstone. Continuing
northward down the path and tracing the crossbedded rock, you'll see that it
ends against a sharp boundary that slopes to the north. Above and north of the
boundary is unlayered sandstone. As the trail rounds a corner, the same massive sandstone crops out as a huge overhanging rock, underneath which twenty
people might easily take shelter.

The thick sandstone in the overhanging rock is a cross section of an ancient
river channel frozen in time. During Pennsylvanian time the river eroded tens
of feet down into the older sand deposits (crossbedded sandstone), cutting the
large, convex-downward shape of the channel. The absence of obvious internal
layering within the channel sandstone indicates that at the end of its life, this
channel probably filled rapidly, unlike the layered, crossbedded sandstone surrounding it. Crossbeds form slowly, as the sand in a river or stream channel
migrates over time. Studies of modern rivers show that such rapid filling can
occur during a major flood. If a natural levee breaks upstream, the flow can
cause a river to abruptly change course and abandon a portion of its channel.
The pebbles protruding from the bottom of the rock overhang are further evidence that this was a river channel. Pebbles are concentrated on river bottoms,
where they tend to roll along in the current. Impressions of fossilized wood
(presumably waterlogged as it was carried downstream) are also present.

Both the 50-foot upper falls (Cherokee Falls) and the 90-foot lower falls (Hemlock Falls) on Daniel Creek developed where the stream crosses from erosion-resistant sandstone into easily eroded shale below. The shale is the dark rock with thin beds in the cliffs below the falls. It forms flat chips as it erodes. The shale cliffs behind the waterfall continually crumble away, removing the support for the tough sandstone ledge at the top. As a result, the position of each waterfall retreats upstream over time. The bed of Daniel Creek is filled with sandstone boulders—some nearly house sized—from ledges that have collapsed in the past.

Lookout Mountain to Carters Dam

The eastern edge of Lookout Mountain is the boundary between the Appalachian Plateau (to the west) and the Valley and Ridge (to the east) physiographic provinces. Outcrops along GA 136 reveal the contrast in the amount of tectonic deformation these two regions experienced. The strata in the Appalachian Plateau are only gently folded, whereas compressive forces affected the Valley and Ridge strata much more severely. The Valley and Ridge strata overlap on thrust faults rather like shingles on a roof, except that strata can be contorted with extreme dip angles, especially where thrust faults reach the surface.

About 3 miles east of the turnoff to Cloudland Canyon State Park, the strata in the outcrops on both sides of the road appear much the same as those at

Vertical beds of Mississippian-age limestone on the eastern slope of Lookout Mountain.

Cloudland Canyon: crossbeds developed within layers that are largely horizontal. Except for the Lookout Valley and Sequatchie Valley anticlines, strata from here all the way to the Rocky Mountains remain horizontal, unaffected by the tectonics that crunched Georgia and much of the East Coast.

In contrast, only 0.3 mile farther east, on the north side of the road just east of the intersection with GA 157, sandstone beds dip westward at more than 45 degrees. This contrast is common as you drive through the Valley and Ridge: you will pass both horizontal beds and those tilted by tectonic forces. For example, 2 miles east of GA 157, the Mississippian-age limestone beds exposed on the left are vertical for about 0.5 mile. The vertical beds lie between two steep, closely spaced faults (not exposed along the road) that parallel the length of Lookout Mountain.

Another 0.25 mile down the road, on the left, there is a small abandoned quarry in Mississippian limestone, in which beds dip gently to the northwest. The Tuscumbia Limestone in the lower part of the quarry has black chert. It is overlain by the crossbedded Monteagle Limestone. The Monteagle Limestone is composed of sand-sized particles called ooids, each of which is composed of concentric layers of calcium carbonate that precipitated around a tiny fragment of shell or other debris.

Near the base of Lookout Mountain the road bends eastward to cross a flat interruption in the slope, which has developed due to the relatively erosion-resistant beds visible in the dirt bank on the north side of the road. These beds are Fort Payne Chert, and beneath it are thin sandstone, siltstone, and shale of the Red Mountain Formation.

At the base of Lookout Mountain, GA 136 crosses McLemore Cove. Like Lookout Valley, this flat stretch formed in the eroded hinge of an anticline. The oldest rocks of the valley are in its hilly and wooded center—cherty carbonate rock of the Knox Group. Younger Chickamauga Group limestone lies both west and east of the Knox Group under flat, fertile farmland. Small outcrops of this gray limestone line both sides of GA 136 about 0.5 mile east of its intersection with GA 341. You can see green chert in the limestone on the north side of GA 136. The chert formed under a layer of altered volcanic ash, or bentonite. Groundwater leached silica from the ash and precipitated it below, replacing the limestone with chert.

Between GA 341 and LaFayette, the highway crosses the north tip of Pigeon Mountain, which is capped by Pennsylvanian-age sandstone. Like Lookout Mountain, Pigeon Mountain is the result of a gentle syncline, in which younger rock has been preserved in its core—the opposite of the anticline valleys. From the air (or the road guide map), Pigeon Mountain's northern tip resembles the bottom of a V. This pattern developed because the hinge line of the syncline plunges southward. The gently dipping sandstone outcrops along the road are of the Red Mountain Formation.

The former mining town of Estelle, site of a blast furnace until about 1910 and Georgia's largest producer of iron from Silurian-age rocks, is about 1 mile southwest of GA 136, along GA 193. Due to the gentle dip of the Silurian-age rocks wrapping around the nose of Pigeon Mountain, a large area of iron-bearing strata outcrops at the surface, and much of it has weathered. Soft, weathered ores had several advantages over hard, unweathered ores: they contained more iron, were easier to dig, and could be stripped from the surface without the expense of underground mining. The deposits here were up to 3 feet thick and contained up to 54 percent metallic iron, a fact that made even underground mining economically feasible once the weathered ore had been exhausted.

Southwest of LaFayette, accessible via GA 193 and Chamberlain Road, is the Crockford–Pigeon Mountain Wildlife Management Area atop Pigeon Mountain. The management area includes Rocktown, a rock-climbing mecca in Pennsylvanian-age sandstone that contains rock pedestals, turtle backs, and massive boulders the size of a three-story building.

Several caves are present in the area. Ellisons Cave on Pigeon Mountain is the deepest cave east of the Mississippi River, with a fantastic 586-foot vertical drop, the deepest vertical drop in the continental United States. Picture a pit beneath the mountain about the height of a 58-story building. Caves develop

A rock pedestal at Rocktown on Pigeon Mountain.
—Courtesy of Bob Tolford

Turtlebacks at Rocktown on Pigeon Mountain. —Courtesy of Bob Tolford

where limestone is dissolved by weakly acidic groundwater flowing through cracks in the rock. Ellisons Cave formed where Mississippian-age limestone dissolved along a fault.

North of LaFayette, GA 136 joins US 27 and follows the Kingston thrust fault briefly before resuming its eastward course and entering the Armuchee Ridges, first climbing Taylor Ridge at Maddox Gap. On the west slope of Taylor Ridge, there are several road cuts in Red Mountain Formation sandstone, which dip to the east. West Armuchee Valley, on the east side of Taylor Ridge, lies along the trough of a syncline floored by Mississippian-age limestone and shale. Dick Ridge, on the east side of the valley, is like a mirror image of Taylor Ridge, with west-dipping Red Mountain Formation sandstone road cuts outlining the syncline. Each time GA 136 crosses the Red Mountain Formation traveling eastward, the outcrops become more prominent because the rock is more resistant to erosion due to an increase in coarser, quartz-rich sediment (sand and gravel). This change in sediment type is evidence that eroding highlands existed to the east during Silurian time.

About 10.9 miles east of LaFayette, GA 136 crosses the Clinchport thrust fault. To the southeast, you can see the northern end of Johns Mountain, part of a 35-mile-long zigzag ridge capped by Ordovician to Silurian sandstone. The ridge extends from the southern end of Johns Mountain, north of the town of

Armuchee, to the northern end of Rocky Face Mountain, north of Dalton. The zigzag ridge pattern is the product of plunging anticlines and synclines. Because the hinges of both structural features plunge south in the area, valleys that are closed at the south end formed along anticlines, with older, mainly Ordovician-age limestone in the center. Valleys that are closed at the north end formed along synclines, which are underlain by younger Mississippian-age shale.

Just east of Johns Mountain (0.4 mile east of GA 201), Pocket Road provides a 7-mile side trip to the Pocket. This valley, closed to the south, developed in the hinge of an anticline tucked away in the zigzag ridge. Five miles south of the turnoff, FS 702 heads west to the parking area for the 1.8-mile loop trail to Keown Falls, two waterfalls that tumble over a 60-foot-tall cliff of Silurian-age sandstone. The falls sometimes dry up in the summertime. About 2.5 miles south of the Keown Falls turnoff is the turnoff (to the left) into the National Forest campground and picnic area at the Pocket, where Ordovician-age limestone underlies a fertile area surrounded on three sides by Horn Mountain.

Satellite image of the Armuchee Ridges. —Courtesy of the U.S. Geological Survey

A large, clear spring wells up in the middle of the picnic area. Because limestone dissolves to form caves, it is frequently associated with large springs. Lime sinks—also associated with limestone—are visible along a 2.5-mile loop trail that explores the local woods.

About 1 mile east of Pocket Road, GA 136 passes amongst small knobby hills within the larger valley. These hills are underlain by the Cambrian-Ordovician-age Knox Group, the oldest rock exposed in the anticline. Along the roadside are fist-sized chunks of white chert, the erosion-resistant mineral responsible for the hilly topography that develops in the Knox Group.

Roughly 2.5 miles east of Pocket Road, GA 136 turns south between Horn Mountain on the west and Mill Creek Mountain on the east, where it crosses the zigzag ridge of sandstone at Snake Creek Gap. From here it parallels Snake Creek for about 5 miles. On May 9, 1864, twenty thousand Union troops moved through this gap on their way to the Battle of Resaca (see the sidebar Geology and the Civil War in Georgia in the preceding road guide).

Leaving Snake Creek, GA 136 turns to the east and emerges into the Great Valley. It crosses the Rome thrust fault less than 1 mile southeast of the intersection with GA 136C. This fault has a more sinuous map pattern than most Valley and Ridge thrust faults, in part because it has been folded along with the rocks above and below it. On the north side of GA 136 at Hyde Road, 5.1 miles east of GA 136C and just west of Resaca, an anticline folded the Rome thrust fault, bringing it and the Clinchport thrust sheet close to the surface. Erosion has cut

Cambrian-age limestone with calcite-filled fractures.

through the Conasauga Group of the overlying Rome thrust sheet and created a "window" through which we can see the underlying Clinchport thrust sheet, exposed as pebbles of black chert in the red soil. Black chert is typical of Devonian- to Mississippian-age rocks not far to the west, where a larger portion of the thrust sheet is exposed at the surface, but not here in the Great Valley.

Much of the Great Valley is underlain by the Conasauga Group. You can see outcrops of this limestone and shale between Resaca and US 411, near the eastern edge of the Great Valley. About 16 miles east of I-75, and 200 yards west of US 411, you can see limestone with calcite-filled fractures on the south side of GA 136. These fractures developed due to tectonic strain and were later filled with the mineral calcite as groundwater passed through them. In general, there is more limestone and less shale to the east, suggesting that the open ocean lay to the east and eroding land to the west during Cambrian time. The highlands that appeared to the east in Ordovician time did not yet exist.

GA 136 leaves the Valley and Ridge and enters the Blue Ridge–Piedmont as it crosses an unexposed thrust fault, part of the Blue Ridge thrust front, 2.2 miles east of US 411. Proterozoic-age rocks were carried more than 100 miles westward along the thrust fault and pushed over younger Paleozoic-age rocks.

Numerous road cuts exposing Proterozoic-age rocks of the Blue Ridge–Piedmont are visible along Carters Dam Road, the turnoff to which is 2.6 miles east of US 411. The road also leads to the dam's visitor center, where on a clear day you can see across the entire Valley and Ridge Province to Lookout Mountain, which is back in the Appalachian Plateau Province.

GA 157
Tennessee State Line—Cloudland
39 miles

See the map on page 134.

GA 157 runs the length of the Georgia portion of the Lookout Mountain plateau, which is capped by erosion-resistant sandstone and conglomerate of Pennsylvanian age. Although the rock layers atop the mountain generally lie flat, they dip inward, toward the mountain along either long edge because Lookout Mountain is a syncline.

Just south of Tennessee, GA 157 provides a scenic view of Missionary Ridge and Chattanooga to the east. The road hugs cliffs of crossbedded sandstone that lie below the Fairyland Country Club, to the west. You can see the overlook at Rock City Gardens up ahead, with its artificially recirculating waterfall.

Rock City Gardens

Rock City Gardens, one block south of GA 157 and less than 1 mile from the Tennessee line, is a commercial attraction that for decades has advertised on barns and billboards across the South.

The Needle's Eye—a path through a north-trending joint at Rock City.

The forces of tectonics and weathering have broken and separated the Pennsylvanian-age sandstone and conglomerate of this locale into isolated blocks the size of buildings. Paths wind between them like streets, hence the term Rock City. Narrow places between the blocks have names such as Needle's Eye and Fat Man's Squeeze. Near the cliffs at the east side of the attraction, isolated boulders have been separated from the others by sliding downslope along a layer of shale, which underlies them.

The vertical joints, or cracks, are oriented in two main directions between the blocks: northeast-southwest, paralleling the trend of Lookout Mountain, and northwest-southeast, at right angles to it. The same two joint orientations are present all along the mountain, and even below it. Cave passages in Mississippian-age limestone, such as at Ruby Falls, 4 miles to the north in Tennessee, have similar orientations because they formed as groundwater, following joints, dissolved the limestone. Groundwater left its mark at Rock City

Liesegang banding on a joint surface at Rock City.

Gardens, as well. As you regain the light north of Fat Man's Squeeze, there is an excellent example of iron oxide staining—a widely observed but poorly understood effect of chemical precipitation called Liesegang banding—left behind by moving groundwater.

The beds of conglomerate are easy to tell apart from the sandstone because they contain white quartz pebbles that stand out in relief. Much of the sandstone and conglomerate is crossbedded. The long, low-angle crossbeds possibly formed on the gentler slopes of a beach or sandbar, and the steeper crossbeds possibly developed in tidal or river channels with stronger currents.

From the overlook known as Lovers Leap, it is said that on an exceptionally clear day you can see seven states: Alabama, Georgia, South Carolina, North Carolina, Tennessee, Virginia, and Kentucky. With even average visibility you can see east across the Valley and Ridge to numerous peaks in the Blue Ridge–Piedmont. Directly east you can see Hawkins Ridge and Missionary Ridge, flanking Happy Valley, a geologic site described in the US 27: Tennessee State Line—Cedartown road guide.

Along Lookout Mountain

From Rock City Gardens to GA 189, the highway climbs to a high ridge crest underlain by the youngest sandstone layers on Lookout Mountain. There are spectacular views in both directions from the campus of Covenant College, which sits on the narrow ridge crest. GA 157 and GA 189 fork about 3 miles south of Covenant College. Continue on GA 157, the east fork. About 5.6 miles past the fork, Old Durham Road leads to the former coal mining community of Durham. Coal mining peaked here in the early twentieth century and ended in the 1950s. Strip mines were dug along the contour of the land to reach the coal, and the overlying shale was removed and piled into spoil heaps, which can still be seen in the area. Pennsylvanian-age plant fossils have been collected from the shale.

In Pennsylvanian time, this area was a swamp. Coal begins as peat, or compressed plant material, preserved from decay in swamp water. Over time the peat was buried and compacted to form coal. The coal's relationships to adjacent rock units tell us that the swamp lay on a delta plain, near sea level, along the western side of a now-eroded mountain chain that developed as Africa collided with North America. The Durham-area swamp was just one of many that existed at the time. The plants in these swamps became the coal that powered the Industrial Revolution (and still supplies much of our electricity today), mined from well-known coal-producing areas, such as West Virginia, Pennsylvania, Nova Scotia, Wales, the Rhineland of Germany, and northern England, including Durham's namesake.

About 5.7 miles farther south, GA 157 intersects GA 136, which leads west to Cloudland Canyon State Park (see the GA 136: Alabama State Line—Carters Dam road guide). About 8.7 miles south of GA 136, you can see a small "mushroom rock" that developed in sandstone on the west side of the road. About 10 miles south of GA 136, GA 157 approaches the east rim of Lookout Mountain, still capped by Pennsylvanian-age sandstone. If you look carefully through the trees on the east side of the road, you may see a pair of irregular, 20-foot-tall rock columns (location: 34°38′51″ N, 85°28′13.41″ W). Called mushroom rocks, pedestal rocks, or hoodoos, they are composed of vertical stacks of sandstone and conglomerate topped by well-cemented conglomerate caprock that gives them their mushroom-like shape. The hard, reddish ridges of rock, about 0.5 inch thick, are lenses of iron oxide cement left behind by iron-rich water passing through the rock. Mushroom rocks are more commonly found in arid areas, where they are sculpted by wind erosion. The mushroom rocks here were sculpted by wind along with freezing and thawing during Pleistocene time, when the climate was colder and drier and there were few or no trees here on the mountain.

Farther from the road, along the eastern rim of the mountain, there are turtlebacks: rounded boulders with a polygonal weathering pattern. On the west side of the road, near the somewhat obscured mushroom rocks, is a sign for Zahnd Natural Area. This locale features much smaller versions of some of the features seen at Rock City. A trail leads west to this "rock town" of huge boulders. The mushroom rocks and turtlebacks are both part of the natural area.

About 0.5 mile farther south, on the east side of the road, is a scenic overlook. Here you can look across McLemore Cove to Pigeon Mountain, a thumb-like projection of the Lookout Mountain plateau. The forested hilly area in the valley below, which developed on Fort Payne Chert and Red Mountain Formation, lies along the hinge line of the McLemore Cove anticline. Pastures that wrap around the forested area to the south are underlain by younger limestone of Mississippian age. McLemore Cove ends to the south, where the anticline

Close-up of one of the mushroom rocks, showing quartz pebbles in the conglomerate and the protruding, iron-cemented layers that resist weathering.

One of the authors with a mushroom rock in the Zahnd Natural Area.
—Courtesy of Miranda Gore

plunges beneath the Pennsylvanian-age sandstone that caps both Lookout and Pigeon mountains. GA 157 continues 13 miles farther south to the town of Cloudland, where it intersects GA 48.

The scenic overlook on Lookout Mountain, facing east across McLemore Cove toward Pigeon Mountain.

US 27
Tennessee State Line—Cedartown
91 miles

Rossville lies in Happy Valley on the west flank of Missionary Ridge, the site of a Civil War battle in which besieged Union troops broke out of Chattanooga in November 1863. Happy Valley is closed at its southern end, like several valleys of northwest Georgia (Lookout Valley, McLemore Cove, and the Pocket). To see the south end, where Happy Valley nestles between converging ridges, head about 2 miles west on GA 2 from its intersection with US 27, 3.4 miles south of the state line.

Happy Valley, like the other valleys, formed as the relatively soft rocks in the core of an anticline eroded. The valley is closed at the south end because the anticline tilts (plunges) southward beneath older Knox Group rocks. As with the other anticline valleys, the oldest rocks (Ordovician-age limestone) are in its center, with slightly younger (Ordovician- to Silurian-age) rocks along its flanks. However, what makes this valley different is that thrust faulting placed older rocks (Knox Group) at the top of the ridges, over the younger Silurian rocks. In the other anticline valleys, the rocks of the ridges that rim them are younger than in the valleys.

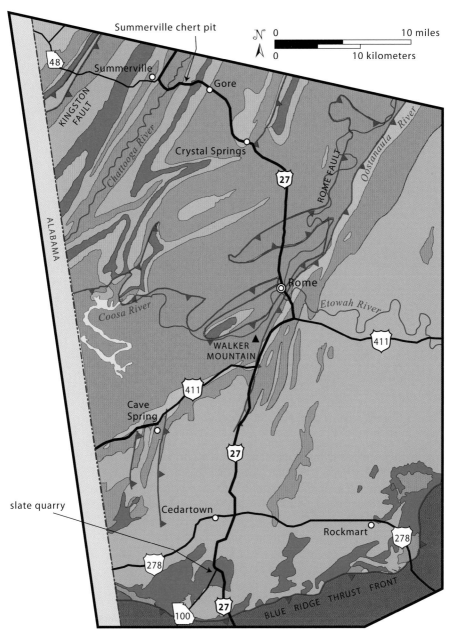

Geology along the southern portion of US 27. See the map on
page 134 for the geology along the northern part of this route.

The following labels appear on the map:

Summerville chert pit

N   0   10 miles
     0   10 kilometers

48   Summerville
                         Gore
KINGSTON FAULT
Chattooga River
Crystal Springs
27
ROME FAULT
Oostanaula River
ALABAMA
Coosa River   Rome
                         Etowah River
WALKER MOUNTAIN
411   411
Cave Spring
slate quarry
Cedartown
                         Rockmart   278
278
100   27   BLUE RIDGE THRUST FRONT

Legend:

Pennsylvanian-age
sandstone and shale

Mississippian-age
limestone, shale, and chert

Silurian- to Devonian-age
sandstone, shale, and chert

Ordovician-age
limestone, shale, and sandstone

Cambrian- to Ordovician-age dolostone,
chert, and limestone (Knox Group)

thrust fault; barbs occur on
the overriding fault block

Cambrian-age limestone, shale,
and sandstone (Conasauga Group)

Cambrian-age sandstone and shale
(Rome Formation)

Precambrian- and Paleozoic-age
metamorphic and igneous rocks

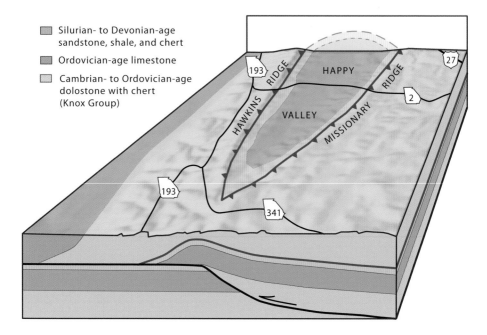

Silurian- to Devonian-age sandstone, shale, and chert

Ordovician-age limestone

Cambrian- to Ordovician-age dolostone with chert (Knox Group)

The plunging anticline at Happy Valley. The Missionary Ridge fault is shown in red.

Although other interpretations have been proposed, the simplest interpretation for the structure of this valley is that the plunging anticline formed after a thrust fault (the Missionary Ridge fault) had placed the Knox Group rocks on top of the Silurian-age strata. Erosion then cut through the folded layers along the hinge of the anticline and through the thrust fault, creating a valley that is open at the north end in Tennessee but closed at the south end.

Chickamauga Battlefield

The oldest and largest military park in the National Park system straddles the Tennessee-Georgia state line, memorializing the Civil War campaign for control of nearby Chattanooga in the fall of 1863. The 5,300-acre Chickamauga Battlefield was the site of the last major Confederate victory of the Civil War. To reach the battlefield, continue straight on LaFayette Highway (old US 27) where US 27 turns west on Battlefield Parkway (GA 2), about 2.9 miles south of the Tennessee state line.

Carbonate rocks underlie the entire battlefield. The hills in the northwest corner of the park are composed of chert-bearing dolostone of the Knox Group. Chickamauga Group limestone forms the gently rolling terrain of the rest of the park. In the eastern half of the park, there are small outcrops of limestone surrounded by thin, gravelly soil. These are called cedar glades because the eastern red cedar is one of the few trees that tolerate the thin, alkaline soil. The largest glade can be accessed from a 500-foot-long path on the north side

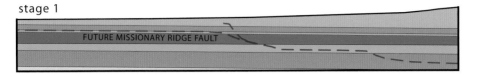

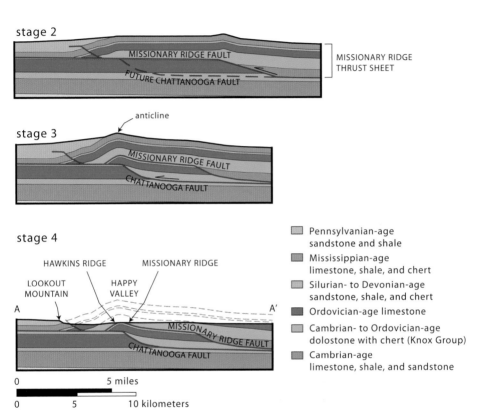

Stages by which Happy Valley may have developed. (1) Sedimentary layers are deposited. (2) Knox Group rocks are forced over Silurian-age and older rocks along the Missionary Ridge thrust fault. (3) Continued tectonic deformation forms the Chattanooga fault. Movement along the fault forms an anticline in the strata of the Missionary Ridge thrust sheet. (4) Erosion removes much of the Missionary Ridge thrust sheet, forming the V-shaped indentation known as Happy Valley and exposing the thrust fault.

of Viniard-Alexander Road, near the park's south border and 0.6 mile east of LaFayette Road. Six football-sized blocks of limestone mark the trailhead.

In a few locations, park streams drain into lime sinks. Along Lytle Road, about 0.5 mile north of Wilder Brigade Monument (built of Chickamauga limestone), there is a historic lime sink north of the railroad tracks and immediately north of the park tablet for Hyndman's division. In 1863 this mowed depression was a cattle pond and a site of fighting. Because of severe drought, it was the only water source on the battlefield, but it quickly became a polluted cesspool known as Bloody Pond.

Union troops controlled a far better water source in the town of Chickamauga, south of the park. Crawfish Spring, like other large springs of the Valley and Ridge, issues from Knox Group carbonate rocks. The weathered Knox Group hills west of town are a mixture of clay and chert, which holds rainwater like a sponge and slowly releases it to fill spaces in the bedrock below. The large and reliable flow of the spring even during severe drought was the reason Union troops were headquartered in the Gordon-Lee Mansion, which still stands across from the spring along GA 341 in Chickamauga.

Chickamauga to Gore

Lafayette Road rejoins US 27 south of the Chickamauga Battlefield. About 4 miles south of Chickamauga, US 27 turns southwest just east of the Kingston fault and follows a valley underlain by Cambrian-age shale. From LaFayette to Summerville, US 27 crosses low hills composed of the Knox Group that offer views of Taylor Ridge to the east, which is topped by Red Mountain Formation sandstone.

About 2 miles east of Summerville, on the south side of US 27, there is a large pit in white chert residuum of the Knox Group. Some of the chert is a banded variety known as agate. Circulating groundwater deposited a layer of tiny quartz crystals along fractures in some of the chert. Called drusy quartz, each mineral crystal is less than 1 millimeter in diameter. Agate and drusy quartz are popular with mineral collectors.

Southeast of the chert pit, US 27 crosses a valley underlain by Ordovician limestone and climbs over Taylor Ridge, where you can see beds of the Red Mountain Formation sandstone. On the west edge of the town of Gore, 4.9 miles east of Summerville, a long road cut on the north side of the road exposes about 15 feet of black Chattanooga Shale overlain by Fort Payne Chert.

Chattanooga Shale

The Devonian-age Chattanooga Shale is an oil shale. A freshly broken piece smells like oil, and some oil shale will burn. Oil shale is considered an "unconventional" energy resource. When oil prices are high enough, it is economically feasible to pulverize the shale and extract the oil. Any shale can contain some natural gas as well, and in the twenty-first century improved drilling technology and increased demand has led to widespread fracking, or fracturing, of underground shale formations to extract gas. Some of the richest deposits are in Devonian-age black shale, where it is thicker than in Georgia, notably from the Marcellus Shale that extends from West Virginia to New York. There is some interest in producing gas from the Chattanooga Shale and other shales in Georgia.

Oil shale's biggest significance is that, over geologic time, it has produced secondary deposits. When oil shale is deeply buried, oil and natural gas are squeezed out of it. These hydrocarbons can seep out at the surface or accumulate in porous rock.

Oil shale begins as a mixture of clay and the remains of marine microbes, such as algae. Although algae are common in the ocean, after they die and settle to the seafloor they are normally quickly oxidized into carbon dioxide or eaten

Fort Payne Chert

Maury Shale

Chattanooga Shale

The Chattanooga Shale outcrop west of Gore.

by organisms, such as burrowing worms. Large quantities of organic remains only accumulate within clay when bottom waters lack oxygen, an environment that is also hostile to worms. The odd chemistry of an oxygen-deprived environment also causes some of the uranium naturally present in seawater to become adsorbed onto clay. Like many black shale formations, the Chattanooga Shale is slightly radioactive and considered a uranium resource. However, commercial extraction of oil, natural gas, or uranium from Georgia's Chattanooga Shale seems unlikely because it is so thin in most places.

Although the Chattanooga Shale is thin, it extends over a large area, and units of equivalent age are recognized as far away as Minnesota. But the Chattanooga Shale is just one of many extensive Middle to Late Devonian–age black shale units in North America and throughout the world. This has led geologists to wonder why oxygen-poor conditions were so widespread at this time. One possibility is eutrophication. When water has excess nutrients, algae populations reproduce rapidly. When they die and begin decomposing, chemical reactions pull dissolved oxygen out of the body of water. This same process occurs today: fertilizer-rich water leads to algal blooms that end with scum-covered ponds and dead zones in the Gulf of Mexico. In Middle to Late Devonian time, land plants with deep roots were taking hold for the first time, extracting nutrients from weathered rock. Dead plant matter washed into waterways, and excess nutrients ultimately accumulated in the sea. These nutrients may have triggered eutrophication episodes, or dead zones, in the world's oceans.

The oxygen-poor seawater that preserved so much oil shale triggered extinctions of marine organisms, such as several groups of trilobites and brachiopods. As a result, one of the five great mass extinction episodes in Earth history occurred in Late Devonian time. The few fossils known from the Chattanooga Shale are floating organisms that drifted in from more hospitable waters, or organisms that survive under low-oxygen conditions.

A bed of green shale overlies the Chattanooga Shale in the Gore road cut. The shale contains both the green mineral glauconite and phosphate nodules, which are made of the mineral apatite. Both minerals are indicators that the shale was deposited slowly. The green shale is the Mississippian-age Maury Shale Member of the Fort Payne Chert, named after Maury County in Tennessee, where it is thick enough that the phosphate is mined for fertilizer. The green shale represents a transition to more-oxygenated conditions. Clearly the overlying Fort Payne Chert was deposited in well-oxygenated water, because in many locations it contains abundant fossils of crinoids, brachiopods, and bryozoans. The fossils were originally calcite, but the calcite was long ago replaced by chert through the movement of groundwater. Fort Payne fossils made of chert are easiest to find in residuum. The groundwater movement that replaced the calcite also left some cavities in the rock. Some are fossil molds in the shape of crinoids or other organisms that dissolved away. Some are also geodes lined with small crystals.

Gore to Cedartown

The broad valley at Gore is underlain by Mississippian-age shale and limestone. To the east, US 27 passes through the Armuchee Ridges, which have a sinuous map pattern caused by plunging anticlines and synclines. Sedimentary rocks of six different ages form ridges here: cherty carbonate rocks of the Ordovician-age Knox Group, Ordovician Sequatchie Formation sandstone, Silurian Red Mountain Formation sandstone, Devonian Armuchee Chert, Mississippian Fort Payne Chert, and Pennsylvanian sandstone. US 27 follows the valleys of Mississippian-age shale and limestone, avoiding the ridges. As a result, the only road cut is in Fort Payne Chert at Crystal Springs.

North of Rome, US 27 crosses the unexposed trace of the Rome fault, which thrust older Cambrian-age shale on top of younger Mississippian-age shale and limestone. Like its namesake in Italy, Rome has seven hills. Some hills, for example, Walker Mountain south of town and west of US 27, developed on the Rome Formation. This is the oldest rock exposed west of the Great Valley, with its distinctive maroon and yellow sandstone, siltstone, and shale. A large exposure is along a rails to trails walking and biking path in downtown Rome, along Silver Creek about 300 yards south of a trailhead at the southwest end of Fourth Street. The beds are mostly vertical, interrupted by a few folds and minor faults, perhaps because of movement that occurred along the nearby Rome fault.

You can see a good roadside exposure of the Rome Formation about 100 yards west of US 27 on Primrose Drive, 4 miles south of Rome and 1.8 miles south of the intersection of US 27 and US 411. Ancient ripple marks, mud cracks, raindrop impressions, and worm burrows indicate that the sediment was deposited on the shore between the high and low tide lines. US 27 separates

The Rome Formation exposed along Primrose Drive. The hand-held specimen has Cambrian-age mud cracks.

from US 411 about 7 miles south of Rome, first crossing hilly country underlain by the Knox Group, then passing around Cedartown, which is underlain by Ordovician-age limestone.

Cave Spring

The town of Cave Spring is named for a spring in a limestone cave. The spring produces 2 million gallons of water daily, flowing into a pond and a swimming pool. To get there, from US 27 in Six Mile head west on US 411 for about 9.5 miles. Located in Rolater Park, about a block south of the town square and US 411, Cave Spring Cave is Georgia's only commercial cave.

The cave is near the base of a hill composed of cherty residuum of Knox Group carbonate rocks, which you can see in the park. Springwater is piped from the back of the cave, allowing tourists to stay dry as they tour the cave, and providing an outlet from which people collect water for personal use.

Caves form when water picks up weak acids from plants and other sources and seeps into cracks in carbonate rock. Over time the rock dissolves. If the water table drops, a cave becomes dry. Water carrying dissolved minerals can evaporate, depositing carbonate minerals such as calcite in the form of stalactites and other features.

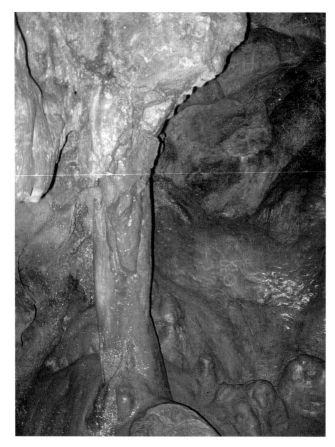

Cave Spring Cave, with stalactites on its ceiling.

Bedrock at the cave entrance has been partially blasted away to provide a level walkway. On the left, about 4 feet above the ground and beginning about 30 feet from the entrance, there are horizontal layers of travertine that were deposited on the face of the dipping strata. The travertine layers, composed of calcite, mark former water levels in the cave.

Access within the cave ends at a concrete wall about 200 feet from the entrance, where the pipes gathering the springwater can be seen through a window in a door. From here a walkway doubles back and climbs about half the distance toward the entrance. Long, cylindrical stalactites hang throughout this part of the cave. These formed as water dripped from the ceiling and precipitated calcite. Also worthy of note is the constant cave temperature of 58°F and the red mud. This residuum forms as the carbonate rocks dissolve.

Slate Quarry near Cedartown

About 4 miles south of Cedartown, a short side trip leads to an abandoned roadside slate quarry, which exposes interesting Ordovician-age rocks. Turn southwest onto GA 100 and continue 0.6 mile to the small quarry on the west side of the road.

The quarry is mostly dug into the Rockmart Slate, a carbonate-rich shale that changed into the metamorphic rock slate through tectonic compression. If you look closely at the slate in the quarry wall, you will see both the original beds, which are bands of differing color, and steeply dipping surfaces called slaty cleavage. The slaty cleavage formed at right angles to the direction of squeezing caused by tectonic compression—the same compression that caused the beds to tilt from their original horizontal position.

Within the slate, and best seen in boulders near the entrance to the quarry, is carbonate-rich sandstone and conglomerate. These are deepwater deposits; they are unlike rocks of the same age exposed to the west and north (along US 27, I-59 and I-75, for example) but similar to those found in several localities along the southeastern edge of the Valley and Ridge, from Alabama to New York.

Rockmart Slate at the GA 100 quarry. The original sedimentary bedding is parallel to the tan, sandy layers, which are up to 2 inches thick in this photo. The slaty cleavage is nearly vertical.

The conglomerate was probably deposited in a submarine landslide called a debris flow. It includes fist-sized pieces of limestone and limy sandstone surrounded by limy sand and silt. The limestone pieces are significant, because unlike the tougher quartz pebbles found in most conglomerate, they would not have survived more than a few miles of transport by rivers or wave action before disintegrating; their presence implies that the conglomerate was derived from eroding limestone cliffs not far from the deepwater where the conglomerate came to rest. This evidence, along with the minimal distances (less than 10 miles) between some eastern Valley and Ridge deepwater deposits and their shallow-water neighbors, has led some geologists to propose that faulting may have lifted some parts of the continental shelf up and pushed neighboring parts down during Ordovician time.

Above the quarry, Mississippian-age Fort Payne Chert rests directly on the Ordovician-age Rockmart Slate. There is no clear explanation as to why nearly 100 million years of sedimentary record is missing at this location.

A boulder of carbonate-rich conglomerate in the GA 100 quarry. A 1-foot-thick bed of sandstone can be seen in the boulder, jutting out on the right. Near the top of the hill in the background is the surface, or unconformity, on which Mississippian-age Fort Payne Chert rests directly on much older Ordovician-age Rockmart Slate.

BLUE RIDGE–PIEDMONT

In this book we discuss the Blue Ridge and Piedmont physiographic provinces as a single geologic region because the metamorphic and igneous rocks that underlie the mountains of the Blue Ridge are quite similar to those underlying the uplands of the Piedmont. The rocks of the Blue Ridge–Piedmont are distinct from those of the rest of Georgia in that they formed at great depths within the Earth. Some are metamorphic, having started out as sedimentary or igneous rocks (now referred to as metasedimentary and metavolcanic rocks), and others are plutonic igneous rock formed from cooled magma.

PREMETAMORPHIC CHARACTER OF THE BLUE RIDGE–PIEDMONT ROCKS

When geologists delve into the history of the Blue Ridge–Piedmont rocks, determining what the rocks were before they were metamorphosed, they find additional ways these rocks differ from those in the rest of Georgia. The rocks of the Valley and Ridge and Coastal Plain were deposited as sediments on a relatively shallow-water continental shelf, or in rivers and swamps near the coast. Rocks of volcanic origin are extremely rare in these provinces, consisting of a few beds of windblown ash from distant volcanoes.

In contrast, the metasedimentary and metavolcanic rocks of the Blue Ridge–Piedmont generally originated in deeper water, on the outer continental shelf and ocean floor or around volcanic island chains. How do geologists know that some of these metamorphic rocks are of volcanic origin? Though not commonly preserved, some rocks retain the textures of lava or ash eruptions. Geologists also use the chemical composition of a metamorphic rock to piece together its premetamorphic story. For example, the chemistry of metamorphic amphibolite resembles volcanic basalt, which is evidence that the amphibolite likely had a volcanic origin.

The metasedimentary rocks of the Blue Ridge–Piedmont have several characteristics that tell geologists they were deposited in deeper water. Mica, a mineral that metamorphism forms from clay, tends to be abundant in these rocks. Clay can be carried long distances from eroding land, whereas sand is more commonly found closer to it. Quartzite, a rock that metamorphism forms from quartz sandstone, such as that found in the Valley and Ridge, is relatively uncommon in the Blue Ridge–Piedmont. Instead of the "clean" quartz sand typical of shallow water, in which waves and other currents have washed the clay away, sand settling in deep water tends to have clay mixed in with it. This "dirty" deposit eventually becomes a type of sandstone called graywacke (pronounced "gray-wacky"). Its metamorphosed form, called metagraywacke, is common in the Blue Ridge–Piedmont.

Metamorphism often erases the characteristics of individual sedimentary layers. However, in the Blue Ridge–Piedmont, metagraywacke commonly lies within what's called a graded bed. In a graded bed, the coarsest material, be it gravel or sand, is at the bottom of a sequence that gradually becomes finer-grained upward, through sand and silt to clay at the top. Graded beds are typically deposited by turbidity currents, and thus are also called turbidites. Triggered by earthquakes or instability in sediment layers, these dense suspensions of mixed sediment travel downslope into deep water. Once a turbidity current reaches a level patch of seafloor, it gradually loses its energy and deposits its sediment in order, from heaviest to lightest. Turbidites are emblematic of deepwater deposits because significant underwater slopes are needed for turbidity currents to gain momentum.

Metamorphism

During metamorphism, minerals change into new minerals due to high temperatures and pressures. The Blue Ridge–Piedmont has a much greater diversity of minerals and rocks than the rest of Georgia because the region contained a wider variety of igneous and sedimentary rocks to start with, and these rocks endured such a range of temperatures and pressures during metamorphism.

For example, a colorful parade of metamorphic minerals documents rising temperatures in clay-rich rocks as they are metamorphosed: first green chlorite, black biotite, and silver muscovite develop, followed by red, globelike garnet; then staurolite (Georgia's state mineral) forms, followed by blade-shaped blue or brown kyanite; and lastly, at temperatures above about 1,100°F, white fibers of sillimanite develop.

As the rocks were metamorphosed, water-based fluids moving through cracks deposited minerals, creating features called veins. If the fluid contained

A garnet crystal on display at the Tellus Science Museum in Cartersville. Penny for scale.

silica, quartz crystallized in the crack; if the fluid contained carbonate, calcite or dolomite formed. Pieces of quartz are the most common stones found throughout the Blue Ridge–Piedmont due to their chemical resistance to weathering. Long after the metamorphic rocks that host them have weathered to soil, the quartz veins remain.

The growth of new minerals and the deformation resulting from tectonic movements creates textures in metamorphic rocks that obliterate the original sedimentary and volcanic textures. Many metamorphic rocks exhibit a texture called foliation. Foliation has two forms: the parallel orientation of long or flat metamorphic mineral grains that were aligned by tectonic stress as they grew, and the segregation of light and dark minerals into layers.

Foliated quartzite (dipping to the right) at East Palisades overlook near Atlanta. The minerals were aligned during metamorphism, creating the visible layering.

In rocks that experienced less-intense metamorphism, such as those found along the western edge of the Blue Ridge, foliation is obvious mainly in finer-grained rocks such as slate and phyllite, which began as clay-rich rocks. These rocks are prone to splitting along parallel surfaces that cut the original sedimentary bedding at an angle. Microscopic metamorphic minerals, segregated into layers, cause the splitting. This foliation is also called slaty cleavage.

In much of the Blue Ridge–Piedmont, you will encounter gneiss and schist, rocks that experienced intense metamorphism. The foliation of gneiss is evident in its alternating light and dark layers, which give it a zebra-stripe appearance. Schist easily splits apart along its foliation, leaving a surface that often glitters with flat, flaky minerals, such as biotite and muscovite.

Often you'll find an outcrop in which the foliation of gneiss or schist was folded (quartz veins can also be folded). Isoclinal folds, in which the opposite sides of a fold parallel each other, are dramatic evidence of the high temperatures and pressures these rocks experienced. They flowed like taffy. Some of the Blue Ridge–Piedmont rocks underwent multiple episodes of folding. Visualize a baker kneading dough when trying to imagine the tectonic processes that repeatedly folded the rock layers of the Blue Ridge–Piedmont.

Gneiss at North Highlands Dam near Columbus.

Isoclinal fold in Arabia Mountain Migmatite near Lithonia.

The taffylike flow of deeply buried metamorphic rocks is also demonstrated in what are called mylonite zones. Like faults, mylonite zones are planar features along which blocks of crust moved, except that instead of having a single break, the entire zone is packed with parallel surfaces along which a block of crust slipped. In general, the mineral grains in a mylonite zone are stretched or flattened and appear to be smeared out. The mineral crystals also tend to be smaller. In Georgia, mylonite rocks are most prominent along the Brevard fault zone and in the southern Piedmont along the Goat Rock, Modoc, and other fault zones.

Igneous Intrusions

Besides metamorphic rocks, the Blue Ridge–Piedmont also contains rocks that developed from melted rock, or magma. When buried in hotter depths, rocks can melt. Rocks containing water begin to melt at around 1,100°F, though dry rocks require temperatures as much as 750°F higher to melt. Quartz and feldspar are among the first minerals to melt. If quartz and feldspar in gneiss melt and then crystallize in the same place, the rock texture called migmatite is created, in which the foliation of gneiss fades out into granitelike patches of quartz and feldspar (the newly crystallized minerals) that lack foliation.

Magma doesn't always stay in place. It is less dense than solid rock, so it can rise through cracks into other rocks and cool and crystallize as what's called an igneous intrusion. Georgia's Blue Ridge–Piedmont contains intrusions of all scales, from pegmatite dikes a few inches across to bodies of granite—formed from massive amounts of magma—many miles in extent. The largest granite intrusions, such as the Elberton Granite, which is the center of Georgia's granite industry, lie in the eastern Piedmont within 70 miles of Augusta. Five distinct granite intrusions of smaller but significant size, including the Stone Mountain Granite, occur in the Atlanta area.

Some of the intrusions were later affected by metamorphism. For example, many large bodies of granitic gneiss, a rock that is dominated by the lighter-colored minerals quartz and feldspar but has relatively thin bands of dark minerals, probably began as granite. In the quarry industry, the term "granite" often includes granitic gneiss.

Mineral collectors are often interested in pegmatite dikes. This igneous rock has a texture similar to granite except it includes relatively large crystals, the growth of which was facilitated by the presence of water in the magma. Pegmatite dikes are mined for muscovite mica, and some contain unusual minerals such as beryl, tourmaline, rose quartz, and amethyst.

Weathering

Many of the rocks in the Piedmont have been deeply weathered, forming a soft "rotten" rock called saprolite. Chemical weathering occurs when rock is exposed to warm, humid conditions at Earth's surface, causing chemical reactions in the rock's minerals. Feldspar in granite, gneiss, and other rocks weathers to white kaolinite clay, which makes the rock soft. Although weathered, the

A backhoe can easily dig through saprolite. The foliation of the original metamorphic rock is visible, dipping toward the lower right.

The foliation in this saprolite, formed from the weathering of gneiss, is clearly visible: the white layers are kaolinite and quartz; the gray layers are dominated by weathered muscovite; and the reddish brown layers are iron oxide and clay, which probably weathered from biotite.

rock still retains some of its original texture, such as foliation. Iron-bearing minerals, such as biotite or amphibole, weather to clay and iron oxide, which stains saprolite a rusty red color. This is the origin of some of Georgia's famous red clay.

THE PATCHWORK OF TERRANES ASSEMBLED IN PALEOZOIC TIME

The most significant boundaries within Georgia's Blue Ridge–Piedmont are faults that separate large pieces of crust called terranes. Collisions between tectonic plates welded the terranes to the edge of North America between about 450 and 265 million years ago, during Paleozoic time. Some of the terranes began as volcanic island chains before they were accreted to North America, adding a great deal of landmass to the continent. The final collision joined Laurentia, the landmass that included most of North America, and Gondwana, a continent composed of Africa and South America as well as other landmasses (for example, Florida and south Georgia). This final joining created the supercontinent Pangaea, which later rifted apart, creating the openings for the Atlantic Ocean and Gulf of Mexico and leaving south Georgia and Florida behind.

Though the term *collision* implies compression, and thus thrust faulting, some Georgia terrane boundaries lie along strike-slip faults. These faults dip steeply into the subsurface. Unlike with thrust faults, the fault blocks on either side of a strike-slip fault slid horizontally past each other, like the halves of a sliding glass door. The terrane southeast of each prominent strike-slip fault in Georgia seems to have slipped southwest, perhaps pushed aside during collisions that began in Virginia and the Carolinas.

Recognizing terranes and their boundaries is not easy because these tectonic collisions happened over a very long period of time, during which many types of rocks were forming. Sedimentary rocks built of Laurentia sediments were being deposited in numerous environments, and the same was occurring with Gondwana. Volcanic islands were forming and eroding, forming their own volcanic and sedimentary rocks. And the rocks were then metamorphosed.

Along the Appalachian chain, geologists recognize the crust of Laurentia where they find radiometric ages of approximately 1.1 billion years. This is the age of igneous and metamorphic basement rock on which the sedimentary rocks of Laurentia continued to be deposited during much of Paleozoic time. Basement rocks of this age have been recovered from deep wells in many parts of the Appalachian Plateau. Exposures of this basement rock are also strung out where faulting has brought them to the surface along the Blue Ridge physiographic province, from Pennsylvania to Georgia.

Georgia hosts not only the southernmost outcrop of basement rock in the Blue Ridge, just east of Cartersville, but also a huge area of basement that lies near the southeastern border of the Piedmont. The Piedmont outcrop is part of the Pine Mountain terrane, which stretches from near Macon into Alabama west of Columbus, Georgia.

An ocean deposited some of the metasedimentary rocks of the Blue Ridge–Piedmont directly on the Laurentian continental crust during the Proterozoic

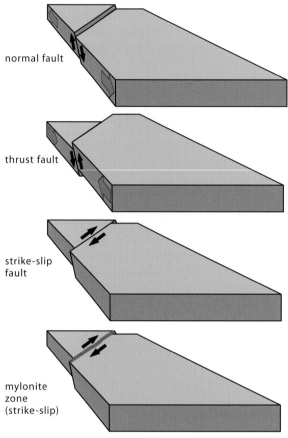

normal fault

thrust fault

strike-slip
fault

mylonite
zone
(strike-slip)

The three fault types, and how mylonite zones differ from faults. Normal faults develop when tectonic forces pull Earth's crust apart and a fault block drops down relative to the other. Thrust faults form when tectonic forces compress the crust and a fault block is forced up and over the other. Strike-slip faults develop when fault blocks slide horizontally past each other. With faults, fault blocks move along brittle fractures. Mylonite zones develop where there is ductile movement, meaning blocks of crust move, in any of the three ways above, along a wide band of rock that separates the blocks and flows like taffy. Arrows represent direction of relative movement.

eon, so we know that these rocks are younger than the 1.1-billion-year-old basement. In addition, geologists are able to use radiometric dating to determine the age of certain mineral grains, such as zircon and monazite, which were deposited in the sedimentary rocks (now metasedimentary rocks). The minerals can tell us the age of the original igneous or metamorphic rock they crystallized within. These ages reveal whether a metasedimentary rock is composed of sediment eroded from Laurentia (1.2 to 1 billion years old), Gondwana (2.2 to 1.8 billion years old), or terranes that developed along the edge of Gondwana between 750 and 600 million years ago. The spatial relationships of the Blue Ridge–Piedmont rocks have helped scientists to further clarify the ages of the metamorphic and igneous events that occurred in the region.

Based on these data and a deep familiarity with field relationships, geologists have created terrane maps of the southern and central Appalachians, with about ten divisions delineated in Georgia. This is a work in progress.

In Georgia, Laurentia includes the Appalachian Plateau, Valley and Ridge, and much of the northwestern Blue Ridge–Piedmont rocks. Exposures of

Laurentia in the Blue Ridge–Piedmont include 1.1-billion-year-old basement rock east and north of Cartersville, as well as younger metasedimentary rocks, but few younger metavolcanic or igneous rocks. Rocks of Laurentia are also exposed in three tectonic "windows" identified with terranes to the east and south. These are known as the Brasstown Bald, Dog River, and Opelika windows.

The Dahlonega gold belt, which crosses northwest Georgia in a northeasterly direction, and the Cowrock and Cartoogechaye (car-TOO-ga-jay) terranes, both in northeast Georgia, contain metasedimentary rocks originally sourced from Laurentia as well as metavolcanic rocks derived from a volcanic island chain. Geologists think the sedimentary deposits formed on the continental crust of Laurentia and adjacent oceanic crust, including sediments derived from and deposited around volcanic islands. The rocks were thrust westward onto Laurentia along thrust faults and deformed and metamorphosed from about 460 to 450 million years ago, during Ordovician time. Called the Taconic event, this tectonic episode built a mountain range that eroded and provided the gravel, sand, and clay that appear in the Ordovician-through-Silurian-age rocks of the Valley and Ridge.

Like the Dahlonega gold belt and the Cowrock and Cartoogechaye terranes immediately to its northwest, the Tugaloo (TOO-ga-loo) terrane contains metasedimentary rocks with a Laurentian source, as well as metavolcanic rocks, especially amphibolite, within the older part of the terrane. In addition, it contains large bodies of granitic gneiss. These former magma bodies formed during Ordovician time, perhaps below the same volcanic island chain that became part of the terranes to the northwest.

The Tugaloo terrane was severely deformed and metamorphosed around 400 to 360 million years ago. Unlike the Taconic event, this mountain building event (called the Neo-Acadian) did not produce a major influx of coarser sediment into Georgia's Valley and Ridge. Sand and gravel eroded from the highlands were spread into New York and Pennsylvania, and some of its clay reached as far south as Georgia. This evidence suggests that the Tugaloo terrane lay farther north along Laurentia when it was accreted to the continent and was later shifted southward along strike-slip faults. Magma intruded the Tugaloo terrane around 300 million years ago, forming the granite seen in the Atlanta area.

The Cat Square terrane was deformed and metamorphosed along with the Tugaloo terrane about 380 to 350 million years ago, but its metasedimentary rocks were derived from a wider variety of sources. A variety of ages taken from zircon minerals indicate that the terrane received sediment from Laurentia and a landmass that had formed in front of Gondwana. A favored interpretation is that the Cat Square terrane represents seafloor that was caught between Laurentia to the west and the Carolina superterrane moving in from the east. It was receiving sediment from both sources until it got caught up in the Neo-Acadian mountain building event. Around 300 million years ago very large magma bodies, including the Elberton Granite, intruded the Cat Square terrane.

The Pine Mountain terrane adjoins the southwest end of the Cat Square terrane in western Georgia. The Pine Mountain terrane consists of

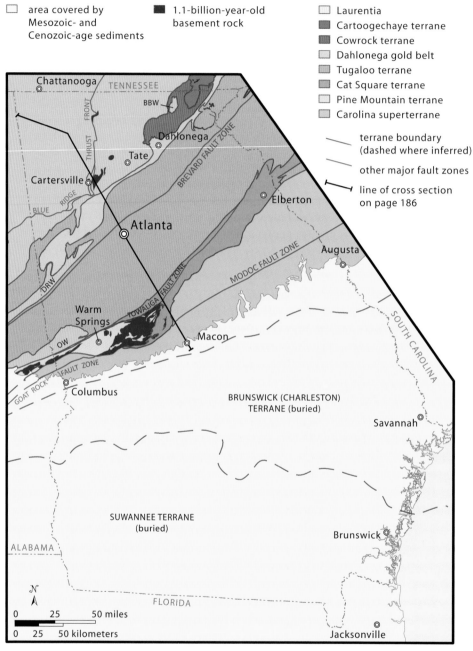

Legend:

- area covered by Mesozoic- and Cenozoic-age sediments
- 1.1-billion-year-old basement rock
- Laurentia
- Cartoogechaye terrane
- Cowrock terrane
- Dahlonega gold belt
- Tugaloo terrane
- Cat Square terrane
- Pine Mountain terrane
- Carolina superterrane
- terrane boundary (dashed where inferred)
- other major fault zones
- line of cross section on page 186

Map labels: Chattanooga, TENNESSEE, BBW, Dahlonega, THRUST FRONT, BREVARD FAULT ZONE, Tate, Cartersville, BLUE RIDGE, Elberton, Atlanta, MODOC FAULT ZONE, Augusta, SOUTH CAROLINA, DRW, TOWALIGA FAULT ZONE, Warm Springs, OW, Macon, GOAT ROCK FAULT ZONE, Columbus, BRUNSWICK (CHARLESTON) TERRANE (buried), Savannah, SUWANNEE TERRANE (buried), Brunswick, ALABAMA, FLORIDA, Jacksonville

0 25 50 miles
0 25 50 kilometers

The interpretations of Georgia's terranes used in this book. The Brasstown Bald (BBW), Dog River (DRW), and Opelika (OW) tectonic windows are labeled. (Based on Hatcher et al., 2007.)

metasedimentary rocks, some of which were probably deposited in shallow water on 1.1-billion-year-old basement rock. This ancient Laurentian rock is Georgia's largest area of exposed basement. Geologists hypothesize that this piece of continental crust was originally a section of Laurentia's continental margin, to the northeast of Georgia. The Cat Square terrane was thrust over it, and both were folded and metamorphosed. Later tectonic activity shifted the basement rock along strike-slip faults to its present anomalous location.

The Carolina superterrane, also called Carolinia, developed in front of Gondwana. The term *superterrane* is used because Carolinia is composed of multiple bodies of rock that had independent histories until they came together and shared a common history. Carolinia's volcanoes were active from about 625 million to 500 million years ago.

Carolinia moved away from Gondwana and collided with Laurentia well ahead of that supercontinent, most likely during the Neo-Acadian mountain building event about 350 million years ago. The northwestern part of Carolinia experienced severe metamorphism, along with the Tugaloo and Cat Square terranes, around 350 million years ago. Around 300 million years ago, magma intruded the terrane, leaving granite and gabbro.

The central and southeastern part of Carolinia underwent metamorphism as Gondwana collided with Laurentia between 325 and 265 million years ago, during Mississippian and Permian time. Called the Alleghany event, this tectonic collision also built a mountain range.

The Brunswick and Suwannee terranes are now entirely covered by Coastal Plain sediments. Based on limited evidence, geologists deduce the Brunswick terrane developed in front of Gondwana, and the Suwannee terrane was part of Gondwana. The arrival of the Suwannee terrane marks the finale of the Alleghany mountain building event some 265 million years ago. The tectonic force behind Laurentia and Gondwana's collision forced the Blue Ridge–Piedmont northwestward over the Valley and Ridge. The thrust faults that moved this rock propagated to the northwest, developing into the thrust faults that contorted the rock layers of the Valley and Ridge, leaving the patterns that erosion would later sculpt into interesting topography.

The igneous and metamorphic rocks of the Blue Ridge–Piedmont were thrust, as a unit, more than 100 miles to the northwest over sedimentary rocks that remain at the surface in the Valley and Ridge. The prominent boundary separating these two physiographic provinces is a complex of thrust faults referred to as the Blue Ridge thrust front. Geologists believe it took more than one generation of thrust faulting to carry the Blue Ridge–Piedmont rocks to their present position. Rocks of the province first moved westward on the north-south-trending faults, and then northward on east-west-trending faults.

EARTHQUAKES DUE TO HUMAN ACTIVITY

Earthquakes are common around reservoirs, both after they are initially dammed and when they experience high water levels. The phenomenon, referred to as reservoir-induced seismicity, happens for two reasons. First, a reservoir raises the water table in the area surrounding it, increasing underground

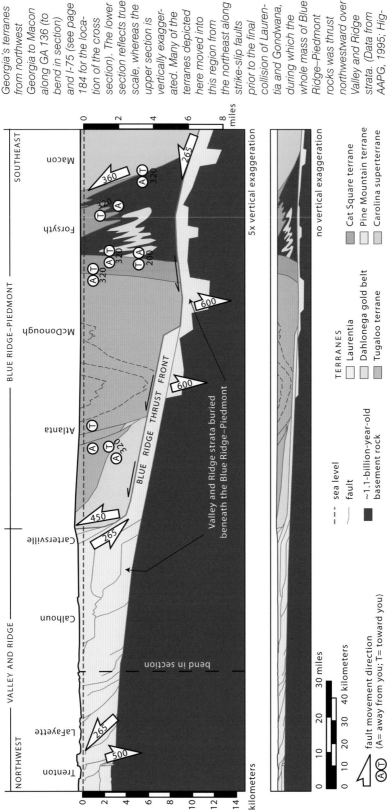

Cross section of Georgia's terranes from northwest Georgia to Macon along GA 136 (to bend in section) and I-75 (see page 184 for the location of the cross section). The lower section reflects true scale, whereas the upper section is vertically exaggerated. Many of the terranes depicted here moved into this region from the northeast along strike-slip faults prior to the final collision of Laurentia and Gondwana, during which the whole mass of Blue Ridge–Piedmont rocks was thrust northwestward over Valley and Ridge strata. (Data from AAPG, 1995; Higgins and Crawford, 2006; Hatcher et al., 2011; and Levin, 2009.)

BREVARD FAULT ZONE:
A TERRANE BOUNDARY THAT WASN'T

One fault zone crossing Georgia has gotten a lot of attention because it coincides with a 375-mile-long furrow in the landscape. Extending from near Montgomery, Alabama, to near Winston-Salem, North Carolina, the Brevard fault zone is so named because it knifes along the Blue Ridge near Brevard, North Carolina, in one of the area's deepest valleys.

The significance of the Brevard fault zone has been difficult to pin down. Like other fault zones in the Blue Ridge–Piedmont, it had a complex history. Geologists have concluded that the fault zone experienced both thrust faulting and strike-slip movement, in which the rocks southeast of the zone slid to the southwest. The strike-slip movement occurred under ductile conditions, meaning the rock body moved along hot rock that flowed like taffy. The ductile flow created a 0.5- to 2.5-mile-wide mylonite zone. When a fault zone is this long and has such a wide mylonite zone, geologists tend to think it represents a major boundary between tectonic plates. However, this assumption has been discarded because the sequences of rock on either side of the zone seem to be identical. The fault zone most likely developed in two stages: first, the area south of the fault slid southwestward for a few tens of miles, and then it was thrust northwestward over the fault zone for a few miles.

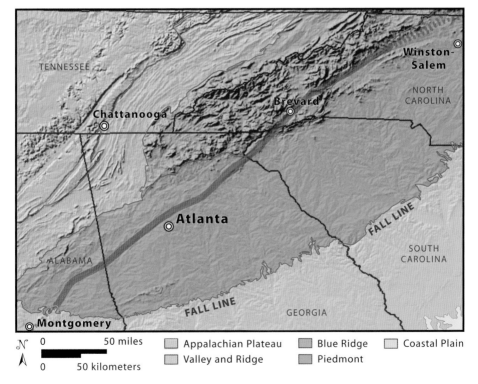

Much of the Brevard fault zone (solid red line) is traceable on a landscape image. Sediments of the Coastal Plain cover the southwestward extension of the fault zone (dashed red line). At its northeast end, two faults (dashed red lines) continue on into Virginia.

water pressure, which can make it easier for rocks to slide along preexisting weaknesses (for example, bedding planes and faults). Since these earthquakes occur where rocks are already primed for movement, the addition of water may just hasten the arrival of an earthquake. Second, the weight of the water adds new stresses to the surrounding rock, possibly triggering earthquakes that otherwise might not have occurred.

Many of the lakes in the Georgia Piedmont have had earthquakes that began about ten years after the reservoir was filled. Most are too small to be felt, but the largest are moderate in scale, between magnitude 4 and 5 on the Richter scale.

PULL-APART ACTIVITY DURING THE MESOZOIC ERA

After about 65 million years the supercontinent Pangaea began to break apart, forming the configuration of continents we recognize today. The tectonic pull-apart process, known as rifting, began during the Mesozoic era, around 200 million years ago. Fault-bounded basins formed in response to the rifting, in

Black diabase (center) in lighter-colored granitic gneiss in Vulcan Materials Norcross Quarry near Atlanta.

which blocks of crust slid down sloping faults. Up to thousands of feet of Triassic- and Jurassic-age nonmarine sediment and basalt lava flows were deposited in them. These basins are found from Georgia to Nova Scotia, Canada, and are a source of reddish brown sandstone used in brownstone apartments and churches. Georgia's basins are among the biggest, but they are buried by Coastal Plain sediments and are known only from deep wells drilled into the subsurface and through other methods.

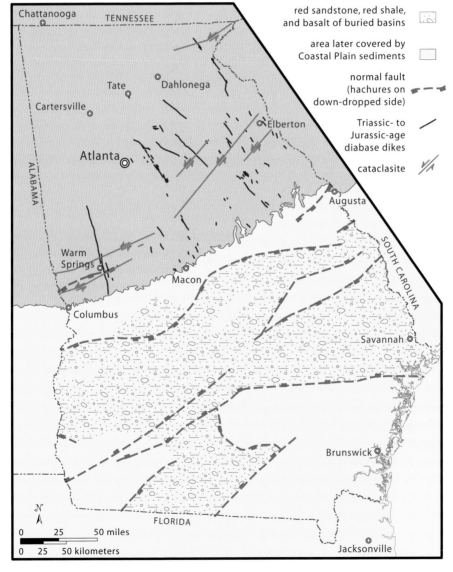

Map of Triassic-Jurassic-age (about 200 million years old) pull-apart features in Georgia. (Compiled from Lawton, 1976; Hatcher et al., 2007; and Huebner et al., 2011.)

Rifting also produced the youngest rocks in the Blue Ridge–Piedmont. As Africa and Europe pulled away from North America, tensional cracks opened within Earth's crust, and hot basaltic magma rose upward to fill them. One of these tensional cracks became dominant, widening to form the Atlantic Ocean. The others remained relatively narrow, forming 200-million-year-old diabase dikes that cross the Piedmont.

THE DAHLONEGA GOLD BELT: A TERRANE DISCOVERED BY PROSPECTORS

Many roads in this book cross the Dahlonega gold belt. This narrow terrane extends 152 miles across northwest Georgia in a northeast-southwest direction, from North Carolina to Alabama. The terrane ranges from less than 1 mile to 25 miles wide and is bounded by faults. Gold was discovered near Dahlonega in 1829, on land that belonged to the Cherokee Nation. Its discovery led to the influx of tens of thousands of fortune seekers and, within a decade, to the notorious Trail of Tears removal of the Cherokee in 1838. With its boomtowns and its transformation of the landscape, Georgia's gold rush cast the mold for gold rushes, such as those that began in 1848 in California and in 1896 in Canada's Klondike region. Many historians think that the phrase "Thar's gold in them thar hills," uttered by a fictional character of Mark Twain's, was inspired by an 1849 speech by Dr. M. F. Stephenson, assayer at the U.S. Mint. He was trying to convince Dahlonega miners leaving for the California gold rush that the Georgia hills still contained millions of dollars worth of gold.

Dahlonega's gold deposits began forming about 500 million years ago, when the area was in deep water near volcanic islands. A variety of sedimentary and volcanic rocks developed around these islands. Turbidity currents deposited sand and clay eroded from the islands. Basaltic lava oozed out onto the seafloor, and magma and gases from below blasted seafloor rock into ash, which settled out of the water forming layers. Seawater circulated downward toward the magma bodies feeding the islands and was heated, which allowed it to dissolve elements such as gold. The element-laden water returned to the seafloor in underwater hot springs, leaving behind deposits of quartz with an elevated gold content. Once the volcanic activity died down, the seafloor gradually was covered with sand and clay eroded from the islands.

About 50 million years later, another volcanic island chain rammed the now-dormant islands from the east. More island chains piled up from the east until the African continent (Gondwana) swept the whole mass of islands onto North America, accreting all these different terranes to North America about 300 million years ago. Over time, overlapping terranes buried the Dahlonega terrane more than 7 miles below the surface, where its original volcanic and sedimentary textures were all but wiped out as the terrane was metamorphosed. The basaltic lava became amphibolite, the hot spring deposits became quartzite, and the ash deposits and other sediments became schist and metagraywacke.

While the rocks were transforming, once again water (as superheated steam) circulated through them, dissolving the metals they contained. The water flowed through cracks generated by faulting, leaving behind gold-bearing quartz veins.

Tiny flecks of gold rim the edge (upper left) of a gold pan. In placer deposits, gold is associated with black heavy mineral grains and coarse quartz sand.

The gold deposits that first attracted hordes of fortune seekers were in mountain streams. Gold is very dense and resists weathering. Once erosion finally brought the Dahlonega rocks to the surface a few million years ago, the gold resisted the chemical weathering that dissolved and changed other minerals in near-surface rocks. This process produced saprolite, or rotten rock, which erosion then washed into streams. Because it is so dense, even small flecks of gold remained with the gravel and coarse sand on stream bottoms, while silt and clay were carried away.

Beginning in the 1830s, the first wave of mining, called placer mining, removed loose gold from stream gravel using gold pans and dredges. After the Civil War hydraulic mining took over, in which miners eroded the saprolite with high-pressure water sprayed from fire hose–like nozzles. Around the turn of the twentieth century, underground and open-pit mining operations removed gold-rich quartz veins directly from the bedrock.

I-20
Alabama State Line—Austell
41 miles

At the Georgia-Alabama state line, I-20 lies just southeast of the Allatoona fault. The fault separates rocks initially deposited on the outer continental shelf of North America (Laurentia), to the northwest, from rocks of the Dahlonega gold belt, to the southeast.

The Dahlonega gold belt is up to 25 miles wide in west Georgia, more than five times its width in most of northeast Georgia, and I-20 crosses it obliquely for nearly 30 miles to just east of Villa Rica.

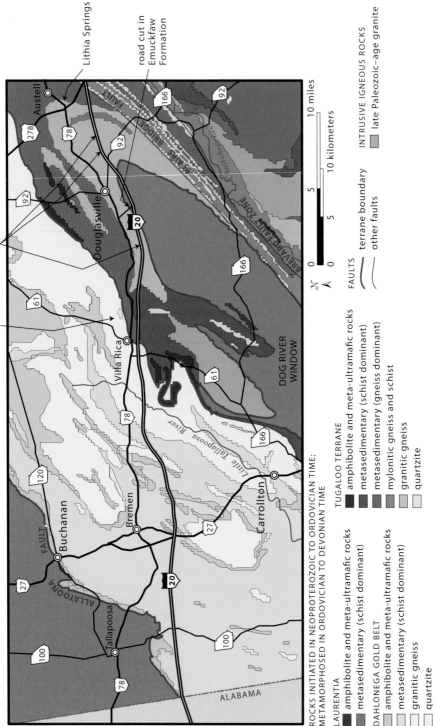

Pine Mountain Gold Museum

road cuts in Austell Gneiss

Lithia Springs

road cut in Emuckfaw Formation

ROCKS INITIATED IN NEOPROTEROZOIC TO ORDOVICIAN TIME;
METAMORPHOSED IN ORDOVICIAN TO DEVONIAN TIME

LAURENTIA
■ amphibolite and meta-ultramafic rocks
■ metasedimentary (schist dominant)

DAHLONEGA GOLD BELT
■ amphibolite and meta-ultramafic rocks
□ metasedimentary (schist dominant)
□ granitic gneiss
□ quartzite

TUGALOO TERRANE
■ amphibolite and meta-ultramafic rocks
■ metasedimentary (schist dominant)
■ metasedimentary (gneiss dominant)
■ mylonitic gneiss and schist
□ granitic gneiss
□ quartzite

INTRUSIVE IGNEOUS ROCKS
■ late Paleozoic–age granite

FAULTS
━ terrane boundary
— other faults

N

0   5   10 miles
0   5   10 kilometers

Geology along I-20 in west Georgia.

Villa Rica (Spanish for "rich town") was named after the wealth of its gold mining district. Some of its eighteen mines were active as early as the 1820s. Unlike the discovery of gold on lands of the Cherokee, the discovery here was initially kept quiet and did not attract as many fortune seekers.

Pine Mountain Gold Museum

Villa Rica's Pine Mountain Gold Museum tells the story of Georgia's forgotten gold rush with a film, maps, artifacts, and an interpretive path among the workings of the Stockmar Gold Mine. From the path you will see veins of quartz, up to several feet thick, which formed in muscovite schist. Pieces of glittery schist litter the path. If there had been significant gold in the veins you see, they likely would have been mined.

Along the self-guided walk you will see the Glory Hole, in which most of the mining occurred, the vats that were once filled with cyanide to process the ore, and mine tunnels framed by weathered quartz veins. The "Unique Geological Formation" stop features a quartz vein, in which small faults provide evidence of the tectonic forces that helped change the rock and provided conduits for the element-laden water to deposit the veins.

To reach the museum, exit I-20 at Liberty Road (exit 26). Head north 1.8 miles to the T-intersection at GA 61 and turn right. After 0.2 mile turn right onto Stockmar Road and continue 1.6 miles to the museum, on your left.

Villa Rica to Lithia Springs

I-20 crosses the southeast boundary of the Dahlonega gold belt south of Villa Rica, between exits 24 and 26, and passes into the Tugaloo terrane. East of exit

A tunnel in a quartz vein of the Stockmar Gold Mine.

*The Emuckfaw Formation at exit 34 of I-20. The brown
spot is probably a metamorphosed concretion.*

30 (Post Road), you can see outcrops of Austell Gneiss (around mile 31 on the
north side of the interstate and around mile 32.5 on the south side). The Austell
Gneiss is composed of black biotite and white quartz and feldspar. Some of the
prominent feldspar crystals are up to 1 inch across. Geologists interpret this
gneiss as being metamorphosed granite.

Metagraywacke of the Emuckfaw Formation is exposed at the west end of
the westbound onramp at exit 34 (GA 5). At first glance the rock looks uni-
formly gray, but the gray color comes from tiny, parallel biotite flakes evenly
distributed throughout the rock, which is dominated by quartz with some feldspar. The oblong rusty patches that are a few inches long are calcium-rich zones
interpreted as metamorphosed carbonate concretions.

The Emuckfaw Formation lies beneath the Austell Gneiss but is exposed
here in a large upfold. This is the northern end of what's called the Dog River
window, where erosion has cut through a thrust sheet and exposed an underlying terrane. The Emuckfaw Formation extends southwest from around exit
34 more than 110 miles to Alabama. Overall, it consists of interlayered schist,
metagraywacke, fine-grained gneiss, very little amphibolite, and characteristic
calcium-rich zones. It is part of Laurentia, the terrane composing the western
Blue Ridge. The Emuckfaw Formation was originally part of the continental
shelf of Laurentia, over which the rocks of the Dahlonega gold belt were thrust
northwestward.

I-20 crosses the Austell Gneiss again on the east side of the Dog River window; there are several exposures of the gneiss on both sides of the interstate for about 4 miles, east of exit 37 (GA 92). At exit 41 (Lee Road), you can see both the Austell Gneiss and the Emuckfaw Formation. The Austell Gneiss may be part of the Laurentian rocks within the window; however, the maps in this book follow a different interpretation, in which the Austell Gneiss is part of the Tugaloo terrane, which was thrust northwestward over both the Dahlonega gold belt and Laurentia.

Exit 41 (Lee Road) leads north to Lithia Springs and Austell, and south to Sweetwater Creek State Park (see Around Atlanta). The town of Lithia Springs, about 3 miles north of I-20, is named after salty tasting springs containing lithium. The medicinal use of the springwater dates back to the Indians, who called it Deer-Lick Springs. The area around the springs was a health resort from about 1880 to 1912. As many as thirty thousand people visited the spring in 1888. Lithium, in the water and in pill form, was used to treat all sorts of disorders in the nineteenth century, among them mania and depression. It fell into disrepute in the twentieth century as evidence of toxic overdoses came to light, but eventually its use in controlled doses revived it as a treatment for bipolar disorder. Lithium occurs in the minerals spodumene and lepidolite, which are sometimes found in pegmatite. An unexposed pegmatite body is likely the source of the lithium at Lithia Springs. The property where the springs are located is not open to the public.

I-75
Cartersville (US 411)—Kennesaw
21 miles

The Tellus Science Museum, just off I-75 in Cartersville at exit 293, has mineral and fossil exhibits (see the appendix for more information). At the same exit, I-75 crosses the White thrust fault, which runs along US 411. On its southeastern side are Cambrian-age Chilhowee Group quartzite and phyllite, Shady Dolomite, and Rome Formation sandstone and shale, which have been thrust northwestward on top of younger Cambrian- and Ordovician-age limestone. All of these rocks are part of the Valley and Ridge Province. However, thickly forested ridges capped by quartzite make the landscape of the White thrust sheet resemble the Blue Ridge. The ridges on both sides of I-75 range 300 to 600 feet higher than the roadway.

Between about miles 292 and 283, I-75 lies less than 1 mile west of the eastern edge of the White thrust sheet and the thrust-faulted northwestern boundary of the Blue Ridge. The road finally crosses the boundary near the south end of exit 283, Old Allatoona Road.

In the Blue Ridge, the oldest basement rocks—among the oldest rocks in Georgia—underlie the area east of I-75 near Lake Allatoona. These rocks, part of Laurentia, include metasedimentary gneiss of the Rowland Springs

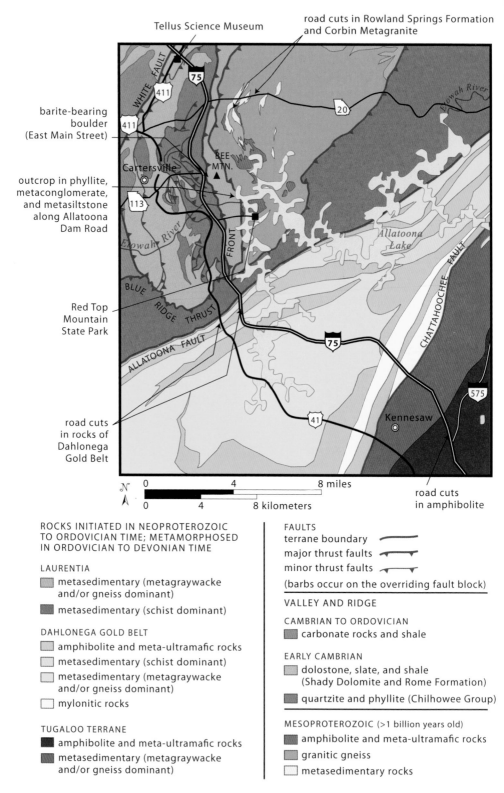

Geology along I-75 between Cartersville and Kennesaw.

Corbin Metagranite at the Vulcan Materials quarry south of GA 20. The white spots are large feldspar crystals.

Formation. This gneiss was metamorphosed at extraordinarily high temperature (above 1,500°F) and pressure more than 1.1 billion years ago. Tan cliffs of the Rowland Springs Formation are exposed on the east side of Vulcan Materials Quarry Road. To reach the road, head east on GA 20 from exit 290 for 2.3 miles and turn right. This gravel road is frequented by quarry trucks, so use caution.

The younger Corbin Metagranite intrudes the Rowland Springs Formation. The Corbin Metagranite has white feldspar crystals more than 1 inch in diameter surrounded by blue quartz grains. Blue quartz grains like these are abundant in some Proterozoic-age metasedimentary rocks in the Blue Ridge, and less commonly in sedimentary rocks in the Valley and Ridge. This suggests that the Corbin Metagranite (or similar rock) served as a sediment source during late Proterozoic and Paleozoic time.

You can see outcrops of Corbin Metagranite on the west side of Vulcan Materials Quarry Road or on the north side of GA 20, 3 miles east of I-75. The Vulcan Materials Company mines the metagranite in a quarry about 1 mile south of GA 20. (Schools and other organized groups can contact Vulcan Materials for tours.)

Cooper's Furnace Day Use Area and Allatoona Dam

I-75 crosses the Etowah River around mile 287, about 1 mile downstream from the Allatoona Dam. The dam was built at the head of a gorge where the Etowah River cuts through erosion-resistant, late Proterozoic–age rocks. The waterpower at this location, used to generate electricity at Allatoona Dam Powerhouse, was a reason for the siting of the Civil War–era industrial town of Etowah. Cooper's Furnace Day Use Area, a park below the dam, is all that remains of the town, now submerged beneath Lake Allatoona. The restored stone structure at the park was part of Mark Anthony Cooper's Iron Works, which processed limonite iron ore dug from pits in Shady Dolomite residuum and breccia in Chilhowee Group rocks. General William T. Sherman destroyed the iron works on his March to the Sea in 1864.

Across the river from Cooper's Furnace, along Allatoona Dam Road, outcrops of late Proterozoic–age rocks have characteristics indicating a low degree of metamorphism. Unlike most of the Blue Ridge–Piedmont rocks, these only reached temperatures of around 575°F. Metaconglomerate, metasiltstone, and phyllite are exposed next to the security fence at the end of the publicly accessible part of the road. Metaconglomerate breaks across pebbles, unlike conglomerate, which breaks around them. In the adjacent metasiltstone and phyllite, you can clearly see both bedding (the contrasting silty and clay-rich layers that were originally laid down in the ocean) and foliation (the slaty cleavage, or lines where the rock splits easily due to the parallel arrangement of microscopic mica flakes that formed during metamorphism).

Both geologic sites are accessible from exit 285. To reach the outcrop near the dam, head west on Red Top Mountain Road for 0.5 mile and turn right (north) onto US 41. Continue for 1.1 miles to Allatoona Dam Road and turn right. The security fence is about 1.6 miles from US 41. To reach Cooper's Furnace, drive 2.2 miles on US 41 and turn right onto Old River Road. The day use area is at the end of the road, 2.4 miles from US 41.

Conglomerate (left) and phyllite (right) at the security gate along Allatoona Dam Road.

Lake Allatoona and Red Top Mountain State Park

On the east side of I-75 at exit 285, Red Top Mountain Road crosses part of Lake Allatoona on its way to Red Top Mountain State Park. Around the lake there are exposures of 1.1-billion-year-old Corbin Metagranite and even older formations.

Red Top Mountain is named for the dark red soils along its crest and western slopes. The soils were derived from the chemical weathering and oxidation of the Red Top Mountain Formation. This dark, iron-rich metamorphic rock was originally gabbro that intruded metasedimentary rocks of the Rowland Springs Formation, which had already undergone extraordinarily high-temperature metamorphism before the gabbro magma arrived. The Red Top Mountain Formation is visible along the lakeshore at the end of Cottage Drive; follow the state park's signs to the rental cottages.

The 1.1-billion-year-old Corbin Metagranite intrudes both the Red Top Mountain and Rowland Springs formations, which means it is younger than both. The exposures of the metagranite in this region, in both outcrops and quarried blocks, are clearly foliated, with feldspar crystals stretched out about four times as long as they are wide. This is interpreted as an effect of ductile deformation that occurred in late Paleozoic time; the minerals were stretched out as rock bodies moved along either side of the metagranite, metamorphosing it in the process. When Lake Allatoona is very low, you can see a weathered exposure of Rowland Springs Formation, intruded by Corbin Metagranite,

Corbin Metagranite near the east end of Bethany Bridge. The feldspar crystals were stretched as the original granite was deformed. The crystal near the center is about 3 inches long.

on the lakeshore about 200 feet north of the west end of Bethany Bridge. The bridge is 1 mile east of I-75 on Red Top Mountain Road.

Allatoona Lake to Kennesaw

Near mile marker 282, I-75 crosses the Allatoona fault, the northern boundary of the Dahlonega gold belt. The section of the gold belt along I-75, now covered by suburban development, was once a busy part of the gold rush.

One of the best exposures of the metavolcanic rocks associated with gold deposits is near mile marker 281, on the east side of I-75. The outcrop is dominated by fine-grained amphibolite and metagabbro, with interlayered phyllite, light-colored gneiss, and quartzite. Some of the rock has a greenish color due to the minerals chlorite and epidote. Chemical analyses indicate that this rock was ocean floor basalt before it was metamorphosed. Coin-sized quartz patches filled with clusters of pea-green epidote are likely amygdules. The minerals were deposited in pockets left by gas bubbles that escaped from the cooling basaltic lava. At the south end of the road cut, you can see the contact between metavolcanic amphibolite and tan layers of the Canton Formation, which is composed of metasedimentary schist and metagraywacke with quartz veins. In the Dahlonega gold belt, gold is concentrated in quartz veins along and near the contact between these two rock units.

Near exit 273 (Wade Green Road), I-75 crosses the Chattahoochee thrust fault, which carried the Tugaloo terrane northward over the rocks of the Dahlonega gold belt. Southeast of Kennesaw is an area dominated by amphibolite. At the south end of exit 269 (Barrett Parkway), on both sides of I-75, you can see outcrops of black amphibolite cut by quartz veins and pegmatite dikes that were contorted by folding.

Metavolcanic rock along east side of I-75 near mile 281.

GA 5 (I-575 and GA 515 in part)
Tennessee State Line—Kennesaw
82 miles

GA 5 begins in McCaysville, just across the state line from Copperhill, Tennessee. McCaysville lies on the southern edge of a copper mining district known as the Copper Basin. Mining, which began in the 1840s and ended in the 1980s, led to one of America's biggest environmental disasters. Fifty square miles of lush forest were laid completely bare, creating a red, martianlike landscape, while the Ocoee River that drains the area was cleared of all life. The trees were cut to fuel open-air smelters that roasted copper sulfide ore. The smelters released sulfur dioxide in the air that combined with water vapor to make sulfuric acid. Extreme acid precipitation killed the remaining plant life. With no plants, erosion removed the topsoil. The effect on the Copper Basin's vegetation lasted into the 1980s, long after the smelters were improved to capture sulfuric acid. Although sulfuric acid was originally just a by-product of the copper mining, later sulfuric acid became the main commercial product of the mining district. Because of aggressive reforestation starting in the 1970s, little of the scar on the landscape remains visible today.

The copper ore deposits probably formed around seafloor hot springs associated with submarine volcanoes, like the gold deposits of the Dahlonega gold

Denuded landscape of the Copper Basin surrounds this mine at the Ducktown Basin Museum in Tennessee.

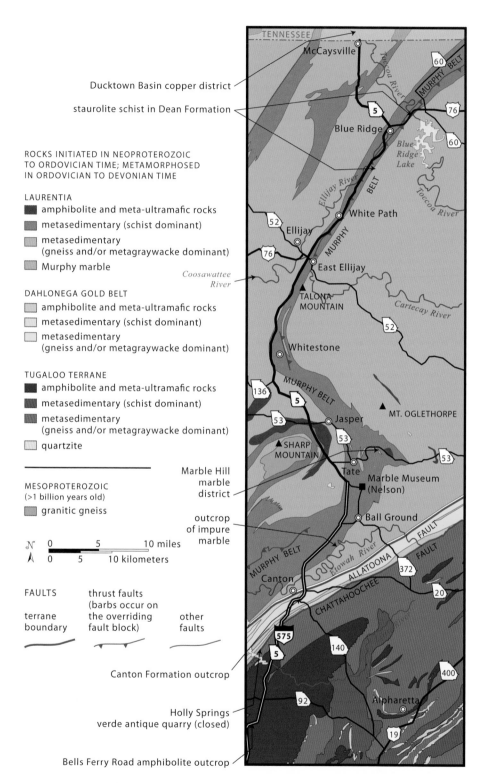

Geology along GA 5, including the Murphy belt. (Modified after Thigpen and Hatcher, 2009.)

belt. Only a minor amount of copper mining occurred in Georgia, west of McCaysville, but the area was also deforested.

Blue Ridge to Jasper

At Blue Ridge, GA 5 turns southwestward to follow the Murphy belt, Georgia's marble region. The Murphy belt is about 100 miles long and up to 8 miles wide, trending southwest from near Andrews, North Carolina, to Canton.

In North Carolina, where it is widest, the Murphy belt has been mapped as a syncline, or concave-up fold. The belt is narrower in Georgia because greater tectonic compression pushed the east side of the syncline westward until the layers on either side of the fold paralleled one another, forming an isoclinal fold.

Between Blue Ridge and Ellijay, GA 5 tracks the core of the syncline, following a valley underlain by light-colored schist. This is the Mineral Bluff Formation, the youngest rock unit in the area. Tusquitee Quartzite and dark phyllite of the Nantahala Formation compose the ridges flanking the valley. The landscape reflects the rock units' differing resistance to erosion. At the village of White Path, 10 miles south of Blue Ridge, there is a crushed-stone quarry on the east side of GA 5. Quarry employees are removing what they call "granite," though geologists likely would consider it metagraywacke. White Path is also the site of one of the few gold mines outside of the Dahlonega gold belt, where four of the five largest gold nuggets ever found in Georgia were mined. The largest weighed more than 4 pounds.

Looking northeast in East Ellijay, you can see a mountain with prominent east-dipping layers just north of the Cartecay River. This is the site of the old Gilmer County prison quarry, in which overturned layers of schist, part of the Murphy belt syncline, are exposed in either the Mineral Bluff or Dean formation. The schist contains chloritoid, a mineral that forms in clay-rich rocks at relatively low metamorphic temperatures and pressures.

Ellijay lies just west of a ridge that runs along the Murphy belt. Four miles south of Ellijay, GA 5 climbs this ridge, with scenic views in both directions. To the east, you can see Talona Mountain (1,851 feet), topped by houses and radio towers and underlain by quartzite.

Between Whitestone and Canton, GA 5 continues as before to follow a valley that coincides with the hinge line of the Murphy belt syncline. However, here the isoclinal syncline has itself been bent over a later anticline and syncline. This distortion results in a 25-mile-long S curve—easily seen on any map—that GA 5 follows.

Around Jasper, views open up to the mountains on either side. Mt. Oglethorpe (3,289 feet), 7 miles to the east, is considered by many to be the southernmost peak of the Blue Ridge Mountains. It was once the southern terminus of the Appalachian Trail. Today the trail terminates on Springer Mountain, 20 miles to the northeast.

Sharp Mountain (2,333 feet), 3 miles to the west, is underlain by quartzite of the Pinelog Formation. The quartzite curves around the northwest edge of the Salem Church anticline, a concave-down fold that exposes 1.1-billion-year-old Corbin Metagranite in its core.

Marble Mining District

Jasper hosts the Georgia Marble Festival one weekend in October, when the world's largest open-pit marble mine and several other mines are open to the public. Marble is metamorphosed limestone, which is mainly composed of the carbonate minerals calcite and sometimes dolomite. The relative softness of these minerals makes marble easier to engrave and carve than other rocks, such as granite. The marble in this region comes in several varieties, including white Cherokee marble, pink Etowah marble, and Creole marble, which is white to gray with dark gray streaks. The dark streaks are clay and other impurities that formed layers when the limestone was metamorphosed.

To visit the heart of the marble mining district, turn east from GA 5 onto GA 53, 4.7 miles south of Jasper. Through Tate, GA 53 follows the Old Federal Road, an early route between Athens and Chattanooga. The U.S. government leased the right-of-way from the Cherokee Nation prior to removing them from the region. One of the travelers along this route was Henry Fitzsimmons, an Irish stonecutter, who noticed an outcrop of marble. He began the marble industry here in 1842 with a mill producing stone monuments. The first public monument produced by the mill, dedicated to soldiers who died battling Indians, stands in the town square of Lawrenceville, Georgia. By the end of the nineteenth century this valley was home to a flourishing industry of independent stonecutters.

Several varieties of Georgia marble are present in the floor tiles at the Michael C. Carlos Museum at Emory University.

Colonel Sam Tate consolidated most of these operations into the Georgia Marble Company in the early twentieth century. Tate House, his pink marble mansion, built in the 1920s, stands on the south side of GA 53, 3.5 miles east of GA 5 and 0.5 mile east of one of Georgia's few marble school buildings. Three miles east of Tate House, on the south side of GA 53, you can see gray, weathered cliffs of marble on the property of Imerys Marble, Inc. South of Tate, in the town of Nelson, the Marble Museum located in the city hall offers a small but comprehensive exhibit covering the genesis, historical and modern uses, and quarrying of Georgia's marble.

Marble quarry near Tate.

Tate House, built of pink Etowah marble.

Marble forms when carbonate sedimentary rocks (limestone or dolostone) recrystallize during metamorphism. Carbonate sedimentary rocks are widespread in the Valley and Ridge and Coastal Plain, so you might expect that marble would be widespread among the metamorphic rocks of the Blue Ridge–Piedmont. So why is nearly all of Georgia's marble concentrated along the narrow trend of the Murphy belt? Geologists might answer that there is a time and place for everything.

The time for carbonate rocks has mainly been since the beginning of the Paleozoic era, when marine animals first began to make carbonate shells and other hard parts. And the place is predominantly in water that is warm, clear, and shallow enough to support a thriving community of organisms. The fact that there is not much marble in Georgia's Blue Ridge–Piedmont must reflect some combination of the following four factors: (1) Its sediments were deposited before the Paleozoic era; (2) the water was too deep, (3) too cold, and/or (4) too muddy for carbonate-producing organisms to thrive. But apparently some or all of these restrictive factors did not apply in the Murphy belt. Why?

One possible scenario is that the Murphy belt carbonate sediments were deposited on the outer continental shelf, farther offshore than the Valley and Ridge rocks to the west, perhaps in Cambrian to Ordovician time, but closer to shore than the majority of the Blue Ridge–Piedmont rocks to the east. Some evidence supports this idea. For example, east of GA 5 between Blue Ridge and Ellijay, metamorphic quartzite, limonite iron ore, and marble (with dolomite) occur in sequence together; to the west, Cambrian-age sandstone, iron ore, and dolostone—the sedimentary equivalents—occur in sequence near

Cartersville. The similarity suggests that these rocks in the Murphy belt are simply the eastward continuation of the Cartersville rocks that happened to get metamorphosed.

A hornblende schist lies adjacent to the marble deposits near GA 53. This rock provides a clue that the origin of the marble belt may be more complicated than simply deposition on the continental shelf. The chemistry of the hornblende schist suggests that it may be metamorphosed basalt, originally spewed from a volcano rooted on the ocean floor. Limestone associated with basalt is more reminiscent of Pacific coral atolls, where thick layers of limestone develop on extinct volcanoes. Coral atolls can develop far from continents. Until more clues are found, the origin of Georgia marble will remain one of the many unsolved mysteries of Blue Ridge–Piedmont geology.

STAUROLITE: GEORGIA'S STATE MINERAL

Staurolite is a brown mineral. "Fairy crosses" can develop in this mineral, in which two crystals grow in such a way that they form the shape of a cross. There are several types of cross, including the St. Andrews, where the two crystals cross at about 60 degrees, and the Maltese, where they cross at nearly 90 degrees. The word *staurolite* comes from the Greek *stavros*, meaning "cross." (Note that some of the fairy crosses offered for sale at souvenir shops are imitations carved from fine-grained rock dyed to a brownish color.) Most of the time staurolite forms individual crystals, generally less than 1 inch long, in the shape of a four-sided prism with flat ends.

Staurolite forms at temperatures of more than 900°F in aluminum-rich metamorphic rocks, such as schist. As they grow, the crystals typically engulf quartz grains, so tiny inclusions of quartz are common in staurolite. Many staurolite crystals will develop with a partial coating of yellowish to orange clay, or silvery gray, fine-grained mica from the surrounding schist that weathered away to leave the crystals behind. In the area north and east of the town of Blue Ridge, staurolite is found in schist in the Dean Formation. It can be found in similar rocks along most of the length of the Murphy belt.

Staurolite crystals from the Murphy belt, on display at the Tellus Science Museum.

Isoclinal folds in impure marble near Canton.

Ball Ground to Kennesaw

An outcrop of impure marble lies a short distance off GA 5/I-575, between Ball Ground and Canton. Contrasts in the carbonate content of the different layers led to an etched appearance in this rock, which has isoclinal folds. Take exit 24 and head south on Airport Drive 0.5 mile to the T-intersection with Ball Ground Highway. Turn right and drive 1.2 miles (0.2 mile past the intersection with Lower Bethany Road). The marble outcrops are atop a small hill about 100 yards south of the road, north of a large industrial building.

Canton lies at the intersection of the southern end of the Murphy belt and the Etowah River valley, which, upstream of here, flows mainly along the Dahlonega gold belt. Near exit 19, in the valley of Canton Creek, GA 5/I-575 crosses the Allatoona fault, which marks the northern edge of the Dahlonega gold belt.

The Canton Formation, one of the rock units that can be traced along the length of the Dahlonega gold belt, is well exposed on the eastern side of GA 5/I-575 near mile marker 18. These tan and white rocks consist of garnet-bearing schist, which probably originated as clay deposited on the ocean floor around the time that seafloor hot springs were giving rise to the gold deposits.

Nearby is the Holly Springs Quarry, one of very few sites in the United States where so-called green marble, or verde antique, has been commercially quarried. Verde antique is a metamorphosed ultramafic igneous rock crossed by green veins of the minerals chlorite and serpentine and light-colored veins of dolomite. Ultramafic rocks are the iron- and magnesium-rich rocks of which

Earth's mantle is composed. The rock may have originated in the mantle beneath the Dahlonega gold belt. The quarry was most active from 1905 to 1915. It is famous among mineralogists as the world-standard locality for a mineral known as hydroxyapatite.

Near exit 14, GA 5/I-575 crosses the Chattahoochee fault, which forms the boundary between the Dahlonega gold belt to the north and the Tugaloo terrane to the south. An amphibolite-dominated part of the Tugaloo terrane is well exposed along the entrance ramp to GA 5/I-575 northbound at exit 4, Bells Ferry Road. The dark rocks at this location are crosscut by dikes of light-colored pegmatite.

GA 52
Chatsworth—Lula
93 miles

Chatsworth lies within the Valley and Ridge in flat country underlain by limestone and shale of Cambrian age. There are splendid views of Cohutta Mountain (4,009 feet) and Fort Mountain (2,848 feet), which are less than 4 miles to the east. At the base of the mountains lies the Blue Ridge thrust front, along which Proterozoic-age metamorphic rocks moved more than 100 miles northwestward over Paleozoic-age sedimentary rocks of the Valley and Ridge Province.

The two industrial plants on either side of GA 52, 0.3 mile east of US 411, once ground talc from Fort Mountain into powder for use as a filler in many different products. Today they grind limestone instead for the same purpose.

About 0.9 mile east of US 411, GA 52 passes a weathered outcrop of Cambrian-age shale on the south side of the road. After crossing the unexposed Blue Ridge thrust front 1.6 miles east of US 411, there are many road cuts in Proterozoic-age metasedimentary rocks for the next 6 miles, all on the north side of the road. This is a busy, curvy highway, so use caution. The first series of road cuts, 1.8 miles east of US 411, expose carbon-rich black slate and metagraywacke, both rust-stained by the weathering of the iron sulfide mineral pyrite. This unit is continuous with similar rocks in a larger road cut 6 miles to the south along US 76, and it is typical of rocks found in the Blue Ridge along the Blue Ridge thrust front between Knoxville, Tennessee, and Birmingham, Alabama. These rocks underwent only a moderate degree of metamorphism, unlike rocks to the east, which experienced a higher degree of metamorphism.

Midway up Fort Mountain, granitic gneiss and lens-shaped bodies (up to 0.2 mile long) of schist are sandwiched between metasedimentary rocks. The schist is green because it contains the mineral chlorite. Both the granitic gneiss and chlorite schist are mylonitic, or "smeared out," due to intense ductile deformation that occurred at great depth along a fault.

Talc, the soft mineral familiar from talcum baby powder, was mined from chlorite schist on Fort Mountain from 1872 until the 1990s. Talc, which contains magnesium, can form by the metamorphism of more than one kind of magnesium-bearing rock. The talc here was long thought to have formed from the metamorphism of dolomite, a carbonate mineral that contains magnesium and is common in sedimentary rocks of the Valley and Ridge. There is some dolomite in the chlorite schist. However, the minerals olivine and serpentine are also present, along with chromium and nickel. These latter minerals are not found in Valley and Ridge rocks but are typical of the ultramafic rocks found throughout the Blue Ridge–Piedmont. Ultramafic rocks are rich in iron and magnesium and poor in silicon, and they normally originate in the Earth's mantle.

About 3.7 miles east of US 411, GA 52 is just downslope of the Lower Bramlett talc mine, and the road passes several more road cuts in mylonitic rocks.

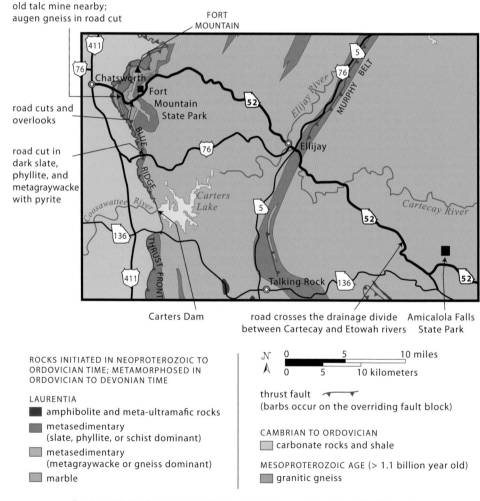

ROCKS INITIATED IN NEOPROTEROZOIC TO ORDOVICIAN TIME; METAMORPHOSED IN ORDOVICIAN TO DEVONIAN TIME

LAURENTIA
- ■ amphibolite and meta-ultramafic rocks
- metasedimentary (slate, phyllite, or schist dominant)
- metasedimentary (metagraywacke or gneiss dominant)
- marble

thrust fault ⤙⤚
(barbs occur on the overriding fault block)

CAMBRIAN TO ORDOVICIAN
- carbonate rocks and shale

MESOPROTEROZOIC AGE (> 1.1 billion year old)
- granitic gneiss

Geology along GA 52 from Chatsworth to Amicalola Falls. (Modified after Thigpen and Hatcher, 2009.) See the road guide map on page 218 for the geology along the eastern portion of the route.

Chlorite schist is not exposed here, but pieces of grayish green phyllonite are present at the base of the first road cut in this series. The granitic gneiss here is augen gneiss (*augen* is German for "eye"), in which dark biotite mica wraps around large crystals of feldspar like eyelashes around an eye. Based on their similarity to deformed Corbin Metagranite and associated rocks near Cartersville, the mylonitic rocks at Fort Mountain are interpreted to be 1.1 billion-year-old basement rocks of Laurentia, on which the metasedimentary rocks of the area were deposited before they were metamorphosed. The granitic gneiss has a radiometric age of 368 million years, which geologists interpret to be the time of the faulting and ductile deformation that produced the mylonitic texture.

About 5.9 miles from US 411, road cuts on both sides of GA 52 expose some of the metasedimentary rocks that form the crest of Fort Mountain. They appear to have been eroded from granitic rock since they have a composition similar to the augen gneiss. Small pebbles of feldspar and quartz (including blue quartz) are present. The presence of the feldspar pebbles suggest the granitic source rock was not far from where these rocks were deposited because feldspar tends to be destroyed by weathering and during transport in water. In marked contrast to the augen gneiss at the previous road cut, this rock doesn't have mylonitic texture. Either the zone of ductile deformation died out over a small distance, or there is a fault between the augen gneiss road cut and this one, which could have placed this metasedimentary rock here after the episode of deformation had ended.

Augen gneiss along GA 52.

Two overlooks, 6.7 miles and 6.9 miles from US 411, provide scenic views to the south and east. Outcrops across the road from the overlooks are black slate and metasandstone. Opposite the second overlook, the contrasting beds of black slate and metasandstone dip moderately toward the west, and then above them a fault cuts across these layers, having placed similar contrasting layers on top of them.

Fort Mountain State Park

The entrance to Fort Mountain State Park on GA 52 is 7.3 miles east of US 411. From the entrance it is 2.3 miles to a parking lot and trailhead for the 1-mile loop trail to the summit of Fort Mountain. If you bear left to take the lower side of the loop first, you will notice that the mountain is littered with boulders. Boulder fields are present on a number of Georgia mountain summits that were probably above timberline during the coldest parts of the ice ages of Pleistocene time. The freeze-thaw action of ice and frost can break up boulders, and without tree cover erosion works more effectively, carrying away loose sediment and leaving boulders behind.

Near the overlook, after about 0.4 mile, you will see exposures of slightly metamorphosed conglomerate and sandstone on the right side of the trail similar to that along GA 52. The wooden overlook platform is one of Georgia's

The wall atop Fort Mountain, with a shallow circular pit (center of picture).

most photogenic spots, with views northward along the line of the Blue Ridge thrust front and westward to Lookout Mountain, which is topped by sandstone and conglomerate deposited at about the time the Blue Ridge thrust front was active. From the overlook, it is a short, steep hike to the summit, which is topped by a tower built of local stone by the Civilian Conservation Corps.

The mysterious stone "wall" that gives Fort Mountain its name is between the tower and the parking lot along the second half of the loop trail. The wall consists of a trail of boulders, about 10 feet wide, a few feet high, and about 900 feet long. The wall crosses the mountain from west to east and then turns to run along the contour of the mountain to the north. In several places circular pits, about 12 feet across and a few feet deep in the center, interrupt the continuity of the wall.

The wall predates European settlement, and according to one Cherokee legend it was built by a fair-skinned race of "moon-eyed people" who inhabited this area before they arrived. Another legend says the wall was built by twelfth-century Welsh seafarers under Prince Madoc, who arrived in Mobile Bay and wandered upriver to this place. Neither legend answers why the structure was built, if indeed the wall is a manmade structure. Some geologists interpret the wall to be the result of natural weathering processes and downslope movement of rock.

Fort Mountain is not the only place in Georgia with stone structures that pre-date European settlement, but elsewhere most of them have been removed. These locales include Stone Mountain east of Atlanta, Rocky Face Mountain near Dalton, Quarry Mountain near Cartersville, and Rock Eagle and Rock Hawk near Eatonton (see the US 441/GA 24: Madison—Milledgeville road guide for more on the Eatonton structures).

Fort Mountain to Amicalola Falls

East of Fort Mountain, GA 52 follows a ridge crest with views on both sides, remaining above 2,000 feet of elevation for more than 6 miles before descending toward Ellijay. The rocks along this stretch are similar to those atop Fort Mountain: metasedimentary rocks, probably of Proterozoic age, that experienced a low to moderate degree of metamorphism. At Ellijay, GA 52 crosses the Murphy belt, a narrow trend of valleys and low ridges, about 100 miles long, following a geologic structure that most geologists agree is a syncline, a concave-up fold. The youngest rocks of the fold are exposed near its core around Ellijay, including marble and schist. See the GA 5 (I-575 and GA 515 in part): Tennessee State Line—Kennesaw road guide.

East of Ellijay, GA 52 follows the Cartecay River upstream, past ridges that mark the east side of the Murphy belt. GA 52 reaches the topographic boundary, or escarpment, between the Blue Ridge and the Piedmont provinces about 13.5 miles east of the GA 5 intersection in East Ellijay. Here, GA 52 crosses a drainage divide, leaving the Cartecay River drainage basin, in which the valley elevations are around 2,000 feet, and dropping into the valley of Amicalola Creek (at about 1,700 feet of elevation), which flows south toward the Etowah River. The escarpment is the setting of Amicalola Falls, about 3 miles to the east.

Amicalola Falls State Park

The entrance to Amicalola Falls State Park is 19.2 miles east of the GA 5 intersection. Amicalola Falls, at 729 feet—four times the height of Niagara Falls—is one of the Seven Natural Wonders of Georgia. It is the third-highest waterfall east of the Rocky Mountains. It is not a sheer drop, but rather a series of seven cascades. Two pathways from different parking areas, one level and the other with stairs, meet at the bridge just below the falls.

At the falls, Little Amicalola Creek descends abruptly from a gradually sloping upper valley, at an elevation over 2,400 feet, to the foot of the escarpment, at 1,700 feet. The falls consists of cascades because, unlike many waterfalls, it has not developed where there is localized contrast in the erosion resistance of bedrock. The rock at the falls is fairly uniform in its resistance to erosion. It seems likely that Amicalola Falls owes its location to stream capture. On a shaded relief map it is easy to see how Little Amicalola Creek might have begun as only one of many short, steep streams that flow down the south side of the escarpment. It seems plausible that it captured the headwaters of a tributary of Anderson Creek, which flows west at a much higher elevation atop the escarpment. This additional source of water increased Little Amicalola's volume of water, eroding the steep-walled gorge below the falls and forming the seven cascades that continue to recede up the valley.

The rocks in the state park are metasedimentary, including gneiss, schist, and metagraywacke, along with small intrusions of pegmatite. Blocks of these rocks litter the area around the creek along the Appalachian Approach Trail.

Two of the cascades near the top of Amicalola Falls.

View from the overlook at the top of Amicalola Falls (and the escarpment separating the Blue Ridge and Piedmont provinces).

Unlike many rocks along mountain streams, they have sharp edges, suggesting that they have come mainly from the collapse of rock as the falls have eroded up the valley. If the boulders had been rolled along in the stream, they would be more rounded. Another clue to active erosion here is the evidence of the gradual movement of rock and soil down the sides of the valley, which geologists call creep. It's especially evident above the West Ridge Falls Access Trail, where trees bend upward to maintain a vertical trunk as their roots ride the soil downslope.

In outcrops along the West Ridge Falls Access Trail, you can see pegmatite and quartz veins cutting across the foliation in the metasedimentary rocks. In addition to tiny garnets, schist in this area contains the minerals kyanite and sillimanite, indicating that these rocks reached metamorphic temperatures of well over 900°F, a higher grade of metamorphism than the rocks of Fort Mountain.

From the top of the falls, you can hike 5 miles to the Len Foote Hike Inn, an ecofriendly wilderness retreat at about 3,050 feet of elevation, or 8 miles to 3,780-foot Springer Mountain, the southern terminus of the Appalachian Trail. Like Amicalola Falls, both are located on the escarpment along the northern edge of the Etowah River drainage basin.

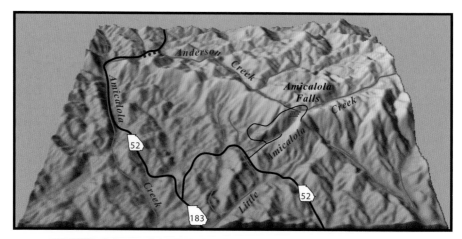

Shaded relief map of the Amicalola Falls area. The red line shows the hypo-thetical path of an Anderson Creek tributary before Little Amicalola Creek cap-tured it. The dashed blue line shows the path of the potential future capture of upper Anderson Creek by another Amicalola tributary, at the gap where GA 52 crosses the divide between the two creeks.

Amicalola Falls to Lula

About 8.9 miles east of the entrance to Amicalola Falls State Park, GA 52 crosses the Hayesville thrust fault, which separates Laurentia from terranes to the east. For 1.5 miles the road passes through the southwest tip of the Cowrock terrane before crossing the Soque River fault to enter the Dahlonega gold belt. Both the Cowrock terrane and the Dahlonega gold belt contain more amphibolite than does Laurentia. Geologists interpret this to mean that the terranes formed on the ocean floor or along a volcanic island chain, east of North America's early Paleozoic-age continental shelf.

Dahlonega was the center of Georgia's gold rush, which began in the 1830s, and is home to a gold museum and two gold mines that offer tours (see the US 19: North Carolina State Line—Cumming road guide). East of Dahlonega (5 miles east of US 19), GA 52 crosses the trace of the Chattahoochee thrust fault, the boundary between the Dahlonega gold belt and the Tugaloo terrane.

After passing through rolling agricultural country with occasional expo-sures of weathered gneiss and schist, GA 52 crosses the Chattahoochee River within the Brevard fault zone, 23 miles east of US 19 and 2 miles west of US 23. Here, mylonitic metagraywacke and schist are exposed on the south side of the river when Lake Lanier is at relatively low levels.

About 2 miles west of the Chattahoochee River bridge, along Flat Creek and on an 8,000-acre private tract known as Glade Farm, gold was mined from stream gravel during the nineteenth century. Several unverified (possibly fraudulent) diamond finds were reported during that time. The farm also con-tains a 125-foot waterfall that has been the site of several fatalities and serious injuries over the years, despite no-trespassing warnings and a fence.

US 19
North Carolina State Line—Cumming
75 miles

Between the North Carolina state line and Blairsville, US 19 is southeast of the Murphy belt. The dominant rock types—schist, metagraywacke, and gneiss— are poorly exposed until the highway begins to climb into the mountains south of Blairsville. US 19 crosses the Hayesville fault about 5.7 miles south of Blairsville, entering the Cowrock terrane.

Brasstown Bald

The highest point in Georgia, 4,784-foot Brasstown Bald, is a short drive from Blairsville. About 7.5 miles south of Blairsville, turn left (east) onto GA 180 and continue 7.3 miles to the GA 180 Spur. Head north on the spur for 2.4 miles to the Brasstown Bald parking area. From the parking area, which is at 4,350 feet of elevation, you can walk to the top along a 0.5-mile-long trail or, for a small fee, ride to the top in a van. The 360-degree view from the visitor center atop the mountain is one of Georgia's most scenic, including Lake Chatuge and the North Carolina's Blue Ridge to the north and the Blue Ridge peaks tracked by the Appalachian Trail to the south.

Surrounded at lower elevations by rocks of the Cowrock terrane, Brasstown Bald is positioned near the center of a tectonic window. The rocks in the window are part of ancient Laurentia, which makes up the western Blue Ridge. The Cowrock terrane was thrust over Laurentia, but erosion cut through it and the underlying thrust fault, exposing the Laurentia terrane.

Lake Chatuge and North Carolina's Blue Ridge in the distance, seen to the north from the top of Brasstown Bald.

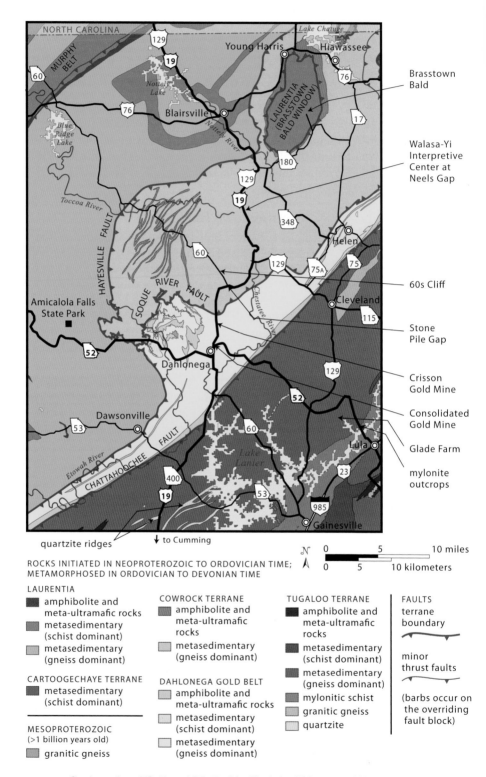

ROCKS INITIATED IN NEOPROTEROZOIC TO ORDOVICIAN TIME;
METAMORPHOSED IN ORDOVICIAN TO DEVONIAN TIME

LAURENTIA

amphibolite and meta-ultramafic rocks

metasedimentary (schist dominant)

metasedimentary (gneiss dominant)

CARTOOGECHAYE TERRANE

metasedimentary (schist dominant)

MESOPROTEROZOIC
(>1 billion years old)

granitic gneiss

COWROCK TERRANE

amphibolite and meta-ultramafic rocks

metasedimentary (gneiss dominant)

DAHLONEGA GOLD BELT

amphibolite and meta-ultramafic rocks

metasedimentary (schist dominant)

metasedimentary (gneiss dominant)

TUGALOO TERRANE

amphibolite and meta-ultramafic rocks

metasedimentary (schist dominant)

metasedimentary (gneiss dominant)

mylonitic schist

granitic gneiss

quartzite

FAULTS

terrane boundary

minor thrust faults

(barbs occur on the overriding fault block)

Geology along US 19 and GA 52. (Modified after Thigpen and Hatcher, 2009.)

The Brasstown Bald window forms an oval about 10 miles long from north to south and about 5 miles wide. Amphibolite and ultramafic rocks of the Lake Chatuge complex define its edges. This complex probably represents the sea-floor on which sediments were deposited that then became gneiss and schist of the Cowrock terrane. Gneiss and schist are exposed in the tectonic window. The schist around the summit, as well as on the west side of the spur road about 250 yards from the intersection with GA 180, is full of red garnet, each crystal a little larger than a pinhead.

Blairsville to Dahlonega

Cowrock terrane rocks are exposed in road cuts all along US 19 as it crosses the mountains south of Blairsville, but the Walasi-Yi Interpretive Center at Neels Gap, where US 19 crosses the Appalachian Trail, is one of the most convenient places to examine them. The Civilian Conservation Corps built the center with native stone (mainly gneiss) in the 1930s. The Appalachian Trail passes through the building, making this the only covered portion of the nearly 2,200-mile-long trail. The rock over the store entrance contains a white zigzag pattern; these are quartz veins that were squeezed and bent into isoclinal folds.

At the intersection with GA 60 the road goes around Stone Pile Gap, a large pile of stones that mark the grave of the Cherokee princess Trahlyta. On GA 60, 1.8 miles north of Stone Pile Gap, is a popular rock-climbing locality along the road. Known as the 60s Cliff, or Pruitt Creek Wall, the south-facing

The 60s Cliff on GA 60.

wall is about 150 to 175 feet wide and 40 to 50 feet high. It is dominated by Cowrock terrane gneiss, migmatite, metasandstone, schist, quartzite, and amphibolite with pegmatite veins. About 3.5 miles north of the 60s Cliff is Woody Gap, where the Appalachian Trail crosses GA 60, providing spectacular views to the south.

About 4 miles south of its intersection with GA 60, US 19 crosses the Soque River fault, leaving the Cowrock terrane and passing southward into the Dahlonega gold belt. Along this stretch you may see signs for Georgia's Wine Highway. There are at least six vineyards within 10 miles of US 19 in the vicinity of Dahlonega, and about fifteen vineyards that lie along the trend of the Dahlonega gold belt from here to the North Carolina state line. There is not a particular bedrock type that determines the red, sandy, clayey soil and hilly topography in which the grapes grow best. Rather, a combination of temperature, wind patterns, and slopes that allow adequate drainage are cited for the success of the area's vineyards.

Dahlonega

Consolidated Gold Mine

At the Consolidated Gold Mine you can pan for gold, polished gemstones, and other minerals (from around the world) in Georgia gravel. Unlike the miners who came before, you are guaranteed to have gold in your pan. The mine is the site of the largest gold mining plant ever built east of the Mississippi River and the only place in Georgia where you can tour an underground mine. Led by an entertaining local guide, you descend into damp, electrically lit passageways dug by miners in the early twentieth century. In the mine, yellow-weathering mica-quartz schist and metagraywacke of the Canton Formation are in contact with darker, orange-weathering amphibolite and magnetite-bearing quartzite of the Univeter Formation. Though miners removed the gold-bearing quartz veins early in the twentieth century, microscopic gold remains scattered throughout the rocks.

When gold rose to over $600 an ounce in the 1980s, entrepreneurs figured the rock that hosted the veins still contained enough gold to turn a profit. They removed the water and mud that had filled the Consolidated Gold Mine, but as mining was to begin, the price of gold dropped, so the owners turned the mine into a tourist attraction. The mine is less than 100 yards south of Dahlonega's Walmart Supercenter parking lot. To get there, turn left (east) onto Consolidated Gold Mine Road about 0.3 mile south of the GA 52 intersection (1.1 miles north of downtown Dahlonega). Continue 0.2 mile to the mine's parking lot.

Crisson Gold Mine

You can pan for gold and hunt for gemstones (including rubies, emeralds, and sapphires from all over the world) at the Crisson Gold Mine, established in 1847 and run as a commercial open-pit mine into the early 1980s. The stamp mill that crushes gold ore mined from bedrock is still operating, which allows the mine to continue to produce small amounts of gold. Visitors can see how gold is extracted from the ore. The mine is on the east side of US 19, 2.5 miles north of downtown Dahlonega and 1.8 miles north of the GA 52 intersection.

Visitors descend the stairway next to outcrops of metagraywacke and biotite schist in the Consolidated Gold Mine.

Dahlonega Gold Museum Historic Site

The Dahlonega Gold Museum, operated by the Georgia Department of Natural Resources, occupies the 1836 Lumpkin County courthouse in the heart of Dahlonega (1 Public Square). This is the oldest courthouse building in Georgia, with bricks made of local clay that have been assayed to contain significant gold. The museum has a gold nugget weighing more than 5 ounces, gold coins produced at a Dahlonega branch of the U.S. Mint in the 1800s, historical photographs, and gold mining equipment.

One block south of the gold museum, the Smith House Restaurant sits upon an old gold mine. You can see the mine shaft through a window in the hall downstairs.

Four blocks southwest of the Dahlonega Gold Museum, along College Circle on the campus of North Georgia College, you can see Price Memorial Hall with its gold steeple. This building is on the site of a U.S. Mint branch that produced $6 million worth of gold coins between 1838 and 1861. In 1959, locals donated 43 ounces of Dahlonega gold, which was ceremoniously taken to Atlanta by mule-drawn wagon train and used to gild the dome of the state capitol.

Dahlonega to Cumming

South of Dahlonega, US 19 joins the GA 400 superhighway to Atlanta and bypasses other towns with a gold rush history, such as Auraria (the name of which is derived from the Latin *aurum*, for "gold") and Dawsonville. North of the intersection with GA 53, US 19 crosses the Chattahoochee fault, the

The Dahlonega Gold Museum with its bricks that contain significant gold.

northern boundary of the Tugaloo terrane. North of Cumming, and 7.9 miles south of the GA 53 intersection, US 19 passes through a gap in a low ridge underlain by quartzite. This is near the northern end of an 80-mile-long belt of northeast-trending quartzite and schist within the gneiss units of the Tugaloo terrane. The erosion-resistant quartzite units are rarely more than a few hundred feet thick, so the ridges they underlie are relatively subdued.

The Sawnee Mountain Preserve, near Cumming, has a visitor center with nature and history exhibits. There is a 0.8-mile-long trail to Indian Seats, where you can stand on huge quartzite boulders on a quartzite ridge. The ridge puts you about 400 feet higher than the surrounding terrain, offering views of Mt. Oglethorpe in the Blue Ridge, 20 miles to the north. There are also two abandoned gold mine shafts on the property. (To get there, from exit 17 head west on GA 306 for 3.6 miles to Bettis Tribble Gap Road. Turn right and continue 1.8 miles to the trailhead for Indian Seats. For the visitor center, continue another 0.9 mile to Spot Road, make a right, and continue 0.1 mile to the parking lot on your right.)

A LaFarge Aggregates quarry, just south of Cumming, mines gneiss and amphibolite of the Tugaloo terrane for crushed stone. Boulders at the quarry entrance include garnet-bearing gneiss and amphibolite, pegmatite with pink feldspar, and clear, pointed quartz crystals lining a cavity in gneiss. A visitor overlook provides a view of the quarry operations. To get there, from exit 14 head east on GA 20 for 0.3 mile and turn right onto Ronald Reagan Boulevard. Continue 1.9 miles to the quarry entrance on the left (east) side of the road.

GA 17 (GA 385 and Alt GA 17 in part)
Hiawassee—Toccoa
67 miles

GA 17 passes through Unicoi Gap (elevation 2,949 feet) on the Tennessee Valley Divide, about 9 miles south of US 76. Here, the divide separates the headwaters of the Hiawassee River, which flows north to the Tennessee River, and Spoilcane Creek, which flows south to the Chattahoochee River. The Hiawassee River valley is in the Cowrock terrane at an elevation of about 2,000 feet, or about 600 feet higher than the Chattahoochee River valley at Helen, which is in the Dahlonega gold belt.

From Unicoi Gap you can hike east along the Appalachian Trail for 5.5 miles to the 4,430-foot summit of Tray Mountain. At the top, purple-flowering rhododendron grows among outcrops of lichen-covered gneiss that has weathered to blocky shapes. The summit offers scenic views of the Piedmont to the south. A large boulder field is present on the north side of the mountaintop. Geologists think it formed when the gneiss was exposed to intense cycles of freezing and thawing during Pleistocene time, when continental glaciers covered New England and much of the Midwest. In that cold climate, the top of Tray Mountain was probably above timberline.

South of Unicoi Gap, GA 17 descends steeply past outcrops of biotite gneiss of the Cowrock terrane on the east side of the road. The road passes southward into the Dahlonega gold belt near the intersection with Alt GA 75, 7.7 miles south of Unicoi Gap and just north of Helen.

Smithgall Woods State Park

Alt GA 75 turns west for a scenic drive through the foothills of the Blue Ridge Mountains, passing Smithgall Woods State Park, the site of one of Georgia's first environmental disasters. In the 1850s gold miners destroyed the land by hydraulic mining, using high-pressure water to strip away the soil and weathered bedrock to extract the gold. (The process was outlawed in the 1880s.) As the land began to recover, lumber companies stripped it bare a second time. Charles Smithgall purchased the land, spent more than $20 million on reclamation, and donated 5,555 acres to the state in 1994.

Dukes Creek, which flows through the park, is reportedly the site of the 1828 discovery that led to Georgia's gold rush. The Martin's Mine Trail leads through the former gold mining area near Dukes Creek, passing vertical mining shafts and a tunnel that followed a gold-bearing quartz vein about 900 feet into the hillside.

Unicoi State Park and Anna Ruby Falls

Less than 1 mile north of Helen, GA 17 intersects GA 356. Unicoi State Park, 1.5 miles east of GA 17 on GA 356, is underlain by metagraywacke, schist, and amphibolite of the Dahlonega gold belt. Just west of the park entrance a 3.3-mile-long Forest Service road leads to the Anna Ruby Falls Scenic Area. The short paved trail that leads to the falls is the most-visited hiking trail in the

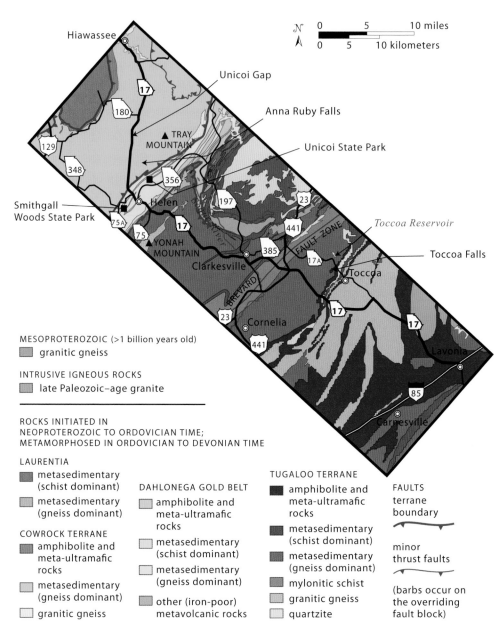

MESOPROTEROZOIC (>1 billion years old)
■ granitic gneiss

INTRUSIVE IGNEOUS ROCKS
■ late Paleozoic–age granite

ROCKS INITIATED IN
NEOPROTEROZOIC TO ORDOVICIAN TIME;
METAMORPHOSED IN ORDOVICIAN TO DEVONIAN TIME

LAURENTIA
■ metasedimentary (schist dominant)
■ metasedimentary (gneiss dominant)

COWROCK TERRANE
■ amphibolite and meta-ultramafic rocks
■ metasedimentary (gneiss dominant)
■ granitic gneiss

DAHLONEGA GOLD BELT
■ amphibolite and meta-ultramafic rocks
■ metasedimentary (schist dominant)
■ metasedimentary (gneiss dominant)
■ other (iron-poor) metavolcanic rocks

TUGALOO TERRANE
■ amphibolite and meta-ultramafic rocks
■ metasedimentary (schist dominant)
■ metasedimentary (gneiss dominant)
■ mylonitic schist
■ granitic gneiss
■ quartzite

FAULTS
terrane boundary
⟶

minor thrust faults
⟶

(barbs occur on the overriding fault block)

Geology along GA 17 between Hiawassee and Toccoa.
(Modified after Thigpen and Hatcher, 2009.)

Chattahoochee National Forest. Two creeks tumble down over gneiss of the Cowrock terrane, forming twin waterfalls. Curtis Creek on the left (west) drops 153 feet, and York Creek on the right (east) drops 50 feet.

Helen to Toccoa

As a gateway from the south to the Blue Ridge Mountains, the Bavarian-themed town of Helen is the third-most-popular tourist destination in the state, after Atlanta and Savannah. Several restaurants and motels overlook the Chatta-hoochee River, flowing southeastward from its headwaters along the Tennessee Valley Divide. Pillow-sized boulders along the river were rounded by abrasion in the riverbed during times of flooding.

South of Helen, GA 17 enters farmland in the Chattahoochee River valley, with fertile soil carried from the mountains and streams and gold carried from the Dahlonega gold belt. Although Helen was not founded until around 1912 (by the logging industry), its county (White County) was second only to Dahlonega's (Lumpkin County) in gold extracted before the Civil War. You can pan for gold and screen for gems (from all over the world) at Dukes Creek Gold and Ruby Mines, about 0.5 mile south of GA 17 on GA 75.

The vineyards on either side of the road in the 5-mile stretch south of Helen are part of Georgia's Wine Highway, which follows the trend of the Dahlonega gold belt and was discussed in the US 19: North Carolina State Line—Cumming road guide.

Two miles south of Helen, GA 17 passes Nora Mill Granary, an authentic, working gristmill. Powered by the Chattahoochee River, its 1,500-pound French burr millstones grind corn and wheat. John Martin constructed the mill in 1876, when he came to the area to mine for gold. About 100 yards south of its intersection with GA 75, GA 17 crosses the Chattahoochee fault, the northern boundary of the Tugaloo terrane. The prominent peak less than 4 miles to the south is Yonah Mountain, which is composed of granitic gneiss similar to that found in many other locations in the Tugaloo terrane.

In the Nacoochee Valley, about 300 yards south of the intersection with GA 75, is a gazebo atop a reconstructed Indian mound. A historical marker states that this is the site of a Cherokee village visited by Hernando de Soto. Excavations in 2004 could not confirm that Cherokee people lived here, although the site was occupied earlier, around the end of the fifteenth century and possibly during the time the Etowah Mounds near Cartersville were occupied (see Ladds Quarry and Etowah Mounds in the I-75: Tennessee State Line—Cartersville road guide in the Valley and Ridge and Appalachian Plateau chapter).

At Clarkesville, you can remain on GA 17 for the partly four lane road that skirts Toccoa, providing the fastest route to I-85 and South Carolina beyond. This road has pretty views as it descends toward Toccoa, but it is outcrop-free. Alternatively, you can follow Alt 17, the old Unicoi Turnpike, for greater scenic, historic, and geologic interest. Follow GA 385 (Historic US 441) northeast from Clarkesville, which connects to Alt GA 17 where the two roads intersect US 23/ US 441. The dark biotite gneiss on the south side of GA 385, 1.5 miles northeast of Clarkesville, is typical of the Tugaloo terrane. The outcrop on the south side of the road 3.7 miles northeast of Clarkesville has a more unusual rock type: light-colored gneiss with coin-sized crystals of muscovite mica.

East of US 23/US 441, the road becomes Alt GA 17. Between about 1 and 3 miles south of US 23/US 441, the road lies within the Brevard fault zone, a prominent 375-mile-long feature within the Tugaloo terrane (see Brevard Fault Zone: A Terrane Boundary That Wasn't). Within this zone, about 1.8 miles south of the intersection, a straight stretch of road crosses Walker Branch valley, which is about 150 feet lower than the hills on either side. This straight valley marks the location of a fault within the Brevard fault zone. It is also the site of continued competition between the Tugaloo and Chattahoochee drainage systems along the Eastern Continental Divide. See Tallulah Gorge State Park in the US 23/US 441 (becoming I-985/US 23): North Carolina State Line—Lake Lanier road guide for more information. Walker Branch drains northeastward to the Tugaloo River and ultimately to the Atlantic, but waters about 1 mile southwest of here drain southwestward to the Chattahoochee and the Gulf of Mexico.

Toccoa Falls.

Alt GA 17 passes the Toccoa Reservoir and begins descending an escarpment about 8 miles south of US 23/US 441 and 3 miles north of Toccoa. Erosion-resistant metagraywacke and quartzite underlie the higher elevations to the north, whereas less-resistant amphibolite underlies the low ground to the south. You can see metagraywacke on the west side of the road, and muscovite-rich quartzite on the east side, as the road continues descending, about 0.5 mile farther south. The resistant escarpment has been caught up in the battle between the Atlantic and Gulf drainage systems. Southeastward-flowing streams, such as Toccoa Creek, have eroded headward into the resistant rock and captured streams at the top of the escarpment that formerly flowed toward the Chattahoochee. The increased streamflow, resulting from the captured stream basin, is part of the reason why Toccoa Falls is one of Georgia's scenic attractions.

Toccoa Falls, just northwest of Toccoa, is one of Georgia's most accessible waterfalls. It is also one of the most beautiful because of its 186-foot sheer drop over a ledge of metagraywacke and biotite gneiss to less-resistant amphibolite below. A 200-foot-long pathway leads to the falls. It was the site of a tragedy in 1977, in which an earthen dam above the falls, built in 1899 to provide hydroelectric power to the Toccoa Falls College, broke after heavy rains. The resulting flood killed thirty-nine people associated with the college, who are remembered in a memorial at the base of the falls. The visitor information center and falls are just north of Toccoa. To get there, turn west onto Forrest Drive into the campus of Toccoa Falls College, 8.7 miles south of US 23/US 441, and follow the signs to the falls.

South of Toccoa Falls, Alt GA 17 passes through the town of Toccoa. A railroad cut on the east side of the road in Toccoa exposes Tugaloo terrane gneiss.

US 23/US 441 (becoming I-985/US 23)
North Carolina State Line—Lake Lanier
83.5 miles

US 23 connects Atlanta to Great Smoky Mountains National Park and the Blue Ridge Parkway, two of the three most popular units of the U.S. National Park system. Highlights along the route include Black Rock Mountain and Tallulah Gorge state parks. All but 1.3 miles of the route lies in the Tugaloo terrane.

US 23 enters Georgia in the long straight valley of the upper reaches of the Little Tennessee River. US 23 begins in the Dahlonega gold belt, but about 1.3 miles south of the state line it crosses a fault and passes onto the Tugaloo terrane, which is dominated by dark metamorphic rocks, including biotite gneiss, schist, and amphibolite. Light-colored bodies of granitic gneiss and granite, up to several miles in extent, are also present. For example, the Rabun Gneiss, which lies east of the Chattahoochee fault, is granite that intruded into surrounding rocks between 374 and 335 million years ago and was later metamorphosed. It forms a body of rock more than 20 miles long and in places more than 2 miles wide. It is exposed west of US 23 at Wolf Fork Road,

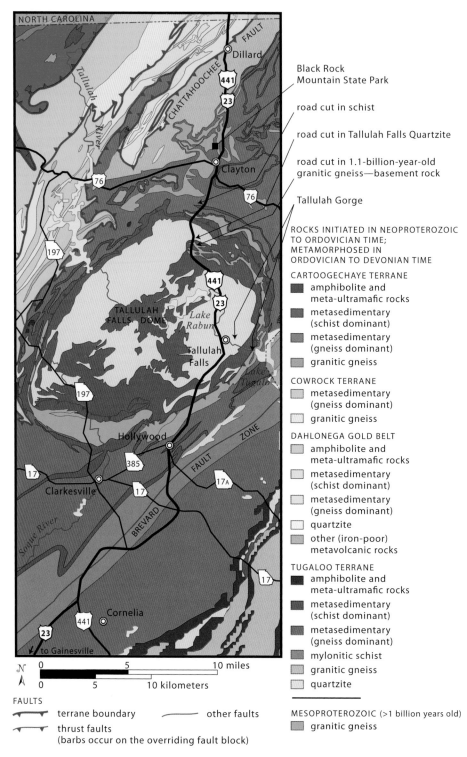

NORTH CAROLINA

Dillard

441

23

CHATTAHOOCHEE FAULT

Clayton

76

76

197

TALLULAH FALLS DOME

Lake Rabun

441

23

Tallulah Falls

Lake Tugalo

197

Hollywood

385

17

17

Clarkesville

17A

BREVARD FAULT ZONE

17

Cornelia

441

23

to Gainesville

Tallulah River

Soque River

Black Rock
Mountain State Park

road cut in schist

road cut in Tallulah Falls Quartzite

road cut in 1.1-billion-year-old
granitic gneiss—basement rock

Tallulah Gorge

**ROCKS INITIATED IN NEOPROTEROZOIC
TO ORDOVICIAN TIME;
METAMORPHOSED IN
ORDOVICIAN TO DEVONIAN TIME**

CARTOOGECHAYE TERRANE
- amphibolite and
  meta-ultramafic rocks
- metasedimentary
  (schist dominant)
- metasedimentary
  (gneiss dominant)
- granitic gneiss

COWROCK TERRANE
- metasedimentary
  (gneiss dominant)
- granitic gneiss

DAHLONEGA GOLD BELT
- amphibolite and
  meta-ultramafic rocks
- metasedimentary
  (schist dominant)
- metasedimentary
  (gneiss dominant)
- quartzite
- other (iron-poor)
  metavolcanic rocks

TUGALOO TERRANE
- amphibolite and
  meta-ultramafic rocks
- metasedimentary
  (schist dominant)
- metasedimentary
  (gneiss dominant)
- mylonitic schist
- granitic gneiss
- quartzite

MESOPROTEROZOIC (>1 billion years old)
- granitic gneiss

N
A

| 0 | 5 | 10 miles |
| 0 | 5 | 10 kilometers |

FAULTS

⊥⊥⊥ terrane boundary ———— other faults

⊤⊤⊤ thrust faults
(barbs occur on the overriding fault block)

*Geology along the northern portion of US 23/US 441.
(Modified after Thigpen and Hatcher, 2009.)*

2.2 miles south of the state line. Here, it is a light gray rock with foliation defined by thin streaks of biotite. You can see different varieties of the Rabun Gneiss in boulders at the entrance to the Vulcan Materials quarry near Dillard. Turn left (east) on Kellys Creek Road, 2.7 miles south of the state line, and continue 1.2 miles to the quarry entrance on your right.

Black Rock Mountain State Park

About 4 miles south of the state line, US 23 passes into darker metasedimentary rocks for which 3,640-foot Black Rock Mountain is named. Black Rock Mountain State Park is Georgia's highest state park. You can see an outcrop of biotite gneiss, showing the effects of multiple folding episodes, at Black Rock Overlook, adjacent to the visitor center. You can also see the drainage basin of the Tallulah River, nearly 1,500 feet below, directing water to the Atlantic Ocean. The paved road in the park follows the Eastern Continental Divide, and only a hundred feet to the north waters flow down a more gradual slope toward the Little Tennessee River, and ultimately the Gulf of Mexico. To reach the overlook turn west onto Black Rock Mountain Parkway, 5.5 miles south of the state line, and continue 2.5 miles to the park visitor center.

Biotite gneiss at the Black Rock Overlook.

Clayton to Tallulah Falls

About 3 miles south of the US 76 intersection in Clayton, US 23 passes onto the Tallulah Falls Dome. On the geologic map, the dome forms an oval about 17 miles long and 11 miles wide. Layers of rock around the margins of the oval seem to double back in an intricate maze, like the edges of leaves in a sliced head of cabbage. This map pattern tells geologists that the rock layers were tightly folded like pleats in fabric. Later, another episode of folding produced the dome shape.

Tallulah Falls Quartzite with a deformed pegmatite dike (white rock).
The vertical lines are drill marks.

Tallulah Falls Quartzite underlies much of the dome. It is more resistant to erosion than the surrounding rocks and forms the mountainous topography surrounding the town of Tallulah Falls. The quartzite is exposed in a road cut on the east side of US 23, 4.7 miles south of Clayton. It is a light gray rock, with little obvious layering, cut by white pegmatite dikes, about 4 inches thick (here mainly composed of the minerals feldspar and quartz), and dark biotite-rich bands, about 2 inches thick, that cross one another. Where the pegmatite dikes encounter the dark bands, they gently bend into alignment with them. The relationship suggests that the dark bands mark zones where the rock flowed at relatively high temperatures—hot enough that the pegmatite dikes, which predated the movement, were able to flow like taffy rather than break. The way in which the dikes are bent shows that the rock above each band was moving in a northward direction relative to the part underneath.

Toward the south end of the outcrop, white patches within the quartzite (about 0.5 inch long and up to 0.25 inch thick) are interpreted as pebbles that were flattened by metamorphism, possibly by the same high-temperature flow that bent the pegmatite dikes.

A few hundred yards south of the quartzite outcrop is a large road cut in granitic gneiss, which has been radiometrically dated at 1.1 billion years. This is probably the most easily accessible outcrop of basement rock in Georgia. In it you will find patches of white feldspar surrounded by biotite, forming augen gneiss. This is a common texture in basement rocks, especially in rocks like granite that have been deformed by faulting at high temperatures.

The association between quartzite and basement rocks is common in Georgia, as it is worldwide. The quartzite was once sand deposited along a coastline as rising sea level gradually flooded the deeply eroded, old landmass Laurentia (the basement rocks). The sand later became quartz sandstone, then metamorphic quartzite. The Tallulah Falls Dome example is a bit unusual because the basement rock layer is only a few hundred feet thick and occurs around the edge of a much thicker body of quartzite. More commonly, the reverse is true: thin layers of quartzite occur on the flanks of a larger outcrop of basement rock. The cause is uncertain, but it may be that only a sliver of basement was brought to the surface on a thrust fault. Here, an unexposed thrust fault may separate the two exposures.

Tallulah Gorge State Park

The entrance to Tallulah Gorge State Park is on the east side of US 23, 10.6 miles south of Clayton. The Jane Hurt Yarn Interpretive Center, with trail access to overlooks and displays on geology, ecology, and history, is 1 mile from US 23. This is the best place to see Tallulah Gorge because stopping is not practical where US 23 crosses the dam impounding Tallulah Falls Lake, 0.4 mile south of the park entrance.

Tallulah Gorge waterfalls as seen from one of the northern overlooks.

Tallulah Gorge is the fourth-deepest canyon east of the Rocky Mountains: 600 feet deep as measured from the rim overlooks to the deepest part of the gorge and deeper if measured from nearby hilltops. The gorge is only about 1,000 feet wide for much of its length, so narrow that it has twice—in 1886 and in 1970—been the setting for a tightrope walk. Because it is so narrow, Tallulah Gorge has steeper slopes than any of the three deeper eastern canyons, which are all more than twice as wide. The narrow gorge limits sun exposure, producing a special environment with unusual green salamanders, wild lily relatives (persistent trillium), and monkey-face orchids.

As you exit the interpretive center, there are overlooks about 300 yards in either direction. To the left, the North Wallenda Tower overlook affords a view of the deepest part of the gorge and marks the northern end of the 1970 tightrope walk by sixty-five-year-old Karl Wallenda. To the right, overlooks provide a view of two waterfalls and access to a staircase that leads to the bottom of the gorge, which is about 350 feet below. Spectacular views reward descent into the gorge, where a swinging bridge crosses the river just above Hurricane Falls. The rocks exposed in the canyon are all Tallulah Falls Quartzite, and the five waterfalls flow over especially resistant quartzite ledges.

The exceptional narrowness of Tallulah Gorge and the steepness of its walls suggest that the gorge was cut relatively recently. If the gorge were as old as the

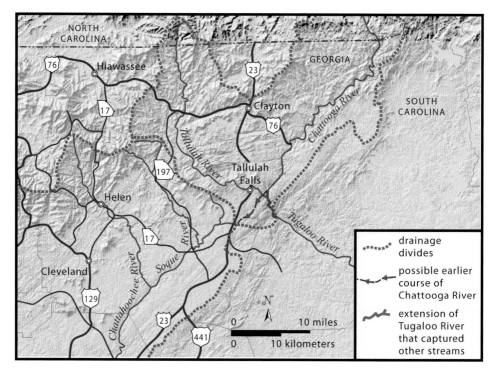

Before the Tugaloo River eroded through the drainage divide separating it from the Tallulah and Chattooga rivers, the flows of those rivers likely drained into the Chattahoochee River. Their capture by the Tugaloo likely caused the carving of Tallulah Gorge.

surrounding Appalachian landscape, gravity-driven erosional processes would have had time to widen it and make its walls less steep. Instead, the steep-walled gorge is likely the result of a stream capture event. The Chattahoochee River basin and the Gulf of Mexico would have been the previous recipients of the waters of both the Tallulah and Chattooga rivers. (Atlanta's water supply issues would be less severe had this remained the case.) At some point, the Tugaloo River to the south, which flows toward the Atlantic, eroded through a ridge separating it from the Chattooga and Tallulah rivers and captured their flows. These rivers began to erode their streambeds down to the level of the Tugaloo, a process that is still going on, and the Tallulah carved its gorge.

Both the Tallulah and Chattooga are renowned for whitewater rapids that result from their steep descent to the level of the Tugaloo River. The Chattooga River, a National Wild and Scenic River, is one of the South's most popular whitewater destinations.

Tallulah Falls to I-85

About 8.7 miles south of the town of Tallulah Falls, US 23 crosses the northwest edge of the Brevard fault zone, a prominent, 375-mile-long band of mylonitic rocks that here is less than 3 miles wide. US 23 splits from US 441 about 1 mile south of the fault zone and heads southwest, paralleling the fault zone until close to its merger with I-985 at Gainesville. From here, US 23/I-985 follows the Brevard fault zone to the intersection with I-85 near Suwanee. The northern edge of the fault zone skirts the southeastern side of Lake Lanier on the Chattahoochee River, which lies about 3 miles to the west. I-985 follows the Eastern Continental Divide. Streams on the west side of I-985 drain to the Gulf of Mexico, and streams on the east side drain to the Atlantic. Off exit 17, just south of Gainesville, the Elachee Nature Science Center has geologic exhibits.

Lake Lanier and Buford Dam

Lake Lanier is impounded behind Buford Dam on the Chattahoochee River. Construction of the dam and powerhouse in the 1950s entailed blasting a channel more than 100 feet deep out of biotite gneiss, schist, and amphibolite and diverting the river's flow into the channel. You can cross the river on a footbridge below the dam to see the walls of dark rock between which the river now flows.

From Buford Dam Road, atop the dam, you can see Lake Lanier, which covers more than 59 square miles and has over 692 miles of shoreline. When lake levels are low, you can see exposures of mylonitic rock in the Brevard fault zone at many places. To get there, from exit 4 on I-985 head northwest on GA 20 for about 3 miles and turn right at Sycamore Road. After 2.6 miles turn left onto Buford Dam Road. Turn left after 1.5 miles at the sign "Lower Pool/ East Powerhouse" to reach the footbridge below the dam, or continue another 0.5 mile to cross the dam.

The Buford Dam Powerhouse and outflow channel as seen from the footbridge at low flow. —Courtesy of Chuck Cochran

Outcrops of mylonitic metagraywacke, schist, and amphibolite in the Brevard fault zone along the south side of Lake Lanier during low water level.

I-85
South Carolina State Line—
I-985 (near Atlanta)
94 miles

See the map on page 242.

I-85 crosses Hartwell Lake on the Tugaloo River at the Georgia–South Carolina border. The Tugaloo River merges with other rivers downstream to form the Savannah River. Hartwell Lake is impounded behind a dam about 17 miles downstream, to the southeast.

Between South Carolina and Atlanta, I-85 passes through metamorphic rocks of the Tugaloo terrane and runs roughly parallel to the Brevard fault zone, which lies about 20 miles to the northwest. Around Atlanta I-85 converges with the fault zone. On a clear day the Blue Ridge Mountains are visible in the distance to the northwest in some places along this route.

At exit 177, signs direct you to Elberton, self-proclaimed Granite Capital of the World. It is about 25 miles south of I-85 on GA 77 (see the GA 72: Athens—South Carolina State Line road guide).

Hurricane Shoals Park

The shoals at Hurricane Shoals Park comprise a rocky area of granitic gneiss along the North Oconee River. The park made the news in 1985 when, on a hot July day, the rock naturally exploded, leaving broken slabs of rock, about 1 inch thick, loose on the surface. This was a dramatic example of the process of exfoliation. The granitic gneiss, like nearly all rocks of the Piedmont, formed many miles below the surface under tremendous pressure. After millions of years of uplift and erosion, the rock reaches the surface, where it naturally expands because the pressure has been removed. Heat can accelerate the expansion and cause the rock to pop up and make a booming sound.

To get there, take exit 140 (Dry Pond Road) and head east, and then north toward Maysville on GA 82 and GA 82 Spur. After crossing over I-85, proceed less than 1 mile to the park entrance on the right.

Nodoroc

Exit 126 leads to Winder, a small town about 10 miles south of I-85. Legend has it that in the late 1700s, about 3 miles east of Winder, there was a 4-acre lake of boiling blue or black mud near the head of Barber Creek, across from what is now the Winder-Barrow Airport. Smoke rose from bubbles at the surface, and bluish flames were occasionally seen. Some sources refer to it as a "mud volcano" or "burning lake of fire," with foul-smelling gases and a plume of black smoke that could be seen for miles. No vegetation grew near the lake, and trees in the vicinity were dwarfed. The Indians called it *Nodoroc*, meaning "hell." Shortly after white settlers arrived, the lake erupted with a loud rumbling noise, and hot mud was thrown up into the air. The lake quieted down after the eruption, was drained, and later was converted to farmland. The soil in the low area is black and soft with high organic content, suggesting the boiling lake was actually a peat bog fire.

Earthquakes at Dacula

About 10 miles south of exit 120, the area just south of the small town of Dacula experienced a swarm of small earthquakes in 1995 and 1996. Residents heard sounds like muffled explosions and felt sporadic rumblings. Agents from the Bureau of Alcohol, Tobacco, and Firearms investigated, suspecting a mad bomber in the woods. Agents contacted seismologists at the Georgia Institute of Technology, who placed seismometers in the area and determined that earthquakes, with magnitudes less than 2 on the Richter scale, were being caused by slight movements along shallow fractures in bedrock that had been weakened by groundwater and weathering. The quakes were heard and felt over only a small area, about 6 miles in diameter, and stopped as suddenly as they started. The only reported damage was a broken window.

US 78
Snellville—Thomson
111 miles

See the map on page 268.

Heading eastward from Stone Mountain into Snellville, US 78 passes through an area of granitic gneiss called the Lithonia Gneiss. It is interrupted by narrow belts of quartzite that underlie ridges such as Lanier Mountain, a 180-foot hill capped by five radio towers just west of Snellville and south of US 78, and Snellville Mountain, just north of US 78 and west of GA 124.

Bakers Rock, 2 miles south of Snellville and off GA 124, is a massive outcrop of Lithonia Gneiss with vernal pools hosting two endangered plant species: the black-spored quillwort and the pool sprite. (Bakers Rock is similar to Arabia Mountain; see Davidson-Arabia Nature Preserve in the Around Atlanta road guide.)

Athens

Athens was founded where an Indian trail crossed a shallow place on the North Oconee River called Cedar Shoals. The shoals and the town are underlain by amphibolite, ultramafic rocks, and biotite gneiss with pegmatite veins. A gneiss outcrop, where according to legend founders of the University of Georgia stood in 1801 when deciding where to build, is on the south side of US 78 Business Route, just downstream of where it crosses the North Oconee River, along the east bank. Several interesting textures can be seen in the outcrop, which are related to the partial melting of the gneiss. The western side of the outcrop is gneiss, with folded light and dark layers, but the eastern side is migmatite, in which the gneiss partially melted and recrystallized as granite. The migmatite still preserves some of the metamorphic, biotite-rich layers. A detailed study of the minerals indicates that the rock developed about 15 miles beneath the surface, where temperatures reached about 1,400°F.

You can see outcrops of Athens Gneiss along the Oconee River on trails at the State Botanical Garden, located at 2450 S. Milledge Avenue in Athens, as well as at the Rock and Shoals State Natural Area, which is 5 miles southeast of Athens. To reach the natural area, take Barnett Shoals Road south from US 78. About 1 mile past the elementary school, look for the Rock and Shoals subdivision on

A core of Athens Gneiss at the University of Georgia in Athens. It was drilled during the construction of an elevator shaft on campus. The core, nearly 8 inches in diameter, has xenoliths and distinct banding, both of which are evidence that it is a metamorphosed igneous rock. —Courtesy of Mike Roden

Athens Gneiss with a dark xenolith and folded dark and light layers, on display in the Geography-Geology Building at the University of Georgia. The block was excavated during construction of the Athens Bypass. —Courtesy of Mike Roden

the right, with a pile of rocks at the entrance. Turn right and go to the end of the paved road. Look for a mailbox at the end of the road containing maps to the outcrop. Go down the stairs and follow the trail about 1,000 feet into the woods, crossing a stream at the shoals. Proceed northward up a steep slope about 300 feet to the flat rock, which is called a pavement outcrop. The outcrop contains weathered granitic gneiss with folds and migmatite, evidence of partial melting. The outcrop is cut by pegmatite veins and has solution pits.

Athens to Washington

East of Athens, US 78 crosses gneiss and schist of the Tugaloo terrane for about 6 miles until it reaches the northwestern edge of the Cat Square terrane near the Clarke-Oglethorpe county line. About 3 miles east of this boundary, you will begin to notice white quartz- and kaolinite-rich soil that has developed on the Elberton Granite, the largest granite body in Georgia.

Shaking Rock Park, just off US 78 on Shaking Rock Road in Lexington, gets its name from a 27-ton boulder balanced atop a granite outcrop. At one time it was possible to shake the boulder by hand. There are a number of unusually shaped granite outcroppings throughout the park that formed by natural weathering processes. The area was once a Creek and Cherokee Indian camping ground. There is small outcrop of granite on the south side of US 78 about 0.4 mile east of the center of Lexington.

In 1773 the botanist William Bartram visited what was called the Great Buffalo Lick, an expanse of white kaolin clay formed from the weathering of feldspar in the Elberton Granite. Buffalo, deer, cattle, and other animals came to lick the clay, which probably contained salts. The location of the lick is uncertain, but the most recent research places it about 0.4 mile south of the Buffalo Creek crossing on GA 22, 8 miles south of Lexington.

US 78 crosses the eastern edge of the Elberton Granite about 6.2 miles east of Lexington, just west of Buffalo Creek, and passes onto rocks of the Carolina superterrane. The contact is marked by a change in slope, with the steeper terrain belonging to the granite. The Carolina superterrane was a volcanic island arc that collided with North America about 350 million years ago and was thrust northwestward over the Cat Square terrane.

The town of Washington, with its colonial-style homes, dates back to 1780. It was here that Jefferson Davis, President of the Confederacy, held his last cabinet meeting and signed the papers that dissolved the Confederacy. Legend has it that the gold from the Confederate treasury (mined in Dahlonega) is buried somewhere on the grounds of the Chennault House, about 14.5 miles north of Washington near the intersection of GA 44 and GA 79.

Graves Mountain

Graves Mountain, 15 miles east of Washington on US 378, is Georgia's best-known mineral collecting locality, with black rutile crystals, kyanite, lazulite, pyrophyllite, iridescent hematite, blue quartz, pyrite, barite, and sulfur. The rock at the top of the mountain is highly weathered, having been leached of iron and several other elements. The result is a white, fine-grained rock that is rich in quartz and kyanite. The iron oxides leached from the uppermost

Rutile crystals from Graves Mountain.

15 cm

portion of the mountain were precipitated in the rocks in the middle zone of the mountain, staining it reddish brown. The lowermost part is unweathered, light bluish gray kyanite-bearing rock. The western side of the mountain was mined for kyanite, which is used in a variety of products to help them withstand high temperatures, such as furnace linings, spark plugs, brake shoes, and molds into which molten metal is poured.

The rock of Graves Mountain was part of the volcanic island chain of the Carolina superterrane. It began as volcanic ash deposited on the seafloor that was altered by hot mineral-rich fluids and then metamorphosed. Prior to metamorphism, quartz, kaolinite, and pyrite were probably the predominant minerals in the rock.

The mountain is visible from US 378, reaching an elevation of about 890 feet and standing about 300 feet above the surrounding terrain. The mine is open to the public for a small fee on certain days during the year, typically in April and October. Check the Georgia Mineral Society website for current information. The entrance is about 0.8 mile past the Lincoln County line, at a gated road. Because of the pyrite that was once mined here, the water in and around the mine is extremely acidic, so visitors are cautioned to avoid standing water.

Lincolnton Metadacite

An outcrop north of Washington is the closest you can come in Georgia to visiting an ancient volcano. This rock, in the Carolina superterrane, was part of a volcanic island chain 570 to 560 million years ago. The outcrop contains what's called the Lincolnton Metadacite, the metamorphosed equivalent of a fine-grained volcanic rock like that erupted from Mount St. Helens in Washington

Lincolnton Metadacite. —Courtesy of Susan Finazzo

State. The Lincolnton Metadacite has relatively large crystals of bluish gray quartz and plagioclase in a finer-grained matrix. The rock was originally deposited as a series of lava and ash flows.

To get there, from Washington head northeast toward Lincolnton on US 378. In Lincolnton, head north on GA 79 for about 3 miles and turn left (west) onto Prater Road, and then left (south) onto Aycock-Norman Road. You can see metadacite on the hillside south of Curry Creek.

Washington to Thomson

About 16 miles south of Washington, US 78 crosses an arm of J. Strom Thurmond Lake. The reservoir is impounded behind Clarks Hill Dam on the Savannah River, about 20 miles to the east. The area around the lake has experienced earthquakes from reservoir-induced seismicity. A number of faults are present near the lake, including one sealed by concrete during dam construction. US 78 enters the Modoc fault zone near where the fault zone crosses part of the lake. The Modoc fault zone is an ancient zone of ductile shearing that is inactive and unlikely to produce earthquakes. However, it also experienced brittle faulting, perhaps as recently as 200 million years ago. Mineral deposits have mostly sealed the brittle faults, but some fractures may remain, which could be producing the small earthquakes as lake waters penetrate down into them. Geologists have been unable to establish a direct relationship between earthquakes and the faults in the area.

US 78 intersects I-20 just north of Thomson. About 3 miles south of I-20, US 78 enters the Atlantic Coastal Plain, where unconsolidated sediments and sedimentary rocks overlap the crystalline rocks of the Piedmont.

GA 72
Athens—South Carolina State Line
48 miles

Athens lies in the Tugaloo terrane, a part of the inner Piedmont underlain by biotite gneiss, mica schist, and amphibolite. About 2 miles east of Comer, you cross a terrane boundary, the northwest edge of the Cat Square terrane, which is composed of schist and gneiss of high metamorphic grade.

Watson Mill Bridge State Park

A good exposure of the rocks in the Cat Square terrane can be seen at Watson Mill Bridge State Park. The covered bridge over the South Fork Broad River is the longest in the state, at 229 feet.

The rock outcrops here consist of biotite gneiss and migmatite. The flat outcrops in the shoals below the covered bridge are excellent exposures. The rock has foliation, which developed as flat biotite flakes were aligned perpendicular to the direction of pressure during metamorphism. Farther downstream, the foliation is folded. Granite dikes intruded into the gneiss near the downstream end of the rock exposure. Granite dikes are also present in the gneiss along the path just below the bridge.

To get there, 2.7 miles east of Comer turn right onto Old Fork Cemetery Road and follow the state park signs for 2.3 miles to the park entrance.

Shoals beneath the covered bridge at Watson Mill Bridge State Park. —Courtesy of Alan Cressler

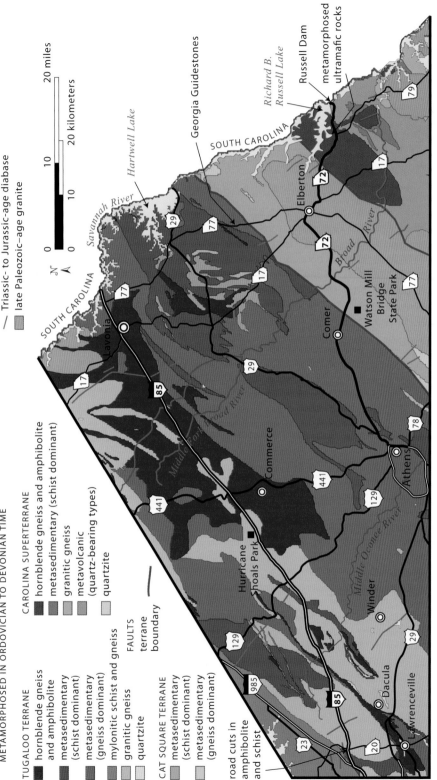

INTRUSIVE IGNEOUS ROCKS

— Triassic- to Jurassic-age diabase

late Paleozoic–age granite

ROCKS INITIATED IN NEOPROTEROZOIC TO ORDOVICIAN TIME;
METAMORPHOSED IN ORDOVICIAN TO DEVONIAN TIME

TUGALOO TERRANE

hornblende gneiss and amphibolite

metasedimentary (schist dominant)

metasedimentary (gneiss dominant)

mylonitic schist and gneiss

granitic gneiss

quartzite

CAROLINA SUPERTERRANE

hornblende gneiss and amphibolite

metasedimentary (schist dominant)

granitic gneiss

metavolcanic (quartz-bearing types)

quartzite

FAULTS

terrane boundary

CAT SQUARE TERRANE

metasedimentary (schist dominant)

metasedimentary (gneiss dominant)

road cuts in amphibolite and schist

Geology along GA 72 and I-85 east of Atlanta.

Elberton

GA 72 crosses the Broad River into Elbert County about 2 miles northeast of Carlton. On the east side of the river you are in Elberton Granite country, and you will occasionally see rounded granite boulders along the road.

Elberton is not a place that you are likely to stumble upon, tucked away as it is in eastern Georgia, off the beaten path. It lies at the center of the Elberton batholith, the largest granite body in the state, and calls itself the Granite Capital of the World. It is impossible to drive through Elberton without seeing granite monuments of all types along the road.

The local granite is bluish gray and prized for its fine-grained, even texture. It is commonly used for tombstones, monuments, curbs, and building exteriors, and for interior uses such as countertops. Look carefully the next time you pass a cemetery in Georgia. The fine-grained, bluish gray stone is probably from Elberton.

The first quarry opened in 1882 near the Broad River to supply stone to construct a railroad. Today in the Elberton area, there are 45 granite quarries, 150 granite processing or manufacturing plants that cut and polish the stone, and 250 separate granite businesses, together employing more than 2,100 people. More than 2 million cubic feet of stone are produced each year and shipped all over the world. At a weight of about 180 pounds per cubic foot, this is a

A large xenolith surrounded by bluish gray Elberton Granite exposed in the side of a quarry. Note the truck for scale. —Courtesy of Paul Schroeder

Dutchy at the Elberton Granite Museum.
—Courtesy of Elberton Granite Association

whopping 180,000 tons of granite each year. The granite industry is the basis of the local economy, with about $175 million in sales each year. As you pass through the Elberton area, you will see quarry entrances and numerous cutting sheds where granite is cut, engraved, and polished.

The Elberton Granite Museum on GA 72 contains historical and educational exhibits on the history of the granite industry as well as methods of quarrying, cutting, and polishing. At the museum you can see the first statue ever made of Elberton Granite. Unfortunately, although it was supposed to be a statue of a Confederate soldier, it was described as looking like a cross between a Pennsylvania Dutchman and a hippopotamus, hence the nickname "Dutchy." Citizens lynched the statue in 1900 by tying a rope around it and pulling it to the ground. They unceremoniously buried the statue in the town square. In 1982, Dutchy was unearthed and put on display in the museum.

Granite is harder than steel because it is composed of quartz, feldspar, and biotite. Granite is so hard that it has to be cut with saw blades impregnated with industrial-grade diamonds or wire saws consisting of steel wire impregnated with diamonds. Wire saws are used to cut granite into intricate shapes. The feldspar in the Elberton Granite, also called Elberton Blue, has minute inclusions of a mineral (probably magnetite) that gives it a bluish gray color.

Elberton Granite. Penny for scale.

Many types of stone of varying colors, both domestic and imported, are cut and polished in Elberton and marketed under the trade name of "granite." The mineral composition of true granite has at least 20 percent quartz and feldspar. "Black granite" and "red granite" are trade names for other types of igneous rock, such as diorite, gabbro, anorthosite, and syenite.

Today, jet burners are used to cut granite out of the bedrock. The burner is a miniature rocket motor attached to a long steel pipe. It combines kerosene or fuel oil with compressed air to form a flame. It generates heat of 2,800°F and causes the rock to expand and disintegrate, or flake off, forming a 4-inch-wide channel. Before jet burner torches were invented, workers removed blocks of granite using a much different method. They drilled lines of closely spaced holes in the granite bedrock and packed them with explosives. The detonation separated the blocks from the bedrock. You may see curbstones with cylindrical drill holes along their edges. This method of stone cutting was used until the 1950s.

Georgia Guidestones

About 8 miles north of Elberton, just off GA 77 at the highest point in Elbert County, you can see the mysterious Georgia Guidestones, sometimes called "America's Stonehenge." The guidestones are massive blocks of Elberton Granite more than 19 feet tall. Text carved into the stones, in twelve languages, gives

The Georgia Guidestones, north of Elberton. —Courtesy of the Elberton Granite Association

ten guidelines, or commandments, to mankind. The stones are aligned to various astronomical coordinates, and the monument also serves as a solar calendar. Sunlight passes through a 7/8-inch hole at noon and shines on a stone, which is inscribed with the days of the year. A group of anonymous donors commissioned the monument in 1979.

Elberton to the South Carolina State Line

About 6 miles east of Elberton, GA 72 enters the Carolina superterrane, which is underlain by metamorphosed volcanic rocks and volcanic sediments. There are a number of small, metamorphosed ultramafic rock bodies on hilltops. These are erosional remnants of a larger, more extensive sheet of meta-ultramafic rocks that was thrust over the Carolina superterrane. Geologists refer to the sheet as the Russell Lake allochthon. The word *allochthon* means that the rocks were transported from somewhere else by a thrust fault. You can see some of these meta-ultramafic rocks near Richard B. Russell Lake at the Elbert Boat Ramp north of GA 72, and along Bobby Brown Park Road to the south.

GA 72 ends at the Georgia–South Carolina state line at the lake, which is impounded by Russell Dam on the Savannah River. Like J. Strom Thurmond Lake just downstream, Richard B. Russell Lake has experienced small earthquakes attributed to reservoir-induced seismicity.

I-185
LaGrange—Columbus
52 miles

I-185 is a 52-mile spur connecting Fort Benning, just south of Columbus, with I-85 just east of LaGrange. Most of the route lies in the Piedmont, but near Columbus I-185 crosses the Fall Line and enters the Coastal Plain. There are few outcrops along the interstate other than two large road cuts north of Columbus.

I-185 crosses three mylonite zones. These ancient fault zones are named, from north to south, the Towaliga, Bartletts Ferry, and Goat Rock. Mylonite zones are areas where rock experienced ductile deformation, in which the rock slipped along thousands of parallel surfaces that smeared it out like hot taffy under high temperatures and at great depths. By examining the way minerals are bent in and near these zones, geologists are able to determine the direction the rock bodies moved. The rocks south of each mylonite zone shifted to the southwest relative to the rest of North America north of the zones. Radiometric dating places the time of movement around 297 to 280 million years ago, toward the end of the collision between North America and Africa that formed part of the supercontinent Pangaea.

From 6 to 20 miles to the west, the Chattahoochee River roughly parallels I-185. Between I-85 and Columbus, six dams were built on the river for hydroelectric power generation. Two dams, the Goat Rock and the Bartletts Ferry, were built on mylonite zones. The question arises, why were dams built on faults? Doesn't this seem like a bad idea? Actually, the mylonite is relatively rich in quartz, a hard mineral that is resistant to weathering and erosion. Therefore, mylonite can make ideal foundations for dams.

I-185 crosses the Goat Rock mylonite zone 1 mile south of GA 315, where there is large road cut on the west side of the road. The Goat Rock mylonite is a dark and fine-grained gneiss with biotite, very-fine-grained quartz, and a scattering of light-colored, fingernail-sized feldspar and muscovite mica crystals. A few miles to the south (0.4 mile north of Smith Road, exit 14), a large road cut, also on the west side of I-185, reveals gneiss that has been partially melted, meaning it's migmatite.

South of the US 80 interchange, I-185 leaves the Piedmont and enters the Coastal Plain. To the west the Chattahoochee River crosses the Fall Line in downtown Columbus, where you can see the rapids from the Riverwalk (see the GA 85: Atlanta—Columbus road guide).

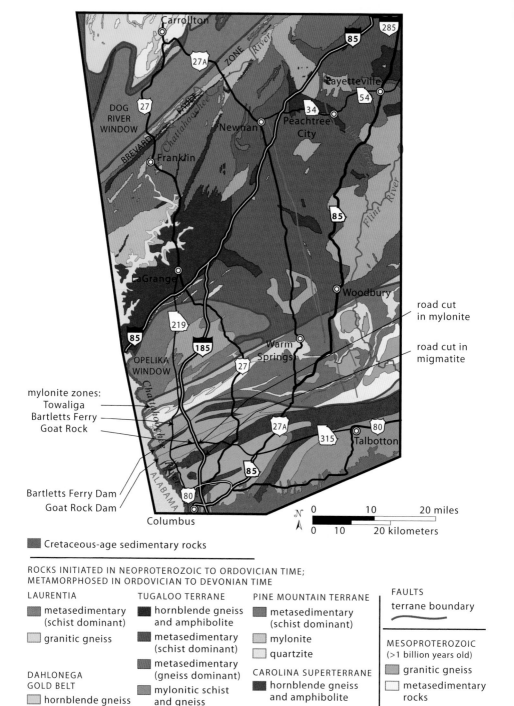

road cut
in mylonite

road cut in
migmatite

mylonite zones:
Towaliga
Bartletts Ferry
Goat Rock

Bartletts Ferry Dam
Goat Rock Dam

Cretaceous-age sedimentary rocks

ROCKS INITIATED IN NEOPROTEROZOIC TO ORDOVICIAN TIME;
METAMORPHOSED IN ORDOVICIAN TO DEVONIAN TIME

LAURENTIA
metasedimentary
(schist dominant)

granitic gneiss

**DAHLONEGA
GOLD BELT**
hornblende gneiss
and amphibolite

metasedimentary
(schist dominant)

granitic gneiss

TUGALOO TERRANE
hornblende gneiss
and amphibolite

metasedimentary
(schist dominant)

metasedimentary
(gneiss dominant)

mylonitic schist
and gneiss

granitic gneiss

quartzite

CAT SQUARE TERRANE
metasedimentary
(gneiss dominant)

PINE MOUNTAIN TERRANE
metasedimentary
(schist dominant)

mylonite

quartzite

CAROLINA SUPERTERRANE
hornblende gneiss
and amphibolite

metasedimentary
(schist dominant)

metasedimentary
(gneiss dominant)

granitic gneiss

mylonite

FAULTS
terrane boundary

MESOPROTEROZOIC
(>1 billion years old)
granitic gneiss

metasedimentary
rocks

**INTRUSIVE
IGNEOUS ROCKS**
Triassic- to
Jurassic-age diabase

late Paleozoic–age
granite

*Geology along I-185 and GA 85 southwest of Atlanta.
(Compiled from Lawton, 1976; and Hatcher, 2011.)*

GA 85
Atlanta—Columbus
97 miles

GA 85 begins from I-75 (exit 237A) just south of the I-285 interchange in Atlanta. The first 48 miles of the route, leading to Woodbury, pass through a region as flat as many parts of the Coastal Plain. The road remains between 700 and 900 feet above sea level for the entire distance. One reason the route is so flat is because much of it lies in the drainage basin of the Flint River, upstream of the resistant ledges of Hollis Quartzite in the Pine Mountain terrane. Other rivers, such as the Chattahoochee River to the west or the Ocmulgee River to the east, gradually descend across the Piedmont from the Atlanta region to the Coastal Plain. They and their tributaries have cut into the whole Piedmont landscape, creating valleys separated by hilly divides. The Flint River drops very little upstream of the Hollis Quartzite of Pine Mountain, and so it has not cut into the landscape much. At Pine Mountain it still has a lot of work to do clearing the wide band of resistant rock out of its way, and at present its drop in elevation is almost entirely in the gorge at that location.

Geologically, this flat area is part of the Tugaloo terrane. Most of the area is underlain by medium to dark gray gneiss, schist, and amphibolite; however, most of the rock outcrops are light gray granite and granitic gneiss, a minor rock type overall in the Tugaloo terrane. About 1.5 miles south of the I-75 interchange, there is a small road cut in granitic gneiss on the west side of GA 85. This is the only hard bedrock you will see between Atlanta and Woodbury.

Within a few miles of the highway there are natural outcrops of granitic gneiss. The most accessible is along GA 18/GA 109, 4.2 miles west of its intersection with GA 85 in Woodbury. The freshest (least weathered) rock is on the right (north) side of the road, but the majority of the natural outcrop appears on the south side of the road, extending for nearly 200 yards to the southwest. The dark bands (foliation) are composed of the mineral biotite, and they are interlayered with white quartz and feldspar. The foliation is folded nearly everywhere it is visible, showing that, as throughout the Tugaloo terrane and most of the Blue Ridge–Piedmont, these rocks flowed like taffy in response to being squeezed at high temperatures.

The Cove

Just southeast of Woodbury, hills rise about 300 feet above their surroundings. A wealth of intriguing geologic features are visible along Cove Road, 0.6 mile south of the GA 18 intersection in Woodbury.

The first feature is a ridge 2.1 miles southeast of the GA 85 intersection. It lies within the Towaliga fault zone, which here is the terrane boundary between the Tugaloo and Pine Mountain terranes. Blocks of fine-grained, silica-rich rock with quartz-filled fractures are evident on the west side of Cove Road. These are examples of cataclasite, or shattered rock (breccia) that was infused with quartz veins. Recent work suggests this cataclasite, found along the Towaliga and other faults, formed as water with dissolved elements surged into fractures after rock shattered during earthquakes. These earthquakes happened during a brittle

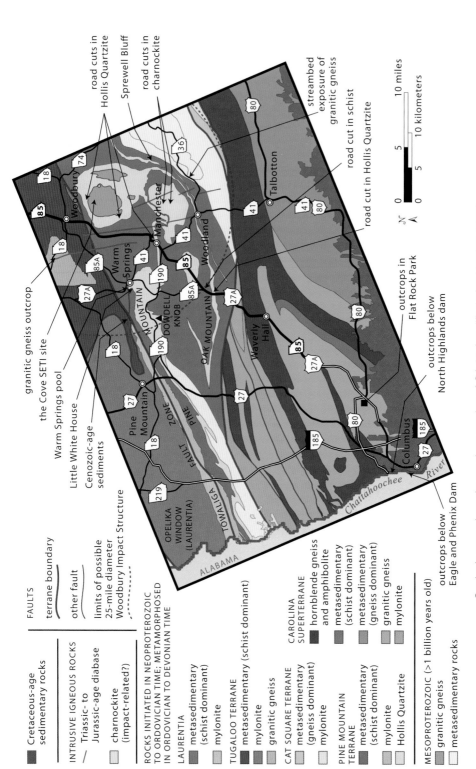

Geologic map along the southern portion of GA 85 (modified after Hatcher, 2011). For the geology along the northern portion, see the map on page 248.

FAULTS
— terrane boundary
— other fault
— limits of possible 25-mile diameter Woodbury Impact Structure

INTRUSIVE IGNEOUS ROCKS
Triassic- to Jurassic-age diabase
charnockite (impact-related?)

Cretaceous-age sedimentary rocks

ROCKS INITIATED IN NEOPROTEROZOIC TO ORDOVICIAN TIME; METAMORPHOSED IN ORDOVICIAN TO DEVONIAN TIME

LAURENTIA
metasedimentary (schist dominant)

TUGALOO TERRANE
metasedimentary (schist dominant)
mylonite
granitic gneiss

CAT SQUARE TERRANE
metasedimentary (gneiss dominant)
mylonite

CAROLINA SUPERTERRANE
hornblende gneiss and amphibolite
metasedimentary (schist dominant)
metasedimentary (gneiss dominant)
granitic gneiss
mylonite

PINE MOUNTAIN TERRANE
metasedimentary (schist dominant)
mylonite
Hollis Quartzite

MESOPROTEROZOIC (>1 billion years old)
granitic gneiss
metasedimentary rocks

granitic gneiss outcrop
the Cove SETI site
Warm Springs pool
Little White House
Cenozoic-age sediments
Woodbury Impact Structure

road cuts in Hollis Quartzite
Sprewell Bluff
road cuts in charnockite

streambed exposure of granitic gneiss
road cut in schist
road cut in Hollis Quartzite
outcrops in Flat Rock Park
outcrops below North Highlands dam
outcrops below Eagle and Phenix Dam

ALABAMA
OPELIKA WINDOW (LAURENTIA)
TOWALIGA
FAULT
PINE
ZONE
Pine Mountain
MOUNTAIN
DOWDELL KNOB
OAK MOUNTAIN
Woodbury
Warm Springs
Manchester
Woodland
Talbotton
Waverly Hall
Columbus
Chattahoochee River

0 5 10 miles
0 5 10 kilometers

N

Cataclasite in the Towaliga fault zone along Cove Road.

period of movement on the fault, thought to have occurred when Georgia was pulling apart and diabase dikes were forming about 200 million years ago.

About 3 miles south of GA 85, on the left side of the road, are outcrops of Hollis Quartzite. Much of the rock is weathered into large, flat pieces about 0.25 inch thick. Quartzite (like sandstone) is made of individual grains, mostly of the mineral quartz. During weathering, or even before the rock is exposed at the surface, fluids can partially dissolve the rock along the boundaries between grains. If the rock is weathered enough, you can easily crumble it into loose quartz grains, otherwise known as sand. Some of the Hollis Quartzite here—and quartzite and sandstone at a few other places worldwide, including Itacolumi, Brazil—has weathered to an intermediate stage, in which the grains have separated but are still held together by their interlocking nature. This allows long, thin blocks of the rock to bend without breaking. This so-called flexible rock is known as itacolumite.

Geologists have not determined why, given that all sandstone and quartzite will gradually weather to loose sand, this flexible transition state is not more common. It's possible that thin layers of mica, which is flexible in small sheets, or special quartz grain shapes hold the rock together, but neither explanation seems to cover all examples of itacolumite.

The outcrop of quartzite lies at the crest of a ridge that surrounds the area known as the Cove. Oddly, seen on a map, these hills and their underlying

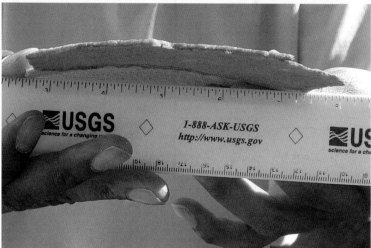

A thin piece of Hollis Quartzite itacolumite, straight and bent.

quartzite form a near-perfect circle that is 3.5 miles in diameter. The geologic structure responsible for the ring of quartzite is called the Cove Dome, a region of rock that was arched upward into a domelike shape. Erosion cut through the less-resistant rocks that existed at the top of the dome, leaving a ring of quartzite hills around the dome's edge. In 2006, some geologists proposed that the Cove Dome is part of a larger structure (the Woodbury Impact Structure) formed by extraterrestrial impact (perhaps a meteorite), like the craters on the moon.

About 1 mile south of the quartzite outcrop is yet another surprise. Because the Cove is ringed with hills that screen out some radio interference, AT&T erected two satellite dishes here as part of its communications network. Later,

Georgia Tech purchased the site to use the receivers as part of the NASA-funded Search for Extraterrestrial Intelligence (SETI). It would be humorous if listening for radio signals of extraterrestrial origin had taken place near the center of a structure of possible extraterrestrial origin.

If you continue on Cove Road an additional 6 miles south, it rejoins GA 85 in Manchester. Manchester lies on the north flank of Pine Mountain, a 21-mile-long ridge underlain by Hollis Quartzite. The quartzite is exposed at several places around Manchester, including along the west side of GA 85 near its intersection with GA 190, 1 mile south of town, and 2 miles east of town near Foster Street along the CSX rail line.

Warm Springs and Little White House Historic Site

Warm Springs, 5.3 miles north and west of Manchester on GA 41, was the retreat of President Franklin D. Roosevelt and the place where he died on April 12, 1945. He came to town in 1924, after he developed polio, because of what he believed were the healing powers of its warm water. Warm Springs is Georgia's largest and most famous warm spring, delivering 914 gallons of water per minute at 88°F year-round.

Springs are present along the base of Pine Mountain because the Hollis Quartzite provides a pathway for underground water movement. Groundwater usually moves in tiny interconnected pores and cracks in rocks, and only rarely as the underground streams in caves that many people imagine. Water enters the cracks and pores at a relatively high elevation, descends through the rock due to gravity, and then flows out of the eroded edges of the rock at the surface. In the case of the Hollis Quartzite, rainwater soaks into the ground high on Pine Mountain and emerges along the base of the mountain in a series of springs. Since quartzite, as a metamorphic rock, contains very few pores, some geologic process has altered the rock to facilitate this groundwater movement. Some of the quartzite may have partially dissolved, forming more-porous rock. It's also possible that the rock cracked during folding and faulting.

The fact that many of these springs are much warmer than normal groundwater means some unusual mechanism is involved in the genesis of these springs. There is no magma below to heat the water, as there is for the famous hot springs and geysers at Yellowstone National Park. The water can only warm up by descending thousands of feet below the surface. In Georgia, as in other parts of the world distant from tectonic plate boundaries, temperature is estimated to rise only about 1°F for every 100 feet of depth. (This temperature rise occurs because Earth's interior is hot, driven by the energy released by naturally radioactive materials.) Based on this figure, the water of Warm Springs must somehow descend about 3,000 feet underground and then return to the surface.

Geologists estimate that this is approximately the depth where descending water, moving in a porous zone near the base of the north-dipping Hollis Quartzite, would encounter the Towaliga fault zone. Geologists presume that mylonite, a nonporous rock type known to be present along the fault zone, blocks the descent of the water and deflects it upward to another porous zone near the top of the Hollis Quartzite. From there it exits the strata at Warm Springs.

rain in recharge zone

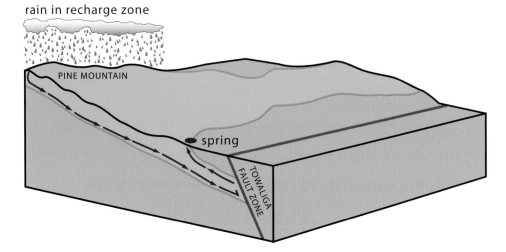

The likely path of water in the Warm Springs area. Water enters the Hollis Quartzite (green) atop Pine Mountain and descends through a porous zone until it encounters a barrier of mylonitic rocks along the Towaliga fault zone (at a depth of about 3,000 feet where temperatures are about 30°F higher than at the surface), which deflects the warmed water back to the surface at Warm Springs.

In the subsurface, the water dissolves mineral salts, including magnesium sulfate, otherwise known as Epsom salt. The water's salinity makes it denser than freshwater, which means that swimmers are more buoyant. This contributes to the relaxation of those who have entered the springs for therapy.

The historic springs and pools on the grounds of the Roosevelt Warm Springs Institute are open to the public as a unit of Roosevelt's Little White House Historic Site. The museum describing their history also includes an exhibit on the geology of the springs. The institute is off GA 41, about 0.5 mile north of the town of Warm Springs.

Roosevelt's Little White House is maintained much as it was the day the president died there, of a stroke, as his portrait was being painted. The house and museum devoted to Roosevelt's life at Warm Springs, 0.4 mile south of town on Alt GA 85, are well worth a visit. The Walk of Flags and Stones, outside the Little White House, is a walkway lined with native stones donated by each of the fifty states. Georgia's is a slab of pink marble cut in the shape of the state.

The Warm Springs area contains a bit of Coastal Plain–like sediment tucked away in the Piedmont. The strata were mined for bauxite (to extract aluminum) in 1916, and they contain kaolin. These deposits, dated as late Paleocene based on fossil pollen, are interpreted as having been deposited on a floodplain. They are similar to bauxite and kaolin deposits of the Nanafalia Formation of the same age near Andersonville.

Overlying the Paleocene-age deposits are younger sediments that resemble the alluvial fan deposits forming adjacent to mountains in desert areas today.

Paleocene-age sediments exposed north of Pine Mountain, with bedding (dashed lines) dipping to the left (north).

Such deposits contain sediments of many grain sizes, from boulders to mud, and are deposited by floods that burst forth from the mountains after infrequent rainstorms. The alluvial fan–like deposits here most likely were eroded from neighboring highlands while climatic conditions in Georgia were dry, perhaps during Pliocene time. The need for highlands to explain the deposits, and the presence of Hollis Quartzite in them, implies that they were deposited once the Hollis Quartzite had become exposed and Pine Mountain existed.

Most remarkably, the strata of both ages have been tilted, dipping as much as 65 degrees to the north. It appears that movement on a nearby fault named the Warm Springs fault shifted them. Although radiometric dating has repeatedly established that faulting and folding in the Piedmont were essentially over early in the Mesozoic era, this sedimentary deposit near Warm Springs shows that faulting happened here as late as Neogene time. Moreover, the pattern of folding near the fault suggests that it was a response to tectonic compression, unlike the closest-known Neogene-age faults in the Coastal Plain of Alabama and other Gulf Coast states, which were mainly caused by gravity: the movement of piles of sediment toward the deep ocean floor of the Gulf. The reason for compression in this area at that time remains a mystery.

To see these sediments, turn left (west) from GA 41 onto GA 194 (Durand Highway) 0.5 mile north of the historic springs. After 0.6 mile, turn left (south) onto Phillips Road and continue for 2.2 miles until the road forks and the pavement ends. Take the left fork onto 7 Branches Road. The outcrop is less than 100 feet past that intersection on the right (north) side of 7 Branches Road. Here, the slightly tilted, younger alluvial deposits of sand and gravel overlie more strongly tilted, fine-grained sand layers of Paleocene age.

F. D. Roosevelt State Park

Extending along Pine Mountain west of Warm Springs is F. D. Roosevelt State Park, Georgia's largest state park. The park contains several stone structures of Hollis Quartzite, built by the Civilian Conservation Corps that Roosevelt founded. Roosevelt's favorite picnic spot, Dowdell Knob, provides panoramic views to the south from an elevation of 1,395 feet.

To reach Dowdell Knob, follow Alt GA 85 south 3.5 miles from Warm Springs. Ascending Pine Mountain, you pass road cuts in Hollis Quartzite on both sides of the road. Turn right (west) along GA 190 and continue 3.1 miles to Dowdell Knob Road on the left (south). The knob is 1.5 miles down the road.

The knob is underlain by Hollis Quartzite, of which several large blocks are present. The ridges visible from Dowdell Knob are underlain by the same erosion-resistant quartzite, including the south edge of Pine Mountain that curves into view 2 miles to the east, and Oak Mountain about 5 miles to the south. The valley between Pine Mountain and Oak Mountain is underlain mainly by schist. To return to Manchester and GA 85, you can retrace your route to Alt GA 85 and continue east on GA 190 along the top of Pine Mountain for 4.9 miles.

View to the southeast from Dowdell Knob at F. D. Roosevelt State Park. The southward bend of Pine Mountain is visible in the distance.

Sprewell Bluff State Outdoor Recreation Area

Along the way to Sprewell Bluff on the Flint River, a scenic gorge and popular canoeing destination, you can visit some of Georgia's oldest rocks and see the rock called charnockite. About 1 mile south of Manchester, head south on GA 41. Continue 5.5 miles and turn left (east) onto GA 36 in Woodland.

About 2.2 miles east of Woodland, the route enters the largest exposed area of 1.1-billion-year-old basement rock anywhere in the Piedmont. The 55-mile-long exposure within the Pine Mountain terrane, up to 15 miles wide, extends from a few miles southwest of Woodland to just east of I-75. You can see an outcrop of 1.1-billion-year-old granitic gneiss in Celeoth Creek, just downstream (south) of the GA 36 bridge 4.1 miles east of Woodland.

About 4.7 miles east of Woodland, turn left onto Chalybeate Springs Road. You soon cross Oak Mountain, underlain by Hollis Quartzite. Continue to weathered outcrops of rounded boulders on the right (east) side of the road, about 4.4 miles north of the GA 36 intersection. Another 0.5 mile farther, much-less-weathered, rounded boulders are on the road shoulder on the right. The boulders are a rock called charnockite. The boulders have developed through a process called spheroidal weathering. In outcrops that lack foliation but have widely spaced joints, the elements attack the edges of the joints first, rounding sharp edges. Over time, rounded boulders remain.

Charnockite is composed mainly of feldspar and quartz and is similar to granite, but it also contains a dark pyroxene mineral called hypersthene (much of which was converted to biotite by later metamorphism). Advocates of the

Charnockite boulders on the shoulder of Chalybeate Springs Road.

hypothesis that Cove Dome is part of an extraterrestrial impact structure think the charnockite is evidence in their favor. They suggest that the rocks in the area were melted by the impact that formed Cove Dome, and that the melted rock crystallized as charnockite. Lab analysis of the charnockite reveals "shocked" forms of quartz and other minerals, which are typical of impact structures.

The name *Chalybeate* (locally pronounced "kal-EE-bit") comes from the Latin *chalybs* (steel) and refers to iron-rich springwater in the region. A resort existed around the springs, on the east edge of Manchester, between 1870 and 1924, and around 1899 there were plans to develop an iron mine in the area.

From the Chalybeate Springs Road turnoff, it is another 5.7 miles along GA 36 to the Flint River, which winds through flatlands underlain by basement rock. About 1.8 miles east of the Flint River bridge, turn left (north) onto Roland Road at the Sprewell Bluff sign and continue 2.1 miles to the next sign at Old Alabama Road. Turn left (west) and continue on the paved road. About 3 miles west of the intersection there is a scenic overlook on the right, from which you can see several miles both up and down the Flint River. The flat-topped ridge about 3 miles upriver is the southeast flank of the Cove Dome, built of Hollis Quartzite. The steep valley slope to the south, at the river bend, and the overlook are both built on Hollis Quartzite. The areas along the river in between are underlain by Manchester Schist.

From the overlook, travel 1.7 miles over partly unpaved road to the Sprewell Bluff State Outdoor Recreation Area. Here, 100-foot-high cliffs of Hollis Quartzite tower above the river where people swim and wade and canoeists drift by.

Boulders of quartzite below Sprewell Bluff along the Flint River.

A beach covered in cobbles of Hollis Quartzite at Sprewell Bluff State Outdoor Recreation Area.

The Hollis Quartzite started as sand deposited in a nearshore environment. It lithified as quartz sandstone and then was metamorphosed to quartzite. Since it is associated with basement rocks here in the Pine Mountain terrane, geologists deduce that the sand was deposited along the shoreline of the large, eroded landmass of Laurentia as sea level rose. Most of the Piedmont rocks, which surround the Pine Mountain terrane, probably were deposited in deep water. So how did the Pine Mountain terrane end up where it is, so far away from other basement rock exposures to the northwest? A currently popular hypothesis is that the Pine Mountain terrane originally was a projecting part of Laurentia that was swept southwestward with other terranes along strike-slip faults. (See The Patchwork of Terranes Assembled in Paleozoic Time in the chapter introduction.)

Manchester to Columbus

About 6 miles south of Manchester, on the west side of the Y-intersection where Alt GA 85 merges with GA 85, you can see an outcrop of quartz-rich schist. GA 85 cuts through Oak Mountain 1.2 miles south of the Y-intersection, and Hollis Quartzite is exposed on the left (east) side of the road. About 4.3 miles south of the Y-intersection, GA 85 crosses the northern boundary of the Carolina superterrane, passing through an area of poorly exposed gneiss, schist, and amphibolite from here to Columbus.

About 11 miles south of Waverly Hall, GA 85 begins crossing a series of low ridges capped by Cretaceous-age sediments of the Coastal Plain. GA 85 crosses one of these ridges at the US 80 interchange. It rises only about 50 feet above the neighboring valleys, which are underlain by Carolina superterrane rocks.

Flat Rock Park

Flat Rock Park, a Columbus city park in the valley of Flat Rock Creek, contains extensive outcrops of granitic gneiss. Dikes of igneous rock cut across the gneiss, including coarse-grained, light-colored pegmatite; gray diorite; and light-colored, fine-grained aplite. The park has many large natural outcrops and a quarry. To reach Flat Rock Park, exit onto westbound US 80 and then immediately exit onto Flat Rock Road (GA 357). Head 0.7 mile north to Warm Springs Road and turn left. Continue 1 mile west to the park entrance on your left.

The Fall Line and Rocky Shoals at Columbus

GA 85 ends in Columbus, where it is called 2nd Avenue, and parallels the Chattahoochee River. Second Avenue (and much of downtown) is built on a thin layer of Quaternary-age sediments; the Chattahoochee River deposited these on Cretaceous-age sediments, which overlie the Carolina superterrane. Less than 0.25 mile west of 2nd Avenue, the Chattahoochee River has eroded into superterrane rocks, including gneiss and amphibolite with folded pegmatite

Gneiss and amphibolite with folded pegmatite dikes along the Chattahoochee River. North Highlands Dam is in the background.

Amphibolite is exposed along the Riverwalk at Rocky Shoals, where a crane is helping to remove the Eagle and Phenix Dam. This Fall Line waterfall was being repurposed as the final drop of a 2.5-mile-long urban whitewater course. —Courtesy of Tom Hanley

dikes, which are well exposed along the Riverwalk 150 yards south of the parking area below North Highlands Dam. This area is reached by a Georgia Power access road from 1st Avenue, one block south of the west end of 35th Street.

Closer to the city center, around 12th Street, the amphibolite that makes up the rapids called Rocky Shoals is also part of the Carolina superterrane. Rocky Shoals is the site of the Eagle and Phenix Dam, which was removed in 2012 as part of a river restoration project.

The rapids on the Chattahoochee provide one of the best illustrations of how geology has influenced the growth of cities along the Fall Line. The Chattahoochee drops 120 feet in 2.5 miles in Columbus, creating a hydropower potential of about 100,000 horsepower (75,000 kilowatts), which has been used by local industry.

Sandbars deposited downstream of the rapids (now covered by impounded water) permitted a river crossing that was the intersection of many Indian trails. Two of the largest Creek Indian villages, Coweta Town and Cusseta Town, grew up on either bank a few miles downstream. The Georgia Legislature established Columbus in 1828, recognizing the location's importance to the burgeoning cotton economy. As the inland limit of river navigation, the town was a natural trading center for cotton. Moreover, the waterpower was harnessed to power textile mills, with three dams built in 1828, 1844, and 1900. By the time of the Civil War, Columbus ranked second (after Richmond, Virginia) among the top four textile production centers in the South (all located along the Fall Line). Columbus was the only one of these centers to have its mills destroyed in the Civil War, burned by Federal troops in a battle that, ironically, took place after the war ended but before word of Lee's surrender reached the area.

The Eagle Mill, rebuilt in 1869 and renamed Eagle and Phenix Mill after the mythical bird that rises from its ashes, became the largest textile producer in the South by 1880. The mill has been renovated as luxury condominiums. The Eagle and Phenix Dam (removed in 2012) was one of a handful of historic structures in Georgia constructed of a material called Rosendale Natural Cement. Natural cement was made from limestone with a high clay content. It was quarried in the Hudson River Valley near the town of Rosendale, New York. The exceptionally durable cement was used in the construction of large masonry and concrete structures in the nineteenth and early twentieth centuries.

The Eagle and Phenix Dam powered the Columbus Iron Works, today renovated as the city's Convention and Trade Center. Built in 1853, it manufactured kettles and ovens, plows, and steam engines (used to power mills, cotton gins, and riverboats), as well as cannons for the Civil War. In 1862, the iron works was leased by the Confederate Navy and converted to the Confederate Naval Iron Works. This factory was the largest manufacturer of naval machinery in the Confederacy and the second-largest iron producer in the Confederacy. Two Confederate ironclad warships, one of them built at the iron works, were recovered from the Chattahoochee River and are on display at the Port Columbus National Civil War Naval Museum. The iron works began producing ice machines in the 1880s and barbecue grills in the 1940s but closed in 1965.

The source of the iron used in the Columbus Iron Works around the time of the Civil War probably included iron from furnaces in Bartow County, near Cartersville. By the 1870s, Birmingham, Alabama, probably became the chief source of iron. Although iron ore is present in the Clayton Formation in the Coastal Plain south of Columbus, it was not mined until the 1950s.

I-75
Atlanta—Macon
75 miles

Heading south out of Atlanta, I-75 crosses a broad expanse of metamorphic rocks, primarily gneiss belonging to four different tectonic terranes, before ending at the Fall Line near Macon. A major inactive fault separates each of the terranes. Outcrops are few, so most of the geology requires short side trips from the interstate.

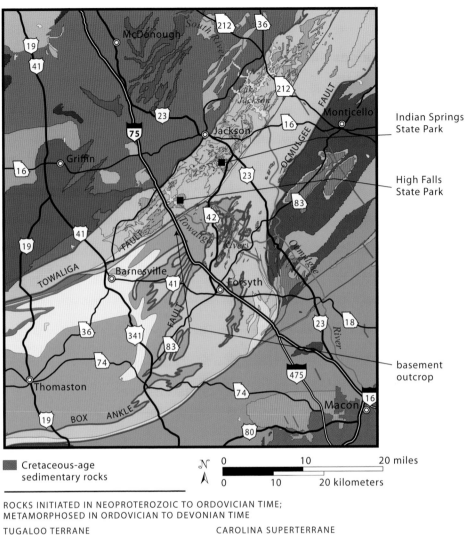

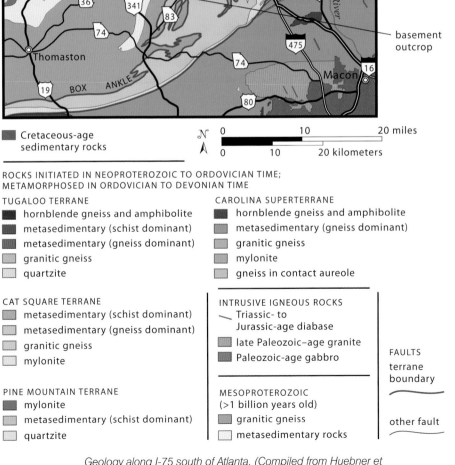

Indian Springs State Park

High Falls State Park

basement outcrop

Cretaceous-age sedimentary rocks

N

0	10	20 miles
0	10	20 kilometers

ROCKS INITIATED IN NEOPROTEROZOIC TO ORDOVICIAN TIME;
METAMORPHOSED IN ORDOVICIAN TO DEVONIAN TIME

TUGALOO TERRANE
- hornblende gneiss and amphibolite
- metasedimentary (schist dominant)
- metasedimentary (gneiss dominant)
- granitic gneiss
- quartzite

CAROLINA SUPERTERRANE
- hornblende gneiss and amphibolite
- metasedimentary (gneiss dominant)
- granitic gneiss
- mylonite
- gneiss in contact aureole

CAT SQUARE TERRANE
- metasedimentary (schist dominant)
- metasedimentary (gneiss dominant)
- granitic gneiss
- mylonite

INTRUSIVE IGNEOUS ROCKS
- Triassic- to Jurassic-age diabase
- late Paleozoic-age granite
- Paleozoic-age gabbro

PINE MOUNTAIN TERRANE
- mylonite
- metasedimentary (schist dominant)
- quartzite

MESOPROTEROZOIC (>1 billion years old)
- granitic gneiss
- metasedimentary rocks

FAULTS
terrane boundary

other fault

Geology along I-75 south of Atlanta. (Compiled from Huebner et al., 2011; Hatcher, 2011; and Lawton, 1976.)

Indian Springs State Park

Indian Springs State Park is the oldest state park in the nation. It's located near Jackson and Flovilla, about 10 miles east of I-75 from exit 205 (GA 16). Follow the signs from the interstate.

The park lies within the Cat Square terrane, much of it within poorly exposed sillimanite schist, but around the park entrance, where the springs are located, there are exposures of the Indian Springs Granite, which is about 300 million years old. (The granite can also be seen at High Falls State Park, where it intruded much older granitic gneiss.)

At the entrance to the park there is a spheroidal boulder of the granite, about 9 feet in diameter, that formed due to natural weathering. Along the stream, you can see outcrops of granite cut by veins of pegmatite with large crystals. Indians believed that a spring in the park had healing properties. The spring is now protected within an open stone building. It flows from a crack in feldspar-rich biotite gneiss, which resembles coarse-grained granite. Some people fill containers with the springwater to drink. The taste is surprising and unpleasant because of the presence of hydrogen sulfide. Early settlers called it Gunpowder Springs because of the taste of the water. The spring flows slowly, with a discharge of less than 1 gallon per minute. This is roughly the rate of the faucet on a bathroom sink, so it takes a minute or so to fill a gallon jug.

Barnesville and Thomaston Area

Exit 201 leads southwest to the Barnesville and Thomaston area, which is known for granitic pegmatite. These lens-shaped or dikelike igneous intrusions have very large crystals of quartz, muscovite, and feldspar, and sometimes tourmaline and beryl. Mining began in the Barnesville area in 1916, and hundreds of thousands of pounds of mica have been extracted from these intrusions. Large sheets of muscovite mica are used for electrical insulation and specialty windowpanes. Smaller pieces are used in a variety of products, including building and fireproofing materials, tires, lubricants, and insulation. This area has been the best source of sheet mica in the Piedmont of the southeastern United States. Although currently no mica is being mined here (or anywhere else in the United States), the potential exists for additional mining in this area.

High Falls State Park

High Falls State Park is located about 1.8 miles northeast of exit 198. High Falls Road crosses the Towaliga River on the way to the park, and you can see a dam on the left (north) side of the road, which impounds High Falls Lake. The dam was built in 1904 to provide electrical power to a gristmill and the city of Griffin to the northwest.

Three rock units of the Cat Square terrane can be seen in a single pavement outcrop in the picnic area on the north side of the road on the east side of the Towaliga River. The main rock type, as throughout the park, is granitic gneiss referred to as High Falls Granite, which has large, elongated crystals of feldspar. Here it contains pieces (xenoliths) of biotite gneiss, the metasedimentary rock of this area that the granite intruded 383 million years ago. The High Falls

Pavement outcrop southwest of the picnic area at High Falls State Park.

Granite was itself intruded by the Indian Springs Granite 300 million years ago. This light-colored, fine-grained granite is also exposed at Indian Springs State Park.

On the right (south) side of the road, High Falls drops approximately 35 feet down over rocky pavement outcrops spanning the river. Hiking trails provide a view of the falls and rocks. The rock exposed along the trails near the falls is the High Falls Granite, with its large, elongated crystals of feldspar.

High Falls State Park lies about 2 miles north of the Towaliga fault, one of the faults separating the Cat Square terrane from the Pine Mountain terrane to the south. The Pine Mountain terrane is the southernmost area in the Appalachians where 1.1-billion-year-old basement rocks of Laurentia are exposed. They consist of garnet-bearing granitic gneiss that formed from the metamorphism of granite.

To see an excellent exposure of the Pine Mountain terrane, head west 0.6 mile from exit 198 and turn left (south) on Unionville/Rocky Creek Road. Continue south 2.4 miles. Much of this road is unpaved, but it is well graded. Cross the Little Towaliga River and make an immediate left onto High Road and park. The outcrop is along the old road down the hill to your left. The rock is granitic gneiss with large feldspar crystals and abundant garnet. Xenoliths of biotite gneiss are also present. To continue south on I-75, proceed south on Rocky Creek Road. After 1.5 miles turn left on Johnstonville Road. Proceed 1.1 miles to exit 193 at I-75.

High Falls on the Towaliga River, downstream of the High Falls Road bridge.
—Courtesy of Chuck Cochran

Basement rock of the Pine Mountain Terrane exposed near the Little Towaliga River. Note the large feldspar crystals in the granitic gneiss and the xenolith of darker rock at the upper left.

Forsyth to Macon

The rocks near Forsyth are biotite gneiss, schist, and amphibolite of Proterozoic age. Southeast of Forsyth, between exits 181 and 177, I-75 crosses the Ocmulgee fault, which is not visible from I-75 but probably runs through the valley north of the rest area. The fault separates the Cat Square terrane to the north and west from the Carolina superterrane to the south and east. The Cat Square terrane rocks are gneiss and schist of high metamorphic grade; they were once marine sediments deposited off the coast of ancestral North America (Laurentia). The Carolina superterrane is composed of rocks that were part of an extensive volcanic island arc chain that existed offshore of Laurentia. The collision between the North American and African tectonic plates pushed the chain of volcanic islands onto the edge of North America about 350 million years ago.

Beginning around mile marker 171, I-75 runs beside the Ocmulgee River for about 7 miles. Imagine being here in July 1994 when Tropical Storm Alberto dropped more than 20 inches of rain on the area and the river reached 500-year flood levels. Levees along the Ocmulgee River burst, covering I-75 with 4 feet of water and stranding motorists on a few patches of higher ground. The flood crested at 35.5 feet on July 10, 1994, 17.5 feet above flood stage. Everything below an elevation of about 315 feet was underwater, including parts of downtown Macon. The Fall Line lies less than 1 mile south of the intersection of I-75 and I-16.

I-20
Atlanta—Augusta
136 miles

Between Atlanta and Augusta, I-20 traverses metamorphic and igneous rocks of the eastern Piedmont. The road is relatively flat, with few visible rock exposures, so you would probably be surprised to learn that the route crosses three separate and distinctive terranes.

Atlanta lies in the part of the inner Piedmont referred to as the Tugaloo terrane. In Rockdale, Newton, and Morgan counties, near I-20, these rocks are mostly granitic gneiss, biotite gneiss, mica schist, and amphibolite. In places you can see a little reddish orange soil exposed, but that is about all.

Just north of Stonecrest Mall (exit 75, GA 124), near Lithonia, are several large rock quarries that mine the Lithonia Gneiss for aggregate (crushed stone), building stone, monument stone, and curbstone. One of these, the Reagin Quarry, is one of the last quarries in Georgia that removes stone by hand and one of only three in the nation that makes "granite" curbstones. Lithonia had come to prominence as a center of quarrying by the 1880s. The name *Lithonia* is derived from the Greek word *litho*, meaning "rock," and *onia*, meaning "place."

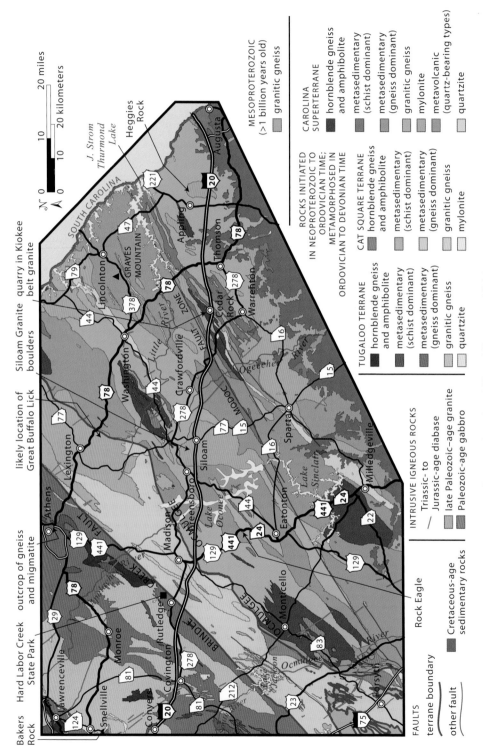

Geology along I-20, US 78, US 441, and GA 24.

The Rockdale County welcome sign is made of Lithonia Gneiss.

At exit 88 (Almon Road) you can see an exposure of biotite gneiss, mica schist, and amphibolite exposed along the entrance ramp to I-20 on the north side of the road. East of exit 98 (GA 11), the terrain becomes slightly hillier and increases in elevation, marking a transition from granitic gneiss eastward into biotite gneiss.

It takes thousands of years for fertile soil to develop on metamorphic and igneous rocks but only a few generations of poor farming practices for it to erode away. About 3 miles north of exit 105 (Newborn Road), Hard Labor Creek State Park was created during the Great Depression to serve as a demonstration project for the reclamation of eroded farmland. The Federal government purchased worn-out farmland and put people to work to restore the soil and waterways. The Civilian Conservation Corps and the Works Progress Administration constructed dikes and terraces and planted thousands of pine trees to stabilize the land and prevent soil erosion. They also dug silt out of streams and lakebeds.

A few miles east of exit 105, I-20 crosses the Brindle Creek fault. Between this point and exit 114, I-20 crosses the Cat Square terrane, which comprises mica schist, biotite gneiss, granitic gneiss, hornblende gneiss, and amphibolite that were metamorphosed between 430 and 400 million years ago, during Silurian and Devonian time.

The most scenic part of the route between Atlanta and Augusta is Lake Oconee, located near mile marker 126. The lake is a reservoir on the Oconee River, formed behind Wallace Dam and operated by Georgia Power for

The muddy water of Lake Oconee south of I-20. Note the saprolite and red clay in the far bank.

electricity generation. It is one of the largest lakes in Georgia. As you drive past, you will see a cliff of orange saprolite, or weathered granitic gneiss, along the southeastern bank of the lake. Note also the muddy, brown character of the lake water. Erosion of the saprolite and red clayey soil that cover the Piedmont causes fine particles of clay, stained reddish brown by iron oxide, to become suspended in the lake and river water. This lack of clarity is typical of lakes and rivers throughout the Piedmont.

Just north of the I-20 bridge over Lake Oconee, the Apalachee and the Oconee rivers merge. Near this juncture was a place called High Shoals of the Apalachee (now flooded by Lake Oconee). The shoals created a shallow, rocky ford where the Hightower Indian Trail crossed the river. This footpath, used by the Cherokee and Creek Indians, connected Cherokee villages along the Etowah River near Cartersville with those in the Augusta area. The name Hightower was derived from the English pronunciation of the Cherokee word *Ita-wa* (Etowah).

The location of the Hightower Indian Trail was controlled by geology. The trail followed ridges and drainage divides, in order to minimize stream crossings and avoid swampy lowlands, and connected rocky places where rivers could be crossed easily. Up until 1802, all of the land west of the Oconee River belonged to the Indians, and the trail served as a dividing line between Cherokee land to the north and Creek land to the south. Pioneers settling the area in the late 1700s and early 1800s also used the trail. The old Georgia Railroad (now CSX) was built along and near this route in the 1830s and 40s. Later, during the Civil War, General Sherman followed part of the trail during his March to the Sea. Engineers found that old Indian trails were the most suitable places to construct roads and railroads .

The portion of Lake Oconee north of I-20 is part of the Oconee National Forest. Unlike many national forests, which are established to preserve woodland areas, this national forest is a monument to environmental degradation. Due to poor cotton farming practices, the soil in this region was heavily eroded. It washed into the Oconee River, and 3 feet of silt now covers its rapids and rocky shoals. After the topsoil had washed away, the land was worthless for farming. During the Great Depression, the government bought the ruined land through the Roosevelt administration's New Deal. It eventually grew up in trees. In 1959 the area became a national forest.

Most of the rocks between exit 114 and Augusta are part of the Carolina superterrane, an ancient volcanic island chain. Between mile markers 133 and 142, near the town of Siloam, I-20 crosses the Siloam Granite, which intruded the surrounding metavolcanic rocks about 269 million years ago. This granite is very different from Georgia's other granites. It has large, pink feldspar crystals. If you look carefully, you can see boulders of the granite in the woods along the interstate. One particularly visible boulder is on the north side of I-20, west of exit 138 to Siloam.

Siloam Granite with pink feldspar crystals and part of a dark gray xenolith (gneiss). Centimeter ruler for scale. —Courtesy of Jon Shuman and Ann Calhoun

East of the Siloam Granite, I-20 is flat because it follows the ridgetops between streams, which flow from west to east, toward the Ogeechee River. The metavolcanic rocks in Taliaferro County were mined in the past. Less than 0.5 mile northwest of the Crawfordville exit (148), gold was mined from the Golucks-Rhodes Mine, and 6 miles farther north, manganese was mined from the Taliaferro magnetite deposit near Springfield.

Just west of exit 160, I-20 enters the Modoc fault zone, which forms the western boundary of the Kiokee belt. These rocks, part of the Carolina super-terrane, experienced a high degree of metamorphism. You know you have entered the Modoc fault zone when I-20 becomes hilly, between mile markers 160 and 165. The fault zone is a 3-to-4-mile-wide zone of schist, mylonite gneiss, and quartzite units that form pronounced ridges. The fault zone dips to the northwest, roughly parallel with the layering in the rocks. The Modoc fault zone is not like a crack in the rock; rather it is what's called a ductile shear zone. This fault zone was so deep within the Earth that temperatures were hot enough to allow the rocks to flow like taffy.

Gold was mined in the Modoc fault zone near the Little River, now flooded by part of Clarks Hill Lake (J. Strom Thurmond Lake), about 5 to 10 miles north of exit 172 (US 78). Most of the gold in this part of Georgia was produced intermittently from the mid-1800s to the early 1900s.

Kiokee belt rocks include gneiss, schist, and amphibolite, some partially melted to migmatite, as well as granite intrusions. There are also a number of small bodies of metamorphosed ultramafic rock in the Kiokee belt northwest of Augusta.

I-20 crosses some of the granite bodies that intruded the Kiokee belt on both sides of the Modoc fault zone some 320 to 300 million years ago, when the fault was active. Notably, I-20 crosses granite at Cedar Rock (exit 165, GA 80), where there is a granite quarry in the Kiokee belt less than 1 mile north of I-20. Also, about 5 miles south of I-20 at this exit, near Warrenton, is the 309-million-year-old Sparta Granite, one of the largest granite batholiths in Georgia. Between mile markers 178 and 181, a few miles east of Thomson, and again around mile marker 192, Coastal Plain sediments overlap the rocks of the Piedmont. Augusta lies at the Fall Line, which separates the Piedmont and the Coastal Plain.

Appling Granite at Heggies Rock

East of Appling (exit 183), less than 3 miles north of I-20 in the middle of the Kiokee belt, is a remarkable 101-acre outcrop of Appling Granite known as Heggies Rock. The rock surface is dotted with roughly circular solution pits, which form as the rock weathers. Some of the pits have thin soil layers in them and form "dish gardens," which are home to a number of rare and endangered plants, including the tiny pool sprite (*Amphianthus pusillus*) and mat-forming quillwort (*Isoetes tegetiformans*), which grows in only one pool on the rock. The dwarf granite stonecrop (*Sedum pusillum*), listed as threatened in Georgia, also lives here. During droughts, many of the plants become dormant, but when the solution pits fill with rainwater the plants are revitalized. Heggies Rock has the

Granite being quarried at the Martin Marietta Aggregates Quarry at Cedar Rock.

Coarse-grained Sparta Granite. The large white feldspar crystals are weathering to kaolin, which has been mined from this rock.

*A solution pit at Heggies Rock containing the endangered mat-forming quillwort (*Isoetes tegetiformans), *which is endemic to Georgia and found only in pools on a few granite outcrops.* —Courtesy of Alan Cressler

highest biodiversity of any of the Piedmont granite outcrops. The public may obtain permission to visit the preserve from the Nature Conservancy.

Augusta

East of exit 196 (I-520), I-20 crosses the easternmost exposure of Piedmont rocks in Georgia. Only small portions of these metavolcanic rocks are exposed at the surface because Coastal Plain sediments mostly obscure them.

Around exit 200 (River Watch Parkway), some of the buildings at one of the largest crushed-stone quarries in the state are visible north of the interstate. The quarry, first opened in the 1850s, produces more than 2 million tons of rock, gravel, and sand each year from mylonite.

Augusta is located on the Fall Line, at the inland limit of navigation. In 1736, General James Oglethorpe ordered the city of Augusta to be built at the junction of the Hightower Indian Trail and other Indian trails on the Savannah River in order to intercept the fur trade.

East of mile 201, the road descends toward the Savannah River, crossing the Augusta Canal. The 7-mile-long canal was constructed in 1845, and it is one of the few industrial canals still in use in the nation. It was built as a transportation route around the rocks and rapids of the Savannah River at the Fall Line. The towpath beside the canal, today part of a trail system, was for the mules that pulled the canal boats. The canal also provided water and hydroelectric power to factories in the city by means of water-driven turbines.

A number of mills and industries were located along the canal, including the Confederate Powder Works. Gunpowder is made from a mixture of sulfur, charcoal, and potassium nitrate, or saltpeter ("stone salt"). The saltpeter was mined from caves in the Appalachian Mountains, particularly the Kingston Saltpeter Cave near Cartersville. During the Civil War, the Confederate Powder Works was the second-largest gunpowder factory in the world, lining the banks of the Augusta Canal for 2 miles. It operated until 1865, producing 7,000 pounds of gunpowder per day. Only the square brick smokestack remains.

Although I-20 crosses a lot of interesting geology, few outcrops can be seen until you reach the Savannah River. During times of low water, from the bridge you can see Piedmont rocks just above the Fall Line, including phyllite and metavolcanic rocks. If you look upstream you may see some of the more resistant granitic gneiss of the Kiokee belt in the distance. Indians were aware of the strategic importance of the Fall Line for transportation and trading. Some of the earliest evidence of human habitation in North America comes from archaeological excavations just downstream of Augusta, where stone tools dating from 18,000 to 16,000 BC have been found.

The Charleston earthquake, the most destructive earthquake to strike the eastern United States, hit Augusta shortly after 9:00 p.m. on August 31, 1886. The earthquake had a magnitude of 6.6 to 7.6 on the Richter scale and was

The Fall Line on the Savannah River. Note the ledges of Piedmont rock in the river.
—Courtesy of Louise Williams Wheeless

felt as far away as Alabama and Ohio. Charleston, South Carolina, experienced extensive damage, but the damage in Augusta was much less severe. A record of the earthquake is preserved in correspondence from the commandant of the Augusta Arsenal, Major J. W. Reilly, to the Chief of Ordnance in Washington, DC. Major Reilly dramatically described the falling plaster as his family fled their crumbling house and noted that brick structures were more severely damaged than wood frame houses.

The Augusta area experiences small to moderate (magnitude 2 to 4.5) earthquakes that have a booming sound with vibrations that rattle windows. Most of these are reservoir-induced earthquakes related to J. Strom Thurmond Lake on the Savannah River, about 10 miles north of town.

US 441/GA 24
Madison—Milledgeville
44 miles

See the map on page 268.

US 441/GA 24 follows part of Georgia's Antebellum Trail, connecting the historic towns of Madison and Milledgeville. About 6 miles south of I-20, US 441/GA 24 crosses a fault that separates the Cat Square terrane to the northwest from the Carolina superterrane, a volcanic island chain that collided with North America about 350 million years ago. A slight difference in topography as you pass through an area with cattle farms marks the transition. Broad, flat hilltops characterize the Cat Square terrane north of the fault, whereas the Carolina superterrane is hilly with narrower hilltops.

Rock Eagle

The Rock Eagle effigy mound is one of the most mysterious pre-Columbian structures in the eastern United States. The mound is constructed of large pieces of white, milky quartz in the shape of a great bird with outstretched wings and a fanlike tail. No one knows who built it or when, but it is at least 1,000 years old. The Rock Eagle is 102 feet from head to tail, with a wingspan of 120 feet, and the mound rises to a height of about 8 feet. It is best viewed from a stone observation tower built in 1937 by the Works Progress Administration. You can see the Rock Eagle at the Rock Eagle 4-H Center, about 12.5 miles south of I-20. Turn west at the entrance and follow the signs. Rock Eagle has a companion stone bird, known as Rock Hawk, about 15 miles to the southeast as the crow flies. Rock Hawk, also constructed of milky quartz, is 78 feet long from its head to its forked tail, with a wingspan of 132 feet. Rock Hawk is located near the entrance to Georgia Power's Lawrence Shoals Recreation Area on Lake Oconee. From Eatonton, turn east on GA 16 and drive approximately 12 miles. Turn left (north) onto Wallace Dam Road and then left onto Lawrence Shoals Road. The driving distance between the stone effigies

*The Rock Eagle effigy mound. The head of the bird is in the
foreground, and the outstretched wings are the larger mound.*

is 22 miles. Rock Eagle and Rock Hawk are the only two bird effigy mounds
known in the eastern United States.

Milledgeville

US 441/GA 24 crosses an arm of Lake Sinclair about 5 miles north of Milledgeville. The Central Georgia Seismic Zone, the most seismically active part
of Georgia, is centered on Milledgeville and has a 50-mile radius. Lake Sinclair and Lake Oconee, both on the Oconee River and within this seismic
zone, have experienced several earthquakes relating to them being dammed
and filled with water (see Earthquakes Due to Human Activity in the chapter introduction). However, significant earthquakes were recorded in this
seismic zone before the dams were built, including a magnitude-4.9 earthquake on March 4, 1914. There are no known active faults within this region
that would have caused the earthquakes. Instead, the earthquakes are probably caused by movement along shallow fractures that have been weakened by water penetrating from the reservoirs or by their weathering.
On the east side of US 441/GA 24 (N. Columbia Street through Milledgeville),
along a busy commercial strip north of downtown Milledgeville, you can see
folded and faulted metamorphic and igneous rock adjacent to a parking lot at
2521 N. Columbia Street. A fine-grained, light gray granite has been thrust-
faulted into a sequence of amphibolite gneiss and schist. The top of the granite

Light gray granite (left) topped by a thrust fault and overlain by folded gneiss with blocks of darker amphibolite.

is cut by a thrust fault, which is topped by folded gneiss and layers and blocks of amphibolite. To the left, behind the building, there is mixture of contorted biotite gneiss, granitic gneiss, and amphibolite exposed. These rocks are part of the Carolina superterrane.

Milledgeville is home to the Georgia College Natural History Museum and Planetarium (231 North Wilkinson Street), with one of the largest exhibits of vertebrate fossils in the Southeast. The Fall Line is southeast of Milledgeville along the Oconee River.

Around Atlanta

Georgia's largest population center sprawls across an area full of geologic interest. Besides Stone Mountain Park, there are half a dozen less-well-known parks in which natural rock outcroppings are part of the attraction. Construction activities, mainly road building and quarrying, have made fresh exposures of a wide variety of bedrock types. The city's rapid rise to prominence in the mid-nineteenth century, and its present status as a world-class transportation center, are connected to geologic factors.

Atlanta is about 1,000 feet above sea level, astride the Eastern Continental Divide. West of the divide the Chattahoochee and Flint rivers flow toward the Gulf of Mexico, and east of the divide the South and Yellow rivers flow toward the Atlantic Ocean. Atlanta's location atop the divide is no accident: it was sited at the junction of railroads that, in turn, followed old Indian trails, which were located along divides to minimize stream crossings. Atlanta was conceived of in 1837 as the meeting point of two rail lines—the Western and Atlantic Railroad and the Georgia Railroad. I-75 to Chattanooga and I-20 to Augusta roughly follow the lines, now operated by CSX. The Georgia State Capitol stands on part of the divide separating the Chattahoochee River and South River drainage networks. The Hartsfield-Jackson International Airport, among the world's busiest, was built in the headwaters region of the Flint River, which has gentle slopes.

The dome of the Georgia State Capitol is covered in gold mined in Dahlonega.

Besides Atlanta's crossroads location, another ingredient of its early growth was waterpower. Several creeks that drain to the Chattahoochee River have significant rapids as they cross the 200-foot drop between the upland and the river. The rapids became the sites of mills to grind corn or wheat, cut lumber, and weave Georgia cotton into cloth. The memory of mills remains in many road names, such as Howell Mill Road and Moores Mill Road along I-75 north of downtown.

Metamorphic and igneous rocks of the Tugaloo terrane underlie the Atlanta area. The landscape is partly determined by rock type. As elsewhere in the

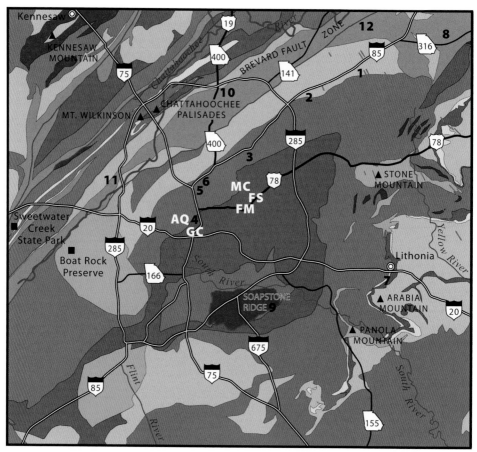

ROCKS INITIATED IN NEOPROTEROZOIC
TO ORDOVICIAN TIME; METAMORPHOSED
IN ORDOVICIAN TO DEVONIAN TIME

TUGALOO TERRANE

- amphibolite and meta-ultramafic rocks
- metasedimentary (schist dominant)
- metasedimentary (gneiss dominant)
- mylonitic schist
- mylonitic gneiss
- granitic gneiss
- quartzite

MUSEUMS AND EXHIBITS

AQ Georgia Aquarium
FM Fernbank Museum of Natural History
GC Georgia Capitol Museum
FS Fernbank Science Center
MC Michael C. Carlos Museum

INTRUSIVE IGNEOUS ROCKS

- Triassic-to Jurassic-age diabase
- late Paleozoic–age granite

faults

Geology of the Atlanta area (modified after Higgins and Crawford, 2006). Parks and museums (see the Museums and Exhibits appendix for more information) are labeled and road cuts and quarries are numbered and keyed to text starting on page 291.

Piedmont, the rocks with higher quartz content, such as quartzite, granitic gneiss, and granite, tend to underlie hills and mountains, whereas rocks with less of this resistant mineral, such as schist and amphibolite, tend to underlie valleys.

Parks of Geologic Interest

Stone Mountain Park

A number of mountains are present in the Atlanta area, including 1,686-foot Stone Mountain and 1,808-foot Kennesaw Mountain. Each rises about 700 feet above the surrounding terrain. These are erosional remnants, or monadnocks.

Both mountains are relatively high because they are made of granite and granitic gneiss, respectively, quartz-rich rocks that are more resistant to erosion than the surrounding terrain. Questions remain, however. Since granite and granitic gneiss are so common around Atlanta, it is odd that there are so few monadnocks. Why do Stone Mountain and Kennesaw Mountain rise more than twice as high above their surroundings as Panola Mountain and Arabia Mountain? And why is Stone Mountain not as large as the Stone Mountain Granite intrusion, which extends several miles eastward of the mountain? The explanations are thought to relate to the spacing of joints, or cracks, in the rock. The less fractured the rock is, the better it will resist weathering. Also, some areas get left behind and escape erosion as streams erode the landscape around them.

Stone Mountain, standing sharply above the surrounding terrain, is visible from a number of highways around Atlanta, including I-85 north, I-285 near

Stone Mountain, viewed here from the east, has an asymmetrical shape, elongated in an east-west direction, with a steep northern face.

the junction with I-85, and US 78. Stone Mountain Park is accessed from the east gate off US 78 and the west gate in the town of Stone Mountain. The Stone Mountain Granite has a fine, uniform texture dominated by white feldspar and gray quartz, along with two types of mica: silvery muscovite and black biotite. The rock formed between 325 and 281 million years ago.

SKYRIDE, CARVING, AND MOUNTAINTOP: As you enter the Skyride parking area you can see a contrast between pavement exposures of granite and the monadnock's steep northern face. Undoubtedly, both erosional history and joints in the bedrock have contributed to the sharp contrast. The immense carving of the Civil War generals Robert E. Lee and Thomas J. "Stonewall" Jackson and Confederate President Jefferson Davis is on the steep north-facing side of the mountain. The Skyride area is the best place to view the rounded profile of Stone Mountain. From the top, you can see more than 100 miles to the Blue Ridge Mountains on a clear day.

Stone Mountain is rounded because it is an exfoliation dome. Exfoliation is the peeling off of layers parallel to the surface of a massive rock body. Rock that formed under great pressure many miles below the surface expands when erosion unearths it. If the rock lacks layers or planes of weakness along which it might split, cracks will develop more or less parallel to the land surface. These are called exfoliation joints. Over time the rock peels off the main rock mass along the joints. Over a large area, the joint pattern curves gently, producing the profile of an exfoliation dome.

As sheets of granite pop up, circular to oval depressions are left beneath them. Most are a few feet in diameter, but they range in size from a few inches to a more than a hundred feet. The smallest pits atop Stone Mountain were once erroneously thought to be the result of lightning strikes. The pits collect rainwater, which causes the rock to chemically weather and provides moisture for plants and organisms, such as lichens and fungi that produce acids that further break down the rock. Because of the dissolving action of acidic water, these pits are called solution pits. Organic matter from plants and other organisms forms soil in the solution pits. On the mountaintop many of the pits host a variety of wildflowers and other plant life, including trees.

Stone Mountain Granite is notable for clusters of black tourmaline crystals surrounded by circular, bleached-white rims about 2 or 3 inches in diameter. Colloquially called "cat's paws," they are common at the top of the mountain and in the East Quarry area.

Raising a Ledge, at the East Quarry area, is an outdoor museum that describes the history of the granite industry as well as the fauna and flora in some of the natural areas of the park. It is located east of the Skyride along Robert E. Lee Boulevard, just before the traffic circle. A stone path beneath the railroad and along a boardwalk leads a short distance through the woods to the exhibit. Granite was quarried here from the 1850s until the 1970s. You can easily identify granite building stone from Stone Mountain by the presence of tourmaline clusters.

SAPROLITE AND DIABASE DIKE: As granite weathers in the humid climate of Atlanta, feldspar weathers to white kaolinite clay, biotite weathers to iron oxide

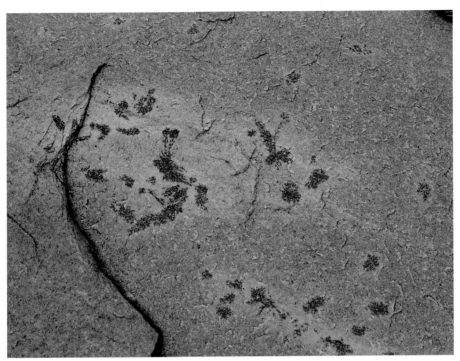

Cat's paws in the Stone Mountain Granite.

Exfoliation in the East Quarry area.

and clay, and quartz weathers to sand, forming a soft, crumbly material called saprolite. Near the intersection of Robert E. Lee Boulevard and Old GA 78, there is an outcrop of saprolite in which the normally hard granite has deeply weathered to predominantly quartz sand and white kaolinite clay.

Near the lake along Old GA 78, there is a diabase dike several feet wide weathering to an orange color. The road is used by the Ride the Ducks attraction, and access is restricted for safety reasons. The diabase dike is about 500 feet east of Robert E. Lee Boulevard on the left side of the road. The orange soil over the weathered dike contrasts sharply with the white saprolite of the granite.

Diabase with a weathering rim of iron oxide and clay.

The orange, weathered diabase dike adjacent to white to gray saprolite weathered from the Stone Mountain Granite in Stone Mountain Park.

The dike is part of a swarm (multiple dikes) of thin, northwest-trending diabase dikes of Mesozoic age (about 200 million years ago) that are by far the youngest rocks in the Blue Ridge–Piedmont Province. They intruded the bedrock as the supercontinent Pangaea began to pull apart prior to the opening of the Atlantic Ocean (see Pull-Apart Activity during the Mesozoic Era in the chapter introduction).

CONFEDERATE HALL AND WALK-UP TRAIL: Confederate Hall, near Stone Mountain's west gate, has a natural history museum that provides an introduction to the geology and life on the mountain, as well as changes in life though geologic time. The exterior walls of the hall are made of migmatite or granitic gneiss, probably Lithonia Gneiss. The entranceway is paved in slate, some of which has a sedimentary structure called graded bedding. Inside Confederate Hall there is a topographic map of Stone Mountain etched into the granite floor tiles.

The 1.6-mile-long Walk-Up Trail leads to the top of the mountain. The lower 600 feet of the trail crosses surfaces that were once hand quarried, and the granite is relatively fresh and unweathered, which makes it easy to see the geologic features. One way geologists establish that a rock was once magma is by finding xenoliths. These are pieces of the rock body that the granite magma intruded

The tapering, curved tip of a pegmatite dike about 250 feet from the start of the Walk-Up Trail. The blocky, raised bumps within the dike are large crystals of feldspar.

that didn't melt and became trapped in the magma before it solidified. At Stone Mountain, xenoliths stand out as dark gray to black patches of rock with metamorphic layering (foliation). A few examples can be seen near the left edge of the quarried surface that exists about 100 feet past the sign at the start of the trail.

About 225 feet past the sign, the trail climbs a ledge about 1 foot high. You may notice occasional strips of lighter rock, less than 1 foot wide and tens of feet long, within the granite. These are dikes. When the granitic magma of Stone Mountain began cooling 300 million years ago, the outside margins of the intrusion, in contact with cooler surrounding rock, solidified first. This outer "shell" of rock contracted as it continued to cool, causing cracks to form. Some of the cracks reached down to the magma and thus filled with magma, becoming dikes.

There are dozens of dikes in the first 600 feet of the trail. The very fine-grained dikes composed of tiny crystals are called aplite dikes. Those made of crystals that are as big as 0.5 inch or more across are called pegmatite dikes. Some dikes cut across each other. For example, 10 feet beyond the ledge, two dikes cross at right angles. If you examine this you will find that one dike was emplaced along a small fault that had earlier shifted the other dike. The displacement along the fault amounts to about 1.5 feet.

Panola Mountain State Park

Panola Mountain, a smaller companion of Stone Mountain, is located in northern Rockdale County along the South River. The mountain is preserved in a more natural state than Stone Mountain because it is protected as part of Panola Mountain State Park. The most ecologically sensitive areas of Panola Mountain can be visited only by guided tour. The park and nearby Davidson-Arabia Nature Preserve are part of the Arabia Mountain National Heritage Area, a National Park Service designation.

The Panola Granite is darker than the Stone Mountain Granite, with larger grains, and is 313 to 293 million years old. The intrusion happened at about the same time as Stone Mountain, but the rock composition indicates a different magma source. The best place to see the Panola Granite is in a small abandoned quarry on the east side of GA 155. The quarry was abandoned because the granite is crosscut by veins and fractures containing the pistachio green mineral epidote. The fractures made the granite less valuable as slabs, and the more easily decomposed mineral epidote made it less useful as crushed rock for gravel.

To get there, from I-285 take exit 48 (GA 155, Flat Shoals Parkway) and follow GA 155 south 7 miles to the quarry on your left, just past the South River bridge, or 8.5 miles to the park entrance on your left.

Davidson-Arabia Nature Preserve

The Davidson-Arabia Nature Preserve, near Stonecrest Mall, is a DeKalb County park encompassing 2,000 acres of granitic gneiss outcrops, wetlands, forests, streams, and a lake. Within the park is Arabia Mountain, which looks like a small version of Stone Mountain but is composed of the Arabia Mountain Migmatite, a rock type within the more widespread Lithonia Gneiss. The rock is about 50 to 100 million years older than that of Stone Mountain, having formed about 375 million years ago.

The preserve is about 2 miles south of I-20. From I-20 take exit 74 (Evans Mill Road, in Lithonia) and head south on Evans Mill Road. Turn left onto Mall Parkway, about a block from the exit, and then right onto Klondike Road. Pass the traffic circle at Rockland Road and take the second driveway on the right to visit the nature center and access pavement outcrops in an old quarry. To walk up the mountain, continue about 1 mile on Klondike Road to the south parking lot, on the left.

Klondike Road divides the park neatly into two parcels. The western side, behind the nature center, has been quarried over a large flat area, so the rock surface is relatively fresh, and structures, such as folded foliation and dikes, are easier to see. On fresh surfaces it is also easier to see the migmatite, a portion of the Lithonia Gneiss that was subjected to such high temperatures and pressures that it partially melted. (The presence of the metamorphic mineral sillimanite indicates that it was heated to temperatures above 1,200°F and buried at a depth of more than 14 miles at the time of metamorphism.) Within the striped, foliated gneiss you see irregular patches that have the salt-and-pepper, non-foliated appearance of granite. These are interpreted as areas where the gneiss melted. Together, they comprise rock type migmatite.

Dikes of light-colored pegmatite, some with tourmaline crystals, cut the gneiss. Within the gneiss there are brownish layers of tiny red garnets. The garnet chemistry suggests the long-ago presence of aluminum-rich clay, indicating that the preserve's portion of the Lithonia Gneiss was once sedimentary.

Folded garnet-rich layers in the Arabia Mountain Migmatite.

On the eastern side of Klondike Road, from the parking lot at the southern edge of the park, you can take the easy 0.6-mile trail to Bradley Peak, the higher of two summits on Arabia Mountain. The peak rises only 150 feet above the surrounding area. The rock here is weathered and pinkish due to the presence of iron oxides.

The area is well-known for its wildflowers, many of which grow in solution pits. The different plants neatly illustrate the concept of biological succession. Lichens grow on bare rock and begin to create soil. Tiny plants and mosses grow where soil has begun to form. Nonwoody flowering plants, such as the yellow Confederate daisy (*Helianthus porteri*), which is only found within a 60-mile radius of Stone Mountain, grow in soil up to a few inches deep. Pine trees are the dominant trees because quartz-rich gneiss makes sandy soil, to which pines are well adapted.

Depending on the time of year, in the gravelly sand that is too shallow even for moss you will either see inch-high, dry stalks bearing seeds, or a lush red carpet (possibly with tiny white flowers), both of the plant called *Diamorpha smallii*. This is a stonecrop that survives in the extremely hot and dry conditions of shallow solution pits. Its inconspicuous stalks hold seeds up high enough so they don't roast on the hot gravel. After rains, the plants survive for days underwater. The plants grow red leaves in January and February and produce white blooms by March or April. The show is over by mid-May, when

Solution pits atop Arabia Mountain. The pit in the foreground is filled with Diamorpha smallii. *Beyond this pit are two vernal pools.*

only the dry seed stalks remain. Around the summit, pits up to tens of feet long contain water except in the driest parts of the summer. These large pits are also called vernal pools. Rare and endangered plants, such as the black-spored quillwort (*Isoetes melanospora*), live in these pits. To avoid stepping on native plants, please tread on bare rock surfaces whenever possible and not in the pits.

Boat Rock Preserve

Boat Rock Preserve, a 7.8-acre park in southwest Atlanta, is underlain by Ben Hill Granite, one of five granite intrusions in the Atlanta area that formed about 300 million years ago. The park contains about a dozen boulders up to 25 feet high, with weathered surfaces that are slightly bumpy due to 0.5-inch or larger feldspar crystals. The boulders discouraged agriculture in the area, so native plants flourish in the relatively undisturbed soil of a mature oak-hickory forest. The rocks lie along a ridge about 150 feet above the elevation of the Chattahoochee River, which has incised the landscape along the Brevard fault zone a little over 1 mile to the west. It is not clear why the Ben Hill Granite has resisted weathering so well here. It is a lovely place to follow paths among boulders, little known except to rock climbers.

To get there, from I-20 west of the city center, head south on Fulton Industrial Boulevard for 3.8 miles and turn left onto Bakers Ferry Road SW. Go 0.5 mile and turn left onto Boat Rock Road SW. After 0.4 mile look for a gravel driveway on the right with a small parking lot, where there is a kiosk with a map of the trail to the boulders.

Boulders of Ben Hill Granite in Boat Rock Preserve.

An exposure of gneiss on Pigeon Hill in Kennesaw Mountain National Battlefield Park.

Kennesaw Mountain National Battlefield Park

Kennesaw Mountain National Battlefield Park, managed by the National Park Service, was the site of a Civil War battle in June and July 1864, in the campaign leading up to the fall of Atlanta to Union forces. The top of Kennesaw Mountain is 1,808 feet above sea level and stands approximately 700 feet above the surrounding terrain. The mountain is composed of light gray gneiss and migmatite surrounded by darker rocks that belong to a large complex of mafic rocks that stretches from Marietta to Canton.

Kennesaw Mountain, the higher peak, has a panoramic overlook near the summit, which is accessible by a park shuttle bus. If you hike to Little Kennesaw Mountain, the lower peak at the west end of the mountain, after 0.2 mile you pass Pigeon Hill, which has house-sized boulders and pavement outcrops with folded gneiss.

To reach the Little Kennesaw Mountain Trailhead, from I-75 (exit 265) head west on GA 120 (North Marietta Parkway) for about 3.6 miles to Burnt Hickory Road. Turn right (north) and continue 1.3 miles to the intersection with Old Mountain Road. Parking is along the west side of Burnt Hickory Road.

Sweetwater Creek State Park

As cotton gained importance in pre–Civil War days, towns grew up around the textile mills. The town of New Manchester no longer exists, but its mill ruins can still be viewed at Sweetwater Creek State Park. Flowing southeast to the Chattahoochee River, Sweetwater Creek has cut a gorge through metagraywacke and gneiss, descending more than 100 feet in elevation over 1.5 miles. The even layering of the southeast-dipping metamorphic bedrock has resulted in straight, parallel ledges that cross the creek. Some of the bedrock contains potholes, circular depressions that stream water has carved out of solid rock. Pebbles that sit in the depressions are swirled in circles by the moving water, and they abrade the bottom of the pothole, causing it to deepen over time. Local amphibolite and metagraywacke are part of the foundation of the five-story brick ruins of the 1849 textile mill that Union troops burned in 1864.

To reach the park, from I-20 (exit 41) follow Lee Road for 1 mile south to Cedar Terrace Road and turn left (east). Continue 0.8 mile and turn right (south) onto Mt. Vernon Road. Drive 0.3 mile to the park entrance on your left (Factory Shoals Road). Follow the signs for 0.6 mile to the park visitor center parking lot and the trailhead for the 0.5-mile hike to the mill ruins.

Quarries and Road Cuts

The rapid growth of Atlanta's population over the past fifty years has led to unprecedented road and building construction. This construction required building materials, such as aggregate (crushed stone, sand, and gravel), so it is not surprising that huge quarries dot the Atlanta area. In addition, road cuts have exposed Piedmont rock throughout the Atlanta area. (The numbers accompanying the following quarries and road cuts are keyed to the road guide map on page 280.)

Vulcan Materials Norcross Quarry as seen from the visitor overlook.

The largest rock quarry is just off I-85, south of Gwinnett Place Mall. The quarried rock is granitic gneiss, which quarry operators refer to as "granite." The Vulcan Materials Norcross Quarry (1) is the largest "granite" quarry in the United States and was the leading granite-producing quarry in the nation in 2010. The quarry is more than 0.5 mile wide and over 600 feet deep. With field glasses you can watch rock-laden, 20-foot-high quarry vehicles toiling up ramps. Here, 200-million-year-old, 5-foot-wide diabase dikes are well exposed (see Pull-Apart Activity during the Mesozoic Era in the chapter introduction), though they are not very visible from the quarry overlook.

To reach the overlook, leave I-85 at exit 102 (GA 378/Beaver Ruin Road) and head south 0.4 mile on Beaver Ruin Road to the quarry entrance on your right, opposite Park Drive. Take an immediate right once inside the gate and proceed to the overlook on your left. (School and other groups can arrange tours of local quarries.) The granitic gneiss of the Norcross Quarry is also exposed along I-85 between Jimmy Carter Boulevard and Pleasantdale Road (2).

The Vulcan Materials Kennesaw Quarry (west of I-75 in Kennesaw) has a large training center that doubles as a museum. Tours can be arranged into the quarry to view black-and-white granitic gneiss. The rock formed from the intrusion of granitic magma into black basalt or gabbro, after which the rocks were metamorphosed into gneiss and amphibolite.

Gneiss and Schist

Gneiss and schist dominate the Atlanta area. There is an excellent exposure of biotite gneiss at the exit 91 off-ramp south of I-85, on the west side of Clairmont Road (3). Pieces of other types of metamorphic rock seem to float in a jumbled, contorted mass in this outcrop. Because the biotite gneiss contains many different types of rock mixed together that don't have continuous layering or foliation, some geologists consider the Clairmont Formation, the map unit to which this outcrop belongs, to be part of a tectonic mélange. A mélange is a mappable body of rock that has been chaotically deformed. Mélanges lack continuous layering and typically include blocks of rock of many different types and sizes. For example, about 10 feet up the outcrop face along Clairmont Road, there is a block of black amphibolite. It is about 3 feet long and is crossed by white veins of quartz. Both the veins and the banding (foliation) within the block are nearly vertical, while the foliation in the enclosing gneiss is horizontal. This suggests a complicated history of deformation, since the compressive forces that produced the foliation in the amphibolite must have happened before this block was broken up and included in the biotite gneiss but before compression from a different direction caused the foliation to form in the gneiss. There is another outcrop of this biotite gneiss 0.3 mile to the south, on the west side of Clairmont Road behind a gas station.

At least one gneiss outcrop in Atlanta is underground. If your introduction to the city is stepping off the MARTA subway platform at Peachtree Center Station (4), you can see gneiss in the walls of the station, though the lighting is not as bright as might be desired and the tracks prevent getting close to the rock.

Aboveground you can examine biotite gneiss, biotite schist, and metagraywacke from the sidewalk on the east side of Spring Street NW, one block south of

A block of amphibolite with nearly vertical foliation occurs within gneiss with nearly horizontal foliation at an outcrop along Clairmont Road south of I-85.

the intersection with Peachtree Street in midtown Atlanta (5), just south of the overpass. If you visit this location, also examine the building stone of Rhodes Hall, the 1904 "castle" just across South Rhodes Court to the north of the Peachtree Street intersection. The hall is open for public tours on specific days. The foundation and basement walls of Rhodes Hall are made of granitic gneiss (Lithonia Gneiss), probably quarried near Lithonia. Above the basement, the walls are made of Stone Mountain Granite, as is clear from the distinctive cat's paws, consisting of clusters of black tourmaline surrounded by white minerals.

You can see large outcrops of gneiss, metagraywacke, and schist from the I-85 northbound off-ramp at Buford Highway (exit 86, and #6 on the geology map), the I-85 northbound on-ramp from Peachtree Street, and on I-75 just north of 17th Street. You can't stop at these locations, but even as you speed by you can see layers of white minerals set in black-and-gray-banded rock. Many of the white bands were pinched into white sausage-like shapes called boudins. As these rocks were deformed at high temperatures and pressures, the white quartz-and-feldspar-rich layers were more brittle than the dark biotite-rich bands, and thus they broke into the boudins.

The steps at the entrance to Rhodes Hall, built of Lithonia Gneiss (below left, with foliation) and Stone Mountain Granite (above left, and in steps) with tourmaline cat's paws.

Light-colored bands and boudins within dark gneiss at exit 86 of I-85.

Quartzite

Quartzite, which underlies narrow ridges, is present in two belts that cross Atlanta. The northwestern belt is 80 miles long and passes through the city of Sandy Springs (named after a spring that issues from quartzite weathering to sand). I-75 crosses a quartzite ridge in this belt, crowned by the Cobb Energy Performing Arts Center, just north of its long descent to the Chattahoochee River. East of I-75, this ridge extends to the cliffs of the Chattahoochee Palisades in the Chattahoochee River National Recreation Area. You can see quartzite at the overlook in the East Palisades area, between I-75 and I-285. The overlook rests atop quartzite outcrops about 160 feet above Devils Racecourse Shoals, rapids named by a nineteenth-century riverboat captain and popular today with whitewater enthusiasts.

The National Park Service manages the Chattahoochee River National Recreation Area. It stretches along 48 miles of river, from the Buford Dam at Lake Lanier southwest to near I-75 at Smyrna. Rocks are exposed along the river in many places, including at natural rock shelters once used by Indians. There are numerous parks and access points.

About 1 mile southwest of the I-75 crossing of the Chattahoochee, a different quartzite ridge underlies Mt. Wilkinson in the town of Vinings. Mt. Wilkinson commands one of the finest views of Atlanta.

The southeastern belt is 45 miles long and lies south and east of Stone Mountain, such as near Snellville (see the US 78: Snellville—Thomson road guide). You can see quartzite of this belt on a ridge that crosses Woodrow Road, south of I-20 near Lithonia. The quartzite is white to yellowish, and sugary to glassy. The glitter in the rock is muscovite mica. To visit this outcrop (7), head south from exit 74 of I-20 (Evans Mill Road). Continue south, crossing the intersection that is 0.1 mile south of the interstate, onto Woodrow Road, and then drive another 0.7 mile to the outcrop. The same ridge continues south, parallel and to the west of Arabia Mountain, from which it is visible.

Amphibolite

Amphibolite dominates the area surrounding Kennesaw, and a smaller area northeast of Lawrenceville. Smaller bodies of amphibolite are also found within many metasedimentary rocks. Most commonly, amphibolite is metamorphosed basalt, diabase, or gabbro, the same rocks that make up oceanic crust. The abundance of amphibolite in the Atlanta area contributes to the idea that its metamorphosed sedimentary rocks were deposited on or close to oceanic crust. Chemical tests suggest that the basaltic magma that became the amphibolite originated in a rift setting such as the Mid-Atlantic Ridge, where oceanic crust forms.

You can see amphibolite near Lawrenceville in the parking lot of a shopping center immediately north of GA 316 on the east side of Lawrenceville-Suwanee Road (8); however, it is not visible from either road. Here, black amphibolite makes up more than 90 percent of an outcrop that also contains several quartzite layers, each a few inches thick. At first the structure appears simple, with straight quartzite layers dipping gently northwest, but high in the outcrop a quartzite layer is folded into a wide, flattened Z, indicating that this rock was severely deformed.

The amphibolite outcrop in a shopping center off Lawrenceville-Suwanee Road. Note the thin, Z-shaped, folded quartzite layer high on the outcrop. The vertical stripes are drill holes.

To get to the outcrop, exit GA 316 at GA 120 and head west 0.4 mile to Lawrenceville-Suwanee Road. Turn left and continue 0.2 mile to the next traffic signal. Turn left into the shopping center and make an immediate right, continuing to the south end of the shopping center.

Soapstone Ridge

Soapstone can be found at Soapstone Ridge, which I-285 crosses near the intersection with I-675. The soapstone was thrust-faulted on top of the other metamorphic rocks in this area. The soapstone is a metamorphosed ultramafic rock that formed within the Earth's mantle or lower crust before it was metamorphosed and thrust to its present position.

Local Woodland Indians had a thriving soapstone industry between 4,200 and 3,000 years ago. Stone cooking bowls carved out of the soapstone were distributed as far away as the Mississippi Valley. The advantages of soapstone cookware include its ease of carving, even distribution of heat, and nonstick properties.

You can see a soapstone boulder in front of the historic DeKalb County Courthouse at the intersection of Ponce de Leon and Clairemont avenues in downtown Decatur. A partly carved soapstone bowl is nearby in the outdoor rock exhibit behind Fernbank Science Center, on Heaton Park Drive near Ponce de Leon Avenue (FS).

Although Soapstone Ridge was once dotted with pits from which soapstone was quarried millennia ago, none of the surviving sites are open to the public. The best roadside exposures are reached by exiting I-285 at Bouldercrest Road (exit 51). Head south on Bouldercrest Road for 2.7 miles and bear left onto Panthersville Road. Outcrops are on the left shoulder of Panthersville Road, from about 400 to 900 feet past the intersection (9). There are several other rock types here, including gneiss and amphibolite, chaotically mixed in with the soapstone. The largest soapstone outcrop is the last large boulder on the left.

Most housing developments on Soapstone Ridge are fairly new, and the region is home to several trucking terminals and landfills. Soapstone's special properties have contributed to this situation. Despite the softness that makes it easy to carve, it is remarkably cohesive and tough. As a result, the soil depth above it varies greatly. This means that, in places, bedrock had to be blasted to make room for homes where only grading had been expected. In addition, the soapstone weathers to make a peculiar type of clay that expands enormously when wet and shrinks as it dries, heaving foundations and cracking concrete slabs. These conditions and the hilly topography kept the area from being developed for housing until well into the 1990s. The low land prices and proximity to the interstates attracted trucking firms and landfills. Soapstone and its clay weathering-products prevent water seepage, which helps prevent landfill waste from contaminating groundwater. The Seminole Landfill, a hill similar in height to Soapstone Ridge near the north end of I-675, did not exist until the late 1990s.

Mylonite

Between Kennesaw and the center of Atlanta there is an area sliced through by several northeast-trending fault zones, each containing mylonite, rocks that have been highly sheared and fractured. The largest of these is the Brevard fault zone. You can see outcrops of mylonite in the Brevard fault zone in two locations along I-285. North of downtown Atlanta, light-colored mylonite gneiss (10) is present on the south side of I-285, 0.6 mile east of the GA 400 interchange (exit 27). West of downtown, 0.3 mile north of exit 13 (Bolton Road), mylonite is exposed on both sides of I-285 (11). Here, a fault brings rock interpreted as mylonitic Ben Hill Granite over dark mylonite gneiss.

If the mylonitic rock exposed along I-285 near Bolton Road is 300-million-year-old Ben Hill Granite, it helps geologists date the movement along the fault zone. The age of the granite and its mylonitic texture indicate that ductile flow along the Brevard fault zone was happening during the final stages of the collision between North America and Africa, possibly the same event that produced thrust faulting in the Valley and Ridge Province. Further lines of evidence that ductile flow occurred this late are the 6-mile-long "tails" that both the Ben Hill Granite and Palmetto Granite have near the Brevard fault zone. It is as if the granite masses were smeared out as they were dragged to the southwest relative to the rocks on the north side of the fault zone.

The most distinctive mylonitic rock type in the Brevard fault zone in the Atlanta area is called button schist. This rock is gray to silvery, with silvery, fine-grained mica and feldspar, quartz, and some chlorite. Shearing has broken the mica, which earlier had formed more or less continuous layers (foliation), into isolated, button-shaped bodies smaller than about 1 inch across. When the rock weathers, these "buttons" are present in the soil. Phyllonite, a rock with fine-grained mica, is also present. It forms through the ductile shearing of coarser-grained schist along a fault zone. You can see these rock types, along with mylonitic quartzite, in road cuts near Duluth. The rocks are visible from the sidewalks on either side of Pleasant Hill Road at the underpass beneath Buford Highway (12).

Mylonite along Pleasant Hill Road near Duluth. The vertical brown stripe is a drill hole.

Appendix

Museums and Exhibits

ALBANY

Flint RiverQuarium
101 Pine Avenue
www.flintriverquarium.com

The aquarium has both freshwater and saltwater tanks, highlighting the ecosystems of the Apalachicola, Chattahoochee, and Flint river basins and revealing the subsurface karst geology of the Ocala Limestone with its sinkholes, aquifers, and underground caves. The building wraps around a 175,000-gallon, 22-foot-deep tank fed by a natural spring; it's designed to look like a blue hole spring. You can examine blocks of limestone all around the aquarium.

ATHENS

Georgia Museum of Natural History
University of Georgia, Natural History Building
101 Cedar Street
http://naturalhistory.uga.edu

The museum houses a geology collection that includes mineral specimens from around the world, more than twelve thousand fossils and casts (including trace fossils and Paleozoic-age fossils from the southeastern United States), and more than twenty thousand specimens from ore deposits and mines. There is an extensive archaeological collection covering 12,000 years of human settlement in Georgia, and also collections of reptiles, amphibians, fish, invertebrates, mammals, birds, plants, and insects.

University of Georgia Science Library
Boyd Graduate Studies Building
210 DW Brooks Drive
www.libs.uga.edu/science/sloth.html

In the library's lobby you can see the skeleton of a giant North American ground sloth (*Eremotherium mirabile*) that roamed coastal Georgia about 14,000 years ago. Discovered during the construction of I-95 near Brunswick, the 13-foot-long skeleton is one of the most complete ever found.

ATLANTA

Fernbank Museum of Natural History
767 Clifton Road NE
www.fernbankmuseum.org

A private nonprofit museum displaying skeletal casts of some of the world's largest dinosaurs. Its "Walk Through Time in Georgia" exhibit interweaves the present landscapes, flora, and fauna of Georgia's physiographic regions with dioramas of the ancient life in Georgia.

Fernbank Science Center
156 Heaton Park Drive
www.fernbank.edu

The exhibit hall has a display of Georgia meteorites and tektites, a cast of a tyrannosaurus skull, several dinosaur reconstructions, and a NASA spacecraft. In the outdoor rock and mineral display behind the building, you can see a collection of Atlanta-area rock types, huge crystals and other mineral specimens, and local soapstone carved thousands of years ago.

Georgia Aquarium
225 Baker Street NW
www.georgiaaquarium.org

The world's largest aquarium, with more than 10 million gallons of water, is located in downtown Atlanta near Centennial Olympic Park. The Georgia Explorer gallery represents Georgia marine habitats, including Gray's Reef. With beluga whales and whale sharks, there are more animals here than in any other aquarium. A fossile whale cast is in the Cold Water Quest gallery.

Georgia Capitol Museum
206 State Capitol
www.libs.uga.edu/capitolmuseum/museum

The museum, located in the Georgia State Capitol building, houses a collection of fossils, gems, and economic minerals of Georgia.

Hartsfield-Jackson International Airport
6000 North Terminal Parkway
www.atlanta-airport.com

The 33-foot-long cast of *Yangchuanosaurus*, one of the largest meat-eating dinosaurs, stands watch over travelers in the airport's atrium. You can also see a number of interesting igneous and metamorphic structures in the polished stone floor tiles.

Michael C. Carlos Museum at Emory University
571 South Kilgo Circle
http://carlos.emory.edu

The museum contains numerous examples of art and antiquities, made of a variety of types of stone and minerals, from the Americas, Egypt, and the Near East. The museum is housed in one of the many buildings on the Emory University campus that are covered in a patchwork of pink, white, and gray Georgia marble. Some slabs of the marble display foliation and folding.

BUFORD

Gwinnett Environmental and Heritage Center
2020 Clean Water Drive
www.gwinnettehc.org

The museum has exhibits on water (including a rivers-to-reef exhibit with freshwater and saltwater aquaria), natural resource conservation, Native American cultures, and sustainability.

CARTERSVILLE

Tellus Science Museum
100 Tellus Drive
http://tellusmuseum.org

Named after the Roman Earth goddess, Tellus Science Museum has an impressive array of large skeletal casts of dinosaurs and other extinct vertebrates, along with many types of Georgia fossils. The Weinman Mineral Gallery highlights Georgia mining and minerals and also has exhibits about rocks, plate tectonics, and minerals from around the world. Outside the museum you can see boulders of different rocks mined near Cartersville.

COLUMBUS

Columbus State University Coca-Cola Space Science Center
701 Front Avenue
www.ccssc.org

Located along the Chattahoochee Riverwalk, the center has interactive exhibits and displays on weather, astronomy, and space exploration. One exhibit includes a seismograph that monitors earthquakes.

DAHLONEGA

Dahlonega Gold Museum Historic Site
1 Public Square
www.gastateparks.org/DahlonegaGoldMuseum

The Museum occupies the 1836 Lumpkin County courthouse, the oldest courthouse in Georgia, with bricks made of local clay that contains gold. Inside you can learn about America's first gold rush and see a gold nugget weighing more than 5 ounces, gold coins produced at the Dahlonega branch of the U.S. Mint in the 1800s, historical photographs, and gold-mining equipment.

ELBERTON

Elberton Granite Museum
1 Granite Plaza
www.egaonline.com/egaassociation/museum

The museum includes exhibits on the history of the granite industry as well as tools used to quarry, cut, sandblast, and polish granite, and a display on the Georgia Guidestones.

GAINESVILLE

Elachee Nature Science Center
2125 Elachee Drive
www.elachee.org

This private nonprofit museum and environmental education center includes an assortment of fossils, including fish, a mammoth tooth and other vertebrate remains, invertebrates of Paleozoic to Cenozoic age, a petrified log, and ferns.

JEKYLL ISLAND

Georgia Sea Turtle Center
214 Stable Road
www.georgiaseaturtlecenter.org

At the center, which is in the old power plant building in the historic district, you can view sea turtles up close and see those being treated and rehabilitated. A huge fossil sea turtle (*Archelon*) hangs from the ceiling.

KENNESAW

Vulcan Materials Kennesaw Quarry
1272 Duncan Road
www.vulcanmaterials.com/vcm.asp?content=results&fac=188

The museum at the quarry showcases minerals, rocks, and fossils and emphasizes the importance of mining to society.

MACON

Museum of Arts and Sciences
4182 Forsyth Road
www.masmacon.com

The museum houses a 40-million-year-old (Eocene-age) fossil whale excavated from a kaolin mine in Twiggs County. Named Ziggy, short for the genus *Zygorhiza*, the whale was found with the vertebrae of a small shark inside, remnants from its last meal.

MILLEDGEVILLE

Georgia College Natural History Museum and Planetarium
221 North Wilkinson Street
www.gcsu.edu/nhm

This museum is dedicated to the earth sciences and houses one of the largest collections of fossils in the Southeast, including plants, invertebrates, fish, an ichthyosaur, a dinosaur, mammals, and birds. Exhibits showcase unique geologic resources of Georgia, including caves, the fossil-rich Hardie Mine site in Wilkinson County, and ice age mammals from Brunswick. It is also an official repository of specimens for the National Park Service.

NELSON

Marble Museum
1985 Kennesaw Avenue

The Marble Museum is located in Nelson City Hall. The exhibit covers the geology of marble, the uses of Georgia marble (from pre-Columbian sculptures found at Etowah Mounds through modern industrial uses), the details and history of quarrying, and examples of famous buildings and monuments that use Georgia marble (such as the seated Lincoln statue at the Lincoln Memorial). The public can view marble quarries at the annual Pickens County Georgia Marble Festival in October.

ROSWELL

Chattahoochee Nature Center
9135 Willeo Road
www.chattnaturecenter.org

This private nonprofit environmental education center has interactive exhibits on the Chattahoochee River watershed. There is a Nature Exchange with thousands of rocks, fossils, shells, and other natural objects that you can examine and even take home with you. You can see birds of prey, reptiles, amphibians, and mammals; examine boulders of several types of rocks; and take a nature hike.

STATESBORO

Georgia Southern University Museum
Rosenwald Building
2142 Southern Drive
ceps.georgiasouthern.edu/museum

The museum displays a number of fossil vertebrate skeletons, including a mosasaur; a mastodon skull; skeletons from modern vertebrates, such as the jaws of a shark, the bill of a sawfish, and the complete skeleton of a bottle-nosed dolphin; and a host of fossil invertebrates. You can also see a 40-million-year-old, 11-foot-long whale skeleton with legs, *Georgiacetus vogtlensis*, the oldest whale ever found in North America. This whale fossil provides a link between land mammals and whales.

SKIDAWAY ISLAND

Marine Education Center and Aquarium
University of Georgia Marine Extension Service
www.marex.uga.edu/aquarium

The center includes many educational exhibits, including the fossil remains of mammoths, mastodons, whales, giant armadillos, and shark teeth, all dredged from the Skidaway River; saltwater tanks of fish, turtles, and invertebrates from the Georgia coast; dioramas of Gray's Reef National Marine Sanctuary (a protected reef east of Sapelo Island); and other exhibits. The staff provides educational programs for groups and schedules tours of the Wilmington River and Wassaw Sound on a research vessel.

STONE MOUNTAIN

Stone Mountain Park
Confederate Hall Historical and Environmental Education Center
1000 Robert E. Lee Boulevard
www.stonemountainpark.com/activities/
history-nature/Confederate-Hall.aspx

Exhibits highlight the unique geological and ecological features of Stone Mountain, including the history of its formation, and also cover plate tectonics and the internal structure of the Earth. You can see fossils characteristic of each of the geological periods.

Stone Mountain Park Quarry Exhibit (Raising a Ledge)
www.stonemountainpark.com/activities/history-nature/Quarry-Exhibit.aspx

At this outdoor quarry exhibit you can explore the granite mining that took place at the mountain from the 1850s to the 1970s. You can examine large boulders of granite and pegmatite, see various geologic features in the granite, and visit exhibits on animals and wildflowers. A carved and polished stone wall lists buildings around the world constructed from Stone Mountain Granite.

TYBEE ISLAND

Tybee Island Marine Science Center
1510 Strand Avenue
tybeemarinescience.org

The center promotes the understanding of the Georgia coast and its marine ecosystem with exhibits of live animals, including fish, amphibians, reptiles, and invertebrates (some in touch tanks). There is also a fossil gallery, and information is provided concerning sea turtles, marine mammals, sharks, and environmental issues. The center also leads beach discovery walks and marsh treks.

VILLA RICA

Pine Mountain Gold Museum
1881 Stockmar Road
www.pinemountaingoldmuseum.com

The museum has exhibits on the history of gold mining in the Villa Rica area. You can pan for gold and tour the ruins of a gold mine.

WARM SPRINGS

FDR's Pools and Warm Springs Museum
401 Little White House Road
www.gastateparks.org/LittleWhiteHouse

Museum and exhibits covering the geology and history of the area's warm springs.

GLOSSARY

agate. Chert with color variations, such as banding. Often prized for jewelry.

ammonites. A group of cephalopods (squid relatives) of the Paleozoic and Mesozoic eras with coiled, chambered shells similar to that of the modern chambered nautilus.

amphibole. A group of minerals ranging from dark green to black that have long crystal shapes. Hornblende is one of the most common.

amphibolite. A black, dark gray, or dark green metamorphic rock that weathers to a dark orange or rusty red color. It is defined by the presence of needle-shaped minerals of the amphibole group, such as hornblende. The other main constituent is white or gray feldspar, which may not be obvious to the unaided eye. Most amphibolite formed from the metamorphism of mafic igneous rock, such as basalt or gabbro.

anticline. A concave-down fold in rock layers that causes older rock strata to be buckled upward along the hinge line of the fold.

aplite. A fine-grained igneous rock, usually with a composition similar to granite, found in dikes.

Appalachian Plateau. The northwesternmost physiographic province in Georgia, composed of flat-topped mountains interrupted by widely separated, straight valleys, all of which are underlain by sedimentary rocks ranging in age from Cambrian to Pennsylvanian.

aquifer. A porous underground body of rock from which water can be obtained.

aragonite. A mineral made of calcium carbonate, of which modern mollusk shells are composed.

arthropod. An invertebrate (for example, crabs, insects, spiders, and millipedes) with an external skeleton, jointed body, and segmented limbs.

augen. A rock texture in which biotite mica wraps around feldspar crystals like eyelashes framing an eye (*augen* is German for "eye"). Usually caused by the ductile deformation of granite in or near a mylonite zone.

barrier island. An island of sand forming part of the outer coastline and remaining out of water at high tide.

barrier island complex. A narrow, linear sand deposit, parallel to the coastline, and a wider, landward deposit predominantly composed of clay. These deposits represent, respectively, a modern or ancient barrier island and its accompanying marsh or lagoon.

basalt. A dark, fine-grained volcanic igneous rock of mafic composition.

basement. The igneous and metamorphic rocks upon which a sequence of sedimentary rocks were deposited.

batholith. An igneous intrusion with an exposed area of more than 40 square miles.

bauxite. A sedimentary rock from which aluminum is derived.

bed. A layer of rock deposited in a discrete episode of sedimentation. The layering of beds is called bedding. The boundary between beds, which once represented an old water bottom or land surface, is a bedding plane.

bedrock. Rock that has not been moved by erosion or human activity from the place where it formed.

bentonite. A mixture of clay minerals dominated by montmorillonite. Usually develops from the alteration of a volcanic ash deposit.

biotite. A black mineral in the mica mineral group with a thin, flat shape.

Blue Ridge. A mountainous physiographic province extending into north Georgia. Underlain by metamorphic and igneous rocks, with some summits higher than 4,000 feet.

brachiopod. A phylum of organisms with a pair of ribbed shells and a fleshy stalk by which they attach themselves to the sea bottom. They were most prevalent in the Paleozoic era and are not related to clams, which they superficially resemble.

breccia. A sedimentary or volcanic rock containing jagged, pebble-sized rock fragments.

Brevard fault zone. A 375-mile-long zone of rock that experienced both brittle and ductile deformation. Extends from near Montgomery, Alabama, to near Winston-Salem, North Carolina, and includes a 0.5- to 2.5-mile-wide mylonite zone.

burrow. A trace left by an organism, such as a worm, as it ate its way through sediment or dug a home.

calcareous. Containing calcium carbonate along with other constituents, such as quartz or clay minerals.

calcite. A light-colored mineral composed of calcium carbonate; the main constituent of limestone, most marble, and many marine fossils.

calcium carbonate. A chemical compound of calcium, carbon, and oxygen that can crystallize either as calcite or aragonite.

carbonate rock. Limestone or dolostone.

Carolina bay. An oval-shaped lake or wetland ranging from a few hundred yards to several miles across within the Coastal Plain.

cement. The substance, usually composed of silica (quartz), carbonate minerals, or iron oxide, that binds the grains of a sedimentary rock.

chert. A sedimentary rock composed of quartz crystals too small to detect except by electron microscope.

chlorite. A green mineral with thin, flat shapes. Develops at low temperatures of metamorphism.

clay. Particles less than $1/256$ millimeter in size; or a group of minerals with thin, flat shapes that are the size of clay grains.

clayey. Clay rich.

Coastal Plain. The southern physiographic province of Georgia, which is characterized by gentle topography because it is mainly composed of soft, easily eroded, unconsolidated sediments.

concretion. A mass of a hard mineral precipitated from a solution within a softer sedimentary or volcanic rock.

conglomerate. A sedimentary rock containing rounded fragments of minerals or rocks (greater than 2 millimeters in diameter) encased in a finer-grained matrix of sediment.

continental shelf. A marine area underlain by continental crust. Today's shelves extend from the coastline to water depths of about 600 feet.

crinoid. A group of animals related to sea stars and sea urchins. Also called sea lilies. Abundant in the Paleozoic era, they left behind cylinder- or donut-shaped fossils from their columnal (stem).

crossbeds. A series of parallel sedimentary layers that are inclined relative to the larger, originally horizontal layer of which they are a part. The downslope direction of the crossbeds relative to the larger layer indicates the direction of the wind or water current that laid them down.

crust. The uppermost layer of Earth. Oceanic crust is made of basalt and rocks of similar composition, ranges from 4 to 7 miles thick, and lies at water depths measured in miles. Continental crust is made mainly of lighter-colored, less dense rock, such as granite or gneiss, that often has a veneer of sedimentary rock. It normally ranges from 15 to 25 miles thick but in rare cases can be over 40 miles thick.

crystalline. A term applied to igneous and metamorphic rocks, especially those made of visible crystals.

delta. A body of sediment deposited where a river enters a standing body of water.

deposition. The process of sediment settling out of water or air.

diabase. A fine-grained, dark gray to black igneous rock. Similar to basalt but has visible crystals and cooled in thin intrusions underground rather than from lava flows.

dike. A thin, flat igneous intrusion that formed when a crack, cutting through older rocks, filled with magma that solidified.

dip. The downslope direction (the direction water would run) on an inclined (or dipping) rock surface, for example, a bed that has been tilted by tectonic movements.

dissected. A description applied to an elevated landscape that has had distinct valleys cut into it by streams of many sizes.

dolomite. A light-colored mineral composed of calcium magnesium carbonate.

dolostone. A rock composed mainly of the mineral dolomite.

dome. A concave-down fold with a roughly circular outcrop pattern.

drainage basin. A watershed on a large scale; the area that drains into a river.

drainage divide. The boundary between two watersheds, such that water flows in different directions on either side of it.

ductile deformation. A process by which rocks and minerals have been severely deformed without breaking; the opposite of brittle deformation.

Eastern Continental Divide. The drainage divide separating waters that flow to the Atlantic Ocean from those that flow to the Gulf of Mexico.

echinoids. A group of sea animals that includes sea urchins and sand dollars.

erosion. The removal of particles of rock from the place where the rock formed by the action of wind, water, gravity, or ice.

erosional surface. A land surface developed by the removal of earth materials by natural agents such as wind, running water, or ice.

escarpment. A narrow zone, sometimes a cliff, that is several miles long and separates two areas of different elevation.

estuary. The part of a river near its mouth that is influenced by tides. Long estuaries are former river valleys that have been flooded by rising sea level.

Fall Line. A geologic boundary where the ancient crystalline rocks of the Piedmont meet the younger sedimentary layers of the Coastal Plain. Waterfalls and rocky shoals in rivers characterize this boundary in Georgia.

fault. A fracture on which there has been relative movement due to tectonic forces.

fault block. A body of rock adjacent to a fault.

fault-bounded basin. An area that becomes low enough to accumulate sediment because of downward movements over time along one or more neighboring faults.

fault zone. A relatively narrow region in which multiple faults, more or less parallel but often interconnected, have developed.

feldspar. The most abundant mineral in Earth's crust; common in igneous and metamorphic rocks. Feldspar is commonly white, gray, tan, or pink, with a blocky shape. When freshly broken, it has smooth, reflective surfaces. Feldspar is not as common as quartz in soil or in sedimentary rocks because weathering alters feldspar to clay.

flint. Black chert.

floodplain. An area adjacent to a riverbed that may lie underwater when the river overflows its banks.

fold. An area in which rock layers were bent, usually by compressive tectonic forces.

foliation. Parallel surfaces or layers in metamorphic rock caused by the growth, flattening, and/or dissolving of mineral grains under stress during metamorphism.

foraminifera (forams). Single-celled organisms with a calcium carbonate exoskeleton. Most are tiny planktonic organisms, 1 millimeter or less in diameter.

Formation. When capitalized, the fundamental formal unit used in geologic mapping to denote a body of similar rock (such as a series of beds) that is interpreted as having been continuous over a relatively large area. Formations are named after places where they are best recognized and can either bear the term *Formation* (for example, Univeter Formation) or the name of a rock type (for example, Suwannee Limestone). A formation can belong to a larger group, which can be part of a supergroup, and it can contain members.

fossil. Any evidence of past life, whether skeletal or a trace (such as a track, burrow, or fecal pellet).

fossiliferous. Fossil bearing.

fuller's earth. A natural clay product composed mainly of the mineral montmorillonite.

gabbro. A dark intrusive igneous rock similar in composition to basalt, but with visible crystals.

garnet. A group of red or brown minerals that form in many-faced, ball-shaped crystals, usually in metamorphic rocks.

geode. A cavity in rock lined with crystals. If the surrounding rock weathers away, the geode may remain as a round rock that reveals its crystals when broken open.

glaciation. A period of time when continental glaciers were extensive (for example, covering parts of North America and Eurasia) compared to the present.

glauconite. A green mineral that sometimes forms on the seafloor when clay mixes with organic matter; for example, in the fecal pellets of seafloor dwellers.

gneiss. A metamorphic rock with a striped appearance due to layers of light minerals (quartz and feldspar) and dark minerals (such as biotite and hornblende).

graded bedding. Bedding in which the grain size gradually gets finer upward through the bed; typically associated with deepwater deposits called turbidites.

grain size. The size of the crystals or mineral fragments that make up a rock.

granite. A white, gray, or pink igneous rock with crystals large enough to be seen by the unaided eye. The main minerals in granite are feldspar, quartz, mica, and sometimes amphibole.

granitic gneiss. Gneiss that is dominated by quartz and feldspar, with relatively thin bands of dark minerals. Most likely formed from the metamorphism of granite.

gravel. Coarse sediment larger than sand (greater than 2 millimeters in diameter).

graywacke. A sedimentary rock similar to sandstone but containing both sand and significant amounts of clay.

groin. A wall-like structure made of stone, concrete, or other durable materials and constructed perpendicular to a shoreline for the purpose of preventing localized longshore currents from eroding a beach.

Group. When capitalized, a formal rock unit consisting of multiple formations.

hammock. An isolated, elevated area within a wetland.

headward erosion. The erosion of rock and soil around the headwaters of a stream that causes the stream to lengthen upstream.

heavy minerals. Minor constituents in rocks. When eroded from a rock, the minerals tend to concentrate as dark layers of sand on beaches because of their high density.

hematite. A mineral made of iron oxide; a main ore of iron. It is chemically the same as rust, and of similar color.

highstand. The maximum sea level reached during a period of rising sea level.

hinge line. The line along which the maximum curvature of a fold occurs.

hornblende. A black mineral usually with a long shape. The most common member of the amphibole group.

ice ages. Popular name for a period of time, mainly during the Pleistocene epoch, in which continental ice sheets periodically advanced and retreated over North America and Europe.

igneous. Rock that cooled from magma either within the Earth (intrusive or plutonic) or at the surface (extrusive or volcanic).

impression. The imprint of an organic or inorganic object (shell, leaf, raindrop, etc.) left in soft sediment that hardens into rock.

interglacial. Related to times of warming and the retreat of continental glaciers within a prolonged period of glaciation.

intertidal. Covered by seawater at high tide and exposed to air at low tide.

intrusion. Any body of igneous rock that cooled underground. Such rock bodies are said to be intrusive.

isoclinal fold. Deformation resulting in the opposite sides of a fold paralleling each other.

joint. A crack in rock along which no movement is known to have occurred (as contrasted with a fault).

kaolin. A soft, white rock made of kaolinite, a clay mineral. Georgia's most important mineral product.

Kaolin Belt. A long, narrow, northeast-southwest-trending region in the Coastal Plain with significant deposits of kaolin.

kyanite. A blade-shaped, often blue or brown metamorphic mineral, usually formed at relatively high temperature and pressure.

lagoon. Open water lying between barrier islands and land.

lava. Magma that erupts at Earth's surface.

leaching. The removal of elements by the passage of chemically active (for example, acidic) water though rock, soil, or other material (such as refuse).

lens. A mass of rock that is thicker in the middle than around the edges.

lime sink. A surface depression caused by the partial collapse of underground caverns in areas underlain by limestone or dolostone.

limestone. A sedimentary rock made primarily of calcium carbonate, usually in the form of the mineral calcite.

limey. Rich in calcium carbonate.

longshore drift. The movement of sediment, mainly sand, along a coastline and in a direction relatively parallel to the coastline. The direction is the result of prevailing winds, because wind drives wave breakers and sand up onto the beach, often in an oblique angle relative to the shoreline, but gravity determines that water and sand in the return flow moves straight downslope.

mafic. Iron- and/or magnesium-rich. Applied to silicate minerals (especially amphibole, pyroxene, or olivine), rocks containing them, or the magma that produces such rocks. Mafic minerals are typically green or black. Mafic magma usually arises from the partial melting of rock within the mantle. Such magma is characteristic of rift zones but is also part of the suite of igneous rocks that form above plates that are being subducted into the mantle.

magma. Molten rock.

magnetite. A silvery to black mineral; a major ore of iron. It is made of iron oxide and is attracted to a magnet.

mantle. The portion of the Earth between the core and the crust.

marble. A metamorphic rock mainly composed of carbonate minerals and derived from limestone or dolostone.

marl. Sediment containing carbonate and clay minerals.

marsh. A wetland dominated by grasses as opposed to trees. Most commonly occurs where salt water prevents tree growth.

massive. Said of a rock layer (such as a thick sandstone bed) or rock type (such as granite) without evident internal layering.

meandering river. A river with somewhat regular and looplike bends, which tend to form in flatlands where there is loose, relatively fine sediment.

meta-. A prefix used to denote the previous identity of a metamorphic rock; for example, a metasedimentary rock was originally a sedimentary rock, and a metagranite started out as granite.

metamorphic. Rock that received its present form by metamorphism.

metamorphism. Change in the texture and often the mineralogy of rock (by the formation of new crystals, or recrystallization) due to heat and pressure but without the rock having melted.

mica. A group of minerals that break into flat sheets with a glittery appearance. The two most common types are muscovite, which is clear or silvery, and biotite, which is black.

mica schist. Mica-bearing schist, generally formed from a clay-rich rock such as shale. *See also* schist.

midden. A mound of refuse left by prehistoric people.

migmatite. A rock with both metamorphic foliation (such as light and dark bands in gneiss) and irregular patches with igneous texture (for example, the randomly oriented grains of quartz, feldspar, and biotite seen in granite). Forms by the partial melting of metamorphic rock.

mineral. A naturally occurring chemical element or compound with a characteristic crystal form.

mold. A cavity in sedimentary rock left behind by the dissolving away of a fossil, such as a shell; the cavity retains the shape of the dissolved fossil.

mollusks. A phylum of animals, usually with a shell or pair of shells, including clams, snails, and squids.

monadnock. An isolated hill or mountain of bedrock that was left behind as erosion decreased the surrounding elevations.

montmorillonite. A mineral in the clay group. It swells greatly when it absorbs water and shrinks by an equal amount when it dries.

mosasaurs. A family of large marine reptile predators of the Cretaceous period.

mudstone. A sedimentary rock made mainly of clay that unlike shale has no tendency to split into thin pieces.

muscovite. A clear to silvery mineral of the mica group.

mylonite. Rock that has undergone "smearing out" (reduction in grain size and flattening and/or stretching of grains) within a relatively broad zone of relative movement between blocks of crust. Depending on the original rock type, rocks with mylonitic texture can be mylonite quartzite, an extremely tough rock; augen gneiss, in which biotite mica wraps along flattened feldspar crystals like eyelashes framing an eye; button schist, which has a braided appearance in outcrop and leaves "buttons" of mica behind in the soil; and phyllonite, which looks like phyllite but formed by a reduction in the grain size of schist.

mylonite zone. A zone of relative movement between two blocks of crust. Unlike a fault there is no single slip surface; rather, the blocks move along rock that flows like taffy.

nannofossils. Fossils of plankton in the diameter range of 0.005 to 0.06 millimeter.

nodule. A rounded concretion, fist sized or smaller, usually in limestone or dolostone.

normal fault. A fault in which one fault block slides down a sloping fault surface relative to the fault block on the other side of the fault. Forms as the result of forces that are pulling an area apart.

ochre. Iron oxide from decomposed rock. Used as a coloring agent.

ore. A mineral or aggregate of minerals from which one or more valuable substances, especially metals, can be profitably extracted.

outcrop. An exposure of bedrock.

oxide. A compound in which oxygen (negatively charged) bonds with a positively charged ion or radical. Geologically, oxide minerals include both metallic ores and gems.

parabolic dune. A sand dune with a concave windward slope and a convex leeward slope.

pavement outcrop. A broad, flat, natural rock outcrop.

peat. Partly decomposed plant material deposited in a swamp or bog. A precursor to coal.

pebbly. Containing pebbles—smooth, rounded sediment in the size range of 2 to 64 millimeters.

pegmatite. A coarse-grained igneous rock, usually with a composition similar to granite, that is found in dikes. Interpreted to have formed from water-rich magma.

period. A formal unit of geologic time. Its boundaries are based on transitions in the world fossil record rather than a specific number of years.

petrified wood. Wood that has turned to stone by the slow replacement of its organic matter with silica. Occurring at the atomic scale, this process preserves the original grain of the wood.

phyllite. A fine-grained metamorphic rock that is similar to slate but contains foliation surfaces that are shiny and silvery due to mica grains that are barely large enough to catch the light.

phyllonite. A metamorphic rock found in mylonite zones that resembles phyllite but formed as the grain size of schist was reduced.

physiographic province/district. A geographic region defined on the basis of differences in the landscape. Georgia includes five physiographic provinces, each of which is divided into districts.

Piedmont. The north-central part of the state that gradually ascends northward from the Fall Line to the mountainous Blue Ridge and Valley and Ridge provinces. It is cut by steep-sided stream valleys and has a few isolated summits. *Piedmont* means "foot of the mountains."

placer deposit. A sedimentary mineral deposit in which water currents and settling have separated valuable, heavier minerals from other sediment.

plagioclase. A mineral in the feldspar group containing sodium and/or calcium.

planktonic. Referring to a life habit in which organisms float and rely mainly on ocean currents for transport.

plate tectonics. The observation that Earth's lithosphere (its crust plus the rigid uppermost mantle) consists of a mosaic of tectonic plates that are in continuous relative motion at speeds on the order of inches per year. Interactions between plates are responsible for a large majority of earthquakes and volcanoes, as well as the gradual formation of ocean basins and mountain ranges. Today's mosaic consists of about seven major plates and many smaller ones.

plateau. An elevated, relatively flat area with at least one distinct edge called an escarpment.

plunging fold. A fold in which the hinge line is not horizontal but rather plunges beneath adjacent rock strata.

precipitate. A solid material left by the evaporation of a solution or a change in solubility.

pyrite. A mineral made of iron sulfide; also called fool's gold.

pyroxene. A group of dark minerals that have stubby shapes and are found in mafic and ultramafic rocks.

quarry. A place where rock is removed for use as is (contrasted with a mine, where the economic product must be extracted from rock).

quartz. A hard mineral, generally light colored, with a wide variety of forms. It is the second-most-abundant mineral in Earth's crust. The main constituent of most sand, it is common in sedimentary rocks and in light-colored igneous and metamorphic rocks.

quartzite. A metamorphic rock made of more than 95 percent quartz. Recognized by its tendency to break across grains, unlike sandstone composed of quartz, which breaks around the grains.

radiometric dating. Obtaining the age of a geologic sample from the known rate at which a naturally occurring radioactive material within it has changed to other substances.

refractory. Resistant to melting at high temperatures.

relict. Existing in different geologic conditions than the ones in which it formed; for example, a relict barrier island that is now inland.

reservoir-induced seismicity. Earthquakes caused or triggered by a lake behind a dam.

residuum. The soil left behind after carbonate minerals have dissolved from limestone or dolostone due to chemical weathering in a humid climate. Composed mainly of clay (typically red due to iron oxide) but may contain significant chert.

rift. A place where Earth's crust is separating.

rifting. The pulling apart of two tectonic plates (or, initially, of one plate into two).

rift zone. A major tectonic plate boundary comprising multiple rifts in which down-dropped areas called rift valleys develop as well as fissures along which lava periodically erupts at the surface.

riverine dunes. Sand dunes that parallel the course of a nearby river. It is believed they formed during drier times, when sand in a dry riverbed was moved by the wind.

rock. A naturally formed, cohesive mixture of minerals (or a single mineral).

salt marsh. Low, muddy areas between barrier islands and the mainland that are threaded with meandering tidal channels and vegetated mainly with cordgrass.

sand. A mineral or rock fragment, most typically made of quartz, between $\frac{1}{16}$ millimeter and 2 millimeters in diameter. Also refers to a deposit composed mostly of sand-sized sediment.

sand dune. A steep-sided mound of sand piled up and gradually shifted by wind.

sand dune field. An array of adjacent sand dunes formed as the result of the same prevailing wind pattern.

sandy. Containing sand.

saprolite. Rock that has weathered to a crumbly state while preserving obvious remnants of its original minerals and texture.

schist. A metamorphic rock that splits easily parallel to the lined up, clearly visible mineral grains it contains. These minerals are most commonly mica, such as muscovite or biotite, but could be chlorite or hornblende. *See also* mica schist.

sediment. Earth material that has settled out of water or air. Can consist of pieces of preexisting rock, body parts of organisms, or mineral crystals that precipitated out of water.

sedimentary rock. Sediment that has been naturally compacted and/or cemented into solid rock.

sedimentation. The process of sediment settling out of wind or water.

serpentine. A hydrated iron- and magnesium-rich mineral with a long fibrous shape. Often develops by the interaction of seawater with hot ultramafic rock.

shale. A sedimentary rock made mainly of clay. Tends to split into thin pieces parallel to its bedding.

silica. Silicon dioxide, the compound that makes up quartz in all its varieties, including chert.

silicate. A member of the largest class of minerals. Contains silicon and oxygen and forms the vast majority of rocks; exceptions include carbonate rocks.

sillimanite. A white, fibrous mineral that formed at the highest metamorphic temperatures.

silt. A mineral or rock fragment, often made of quartz, between $\frac{1}{256}$ and $\frac{1}{16}$ millimeter in diameter.

sinkhole. A surface depression that resulted from the collapse of an underlying cavity for any reason; for example, a lime sink.

slate. A fine-grained metamorphic rock formed from the weak metamorphism of shale. It is tougher than shale because it contains microscopic mica in place of clay minerals, and it splits along the foliation produced by the lining up of the mica, which is also called slaty cleavage.

slaty cleavage. Parallel planes in slate or phyllite along which the rock tends to split, often at an angle to the original sedimentary bedding. A type of foliation.

soapstone. An informal term for an easily carved but cohesive rock. The minerals that give soapstone its typical green color and make it soft enough to carve are talc, chlorite, and/or serpentine. Most soapstone is metamorphosed ultramafic rock, so an initial history in the mantle is often inferred.

solution pit. A depression in a rock outcrop that has been deepened by the dissolving action of plant acids over time. It may be filled with soil and plants, or seasonally with water (a vernal pool).

sound. A relatively long arm of the ocean separating one or more islands from the mainland.

spoil. Piled-up earthen material left behind after dredging or mining.

staurolite. A brown metamorphic mineral that sometimes forms "fairy crosses," in which two crystals have grown together in the shape of a cross. The state mineral of Georgia.

stratum (pl. strata). A single bed or series of beds of like material.

stream capture. The event in which a stream, usually through headward erosion, cuts into the valley of another stream, causing the upper part of the other stream to become its tributary.

strike-slip fault. A fault on which a fault block moves or has moved laterally relative to the fault block on the other side of the fault.

sulfide. A compound containing sulfur. Geologically, sulfide minerals have a metallic luster and are relatively easily smelted.

supercontinent. An ancient continent that contained all or most of the world's continental landmass.

swale. A low area that is often damp; more specifically, a trough on a beach that parallels the coastline.

syncline. A concave-up fold that causes younger rock layers to be buckled downward along the hinge line of the fold.

tabby. A mixture of equal volumes of oyster shells, lime (burned oyster shells), sand, and water used in construction mainly during Georgia's colonial period.

talc. A soft, gray to green metamorphic mineral rich in magnesium.

tectonic. Relating to the deformation of masses of rock by forces within the Earth.

tectonic plate. A rigid section of the Earth's lithosphere that moves independently of adjacent plates and interacts with them at its boundaries.

terrace. A flat area bounded by an upslope escarpment on one side and a downslope escarpment on the other. It was formerly a local lowland that now lies adjacent to land that has eroded to a lower level.

terrane. A geologic region interpreted to have developed separately from neighboring regions for at least part of its history, implying that it existed a considerable distance from the neighboring regions as it developed. Boundaries between terranes are major faults and possibly former tectonic plate boundaries.

thrust fault. A fault along which one mass of rock (called a thrust sheet) slid up and onto the top of an adjacent rock mass.

thrust sheet. A sheetlike mass of rock that was thrust up onto another mass of rock along a thrust fault.

tidal creek. A winding channel of open water within a salt marsh in which the direction of flow is governed by the tides.

tidal inlet. A break in a chain of barrier islands through which the direction of flow is determined by the tides.

trace fossil. A fossil, such as a burrow or a fecal pellet, that preserves evidence of past life but not the actual skeletal remains.

travertine. A form of calcite deposited from solution by groundwater or surface water; for example, in cave formations.

trilobites. An extinct class of arthropods that lived in the Paleozoic era. Their bodies were divided into three major sections.

turbidite. A sedimentary unit displaying graded bedding and deposited by a turbidity current. Interpreted as having formed in deep water due to its sedimentary characteristics. *See also* graded bedding.

turbidity current. A slurry of sediment that moves down an underwater slope.

turtleback. A rounded boulder with a polygonal weathering pattern.

ultramafic. Rock composed of more than 95 percent mafic (iron- and/or magnesium-rich silicate) minerals, such as olivine and pyroxene, without quartz or appreciable feldspar. The mantle is made of ultramafic rock, and most ultramafic rocks in the crust are thought to have originally formed in the mantle.

unconformity. A surface separating rock strata of significantly different ages that represents a time during which rock layers were eroded or never deposited.

Valley and Ridge. A physiographic province in northwest Georgia that consists of long, parallel ridges separated by flatlands, both of which are underlain by sedimentary rocks ranging in age from Cambrian to Pennsylvanian.

vein. A mineral deposit formed along a crack in rock by the precipitation of minerals from water. Quartz and calcite are the most common vein minerals.

volcanic ash. Fine rock fragments from an explosive volcanic eruption.

watershed. The area that drains to a point on a stream.

weathering. The process by which rocks break down near the surface due to exposure to air, water, and the action of organisms. Weathered rock can be recognized by chemical changes to the minerals that make them dull, stained, or soft as compared to "fresh," unweathered rock.

xenolith. A body of rock that was surrounded by magma and now lies within an intrusion.

zircon. A mineral found in tiny crystals in minor amounts in many rocks. Useful for radiometric dating.

REFERENCES

Unless otherwise noted, geologic maps were modified from a digital version of the Georgia State Geologic Map (Lawton, 1976), obtained in 2007. It can be downloaded in a variety of formats at http://tin.er.usgs.gov/geology/state/state. php?state=GA. The landscape images were derived from the National Elevation Dataset, available via the National Map Viewer at http://viewer.nationalmap.gov/ viewer. The 3D-perspective views were developed using 3DEM software from U.S. Geological Survey Digital Elevation Model data, made available via Geo-Community at http://data.geocomm.com/dem/demdownload.html.

Abrams, C. E., and K. I. McConnell. 1977. *Geologic guide to Sweetwater Creek State Park.* Georgia Geologic Survey geologic guide 1.

Albin, E. F. 1991. Georgiaites: Tektites in central Georgia. *Fernbank Quarterly* 16 (2):15–20.

Alexander, C. R., and V. J. Henry. 2007. Wassaw and Tybee islands: Comparing undeveloped and developed barrier islands. In *Guide to fieldtrips: 56th annual meeting of the southeastern section of the Geological Society of America,* eds. F. Rich and C. Alexander, p. 187–98.

Allard, G. O., and J. A. Whitney. 1994. *Geology of the inner Piedmont, Carolina terrane, and Modoc zone in northeast Georgia.* Georgia Geologic Survey project report 20.

American Association of Petroleum Geologists. 1995. *Southeastern region geological highway map.*

Anderson, J. R., et al., 1999. *Tertiary/Cretaceous stratigraphy and paleontology of the north-central Coastal Plain of Georgia.* Geological Society of America, southeastern section, field trip 8.

Atkins, R. L., and L. G. Joyce. 1980. *Geologic guide to Stone Mountain State Park.* Georgia Geologic Survey geologic guide 4.

Bently, R. D., et al. 1966. *The Cartersville fault problem.* Georgia Geologic Survey guidebook 4.

Bergström, S. M., et al. 2004. The greatest volcanic ash falls in the Phanerozoic. *Sedimentary Record* 2 (4):4–8.

Big Ten, Inc. 1982. *Big Ten's map of Georgia gold.* Big Ten, Inc.

Boyd, B. 2001. *Waterfalls of the southern Appalachians.* Fern Creek Press.

Brown, F., and S. M. L. Smith. 1997. *The riverkeeper's guide to the Chattahoochee.* Chattahoochee Riverkeeper.

Brown's Guide to Georgia. Where to go, things to do in Georgia. www.browns-guides.com.

Butts, C., and B. Gildersleeve. 1948. *Geology and mineral resources of the Paleozoic area in northwest Georgia.* Georgia Geologic Survey bulletin 54.

Carpenter, R. H., and T. C. Hughes. 1970. *A geochemical and geophysical survey of the Gladesville Norite.* Georgia Geologic Survey information circular 37.

Carter, B. D., ed. 1995. *Paleogene carbonate facies and paleogeography of the Dougherty Plain region.* Georgia Geological Society guidebooks 15 (1).

Case, G. R., and D. R. Schwimmer. 1988. Late Cretaceous fish from the Blufftown Formation (Campanian) in western Georgia. *Journal of Paleontology* 62 (2):290–301.

Chowns, T. M., ed. 1972. *Sedimentary environments in the Paleozoic rocks of northwest Georgia.* Georgia Geological Society 7th annual field trip. Georgia Geologic Survey guidebook 11.

Chowns, T. M., ed. 1976. *Stratigraphy, structure, and seismicity in Slate Belt rocks along the Savannah River.* Georgia Geological Society 11th annual field trip. Georgia Geologic Survey guidebook 16.

Chowns, T. M. 1977. *Stratigraphy and economic geology of Cambrian and Ordovician rocks in Bartow and Polk counties, Georgia.* Georgia Geological Society 12th annual meeting and field trip.

Chowns, T. M. 1989. Stratigraphy of major thrust sheets in the Valley and Ridge Province of Georgia. In *Excursions in Georgia geology,* Georgia Geological Society guidebooks 9 (1), ed. W. J. Fritz, p. 211–38.

Chowns, T. M., Holland, S. M., and W. C. Elliott. 1999. *An introduction to sequence stratigraphy: Illustrations from the Valley and Ridge Province in Georgia and Alabama.* Georgia Geological Society guidebooks 19 (1).

Chowns, T. M., and R. L. Kath, eds. 2004. *Paleozoics, northwest Georgia: Structure, seismicity, geomorphology, hydrology, and economic geology.* Georgia Geological Society guidebooks 24 (1).

Chowns, T. M., and B. J. O'Connor, eds. 1992. *Cambro-Ordovician strata in northwest Georgia and southeast Tennessee: The Knox Group and the Sequatchie Formation.* Georgia Geological Society guidebooks 12 (1).

Chowns, T. M., and C. T. Williams. 1983. Pre-Cretaceous rocks beneath the Georgia Coastal Plain: Regional implications. In *Studies related to the Charleston, South Carolina, earthquake of 1886: Tectonics and seismicity,* U.S. Geological Survey professional paper 1313, ed. G. S. Gohn.

Churnet, H. G. 1997. *Seeing southeastern geology through Chattanooga.* HGC Publishers.

Clark, W. Z., Jr., and A. C. Zisa. 1976. *Physiographic map of Georgia.* Georgia Department of Natural Resources. http://georgiainfo.galileo.usg.edu/physiographic/physio-dist.htm.

Cocker, M. D., and J. O. Costello. 2003. *Geology of the Americus area, Georgia.* Georgia Geological Society guidebooks 23 (1).

Cook, R. B. 1978. *Minerals of Georgia: Their properties and occurrences.* Georgia Geologic Survey bulletin 92.

Costello, J. O., ed. 2002. *Geologic features of eastern Pickens, Dawson, and western Lumpkin counties, Georgia.* Georgia Geological Society guidebooks 22 (1).

Crawford, T. J., Gillespie, W. H., and J. H. Waters. 1989. The Pennsylvanian system of Georgia. In *Excursions in Georgia geology,* Georgia Geological Society guidebooks 9 (1), ed. W. J. Fritz, p. 1–20.

Crawford, T. J., et. al. 1999. *Revision of stratigraphic nomenclature in the Atlanta, Athens, and Cartersville 30′ x 60′ quadrangles, Georgia.* Georgia Geological Survey bulletin 130.

Cressler, C. W. 1963. *Geology and groundwater resources of Catoosa County, Georgia.* Georgia Geologic Survey information circular 28.

Cressler, C. W. 1964. *Geology and groundwater resources of Walker County, Georgia.* Georgia Geologic Survey information circular 29.

Cressler, C. W. 1970. *Geology and groundwater resources of Floyd and Polk counties, Georgia.* Georgia Geologic Survey information circular 39.

Cressler, C. W. 1974. *Geology and groundwater resources of Gordon, Whitfield, and Murray counties, Georgia.* Georgia Geologic Survey information circular 47.

Croft, M. G. 1964. *Geology and groundwater resources of Dade County, Georgia.* Georgia Geologic Survey information circular 26.

Department of Geology and Geography, Georgia Southern University. 2004. *Tybee Island Field Trip.* http://cosm.georgiasouthern.edu/geo/tybee.pdf.

DePratter, C., and J. D. Howard. 1977. History of shoreline changes determined by archaeological dating, Georgia coast, USA. *Transactions of the Gulf Coast Association of Geological Societies* 27:252–58.

Digital Library of Georgia. *"Thar's gold in them thar hills": Gold and gold mining in Georgia, 1830s–1940s.* http://dlg.galileo.usg.edu/dahlonega.

Duncan, M. S., and R. L. Kath, eds. 2009. *Fall Line geology of east Georgia: With a special emphasis on the upper Eocene.* Georgia Geological Society guidebooks 29 (1).

Frazier, W. J. 2007. Coastal Plain geologic province. *New Georgia encyclopedia.* www.georgiaencyclopedia.org.

Frazier, W. J., and T. B. Hanley, eds. 1987. *Geology of the Fall Line: A field guide to structure and petrology of the Uchee Belt and facies stratigraphy of the Eutaw Formation in southwestern Georgia and adjacent Alabama.* Georgia Geological Society guidebooks 7 (1).

Fritz, W. J., ed. 1989. *Excursions in Georgia geology.* Georgia Geological Society guidebooks 9 (1).

Fritz, W. J., and R. D. Hatcher, Jr., eds. 1989. *Geology of the eastern Blue Ridge of northeast Georgia and the adjacent Carolinas.* Georgia Geological Society guidebooks 9 (3).

Fritz, W. J., and T. E. La Tour, eds. 1988. *Geology of the Murphy Belt and related rocks, Georgia and North Carolina.* Georgia Geological Society guidebooks 8 (1).

Furcron, A. S. 1948–52. The Georgia Story (the geological history of Georgia). *Georgia Mineral Newsletter 1–5.*

Georgia Conservancy. 1998. *Longstreet highroad guide to the Georgia mountains.* Taylor Trade Publishing.

Georgia Geologic Survey. 1969. *Mineral resource map of Georgia.* Map SM-1, scale 1:500,000.

Georgia Humanities Council. 2011. *The new Georgia encyclopedia.* www.georgiaencyclopedia.org.

German, J. M. 1985. *The geology of the northeastern portion of the Dahlonega gold belt.* Georgia Geologic Survey bulletin 100.

German, J. M. 1988. *The geology of gold occurrences in the west-central Georgia Piedmont.* Georgia Geologic Survey bulletin 107.

Gore, P. J. W. 1999. *Geology of Stone Mountain, Georgia.* http://facstaff.gpc. edu/~pgore/stonemtn/stonemountain.html.

Grant, W. H. 1962. *Field excursion: Stone Mountain–Lithonia District.* Georgia Geologic Survey guidebook 2.

Grant, W. H., Size, W. B., and B. J. O'Connor. 1980. Petrology and structure of the Stone Mountain Granite and Mount Arabia Migmatite, Lithonia, Georgia. In *Excursions in southeastern geology,* Geological Society of America, ed. R. W. Frey.

Griffin, M. M. 1982. *Geologic guide to Cumberland Island National Seashore.* Georgia Geologic Survey geologic guide 6.

Griffin, M. M., and R. L. Atkins. 1983. *Geologic guide to Cloudland Canyon State Park.* Georgia Geologic Survey geologic guide 7.

Griffith, G. E., et al. 2001. *Level III and IV ecoregions of Georgia.* http://www1. gadnr.org/cwcs/PDF/ga_eco_l3_pg.pdf.

Hamilton, N., ed. 1993. *Natural selections: A guide to great science and nature outings in Georgia.* APPLE Corps, Inc.

Hanley, T., and M. G. Steltenpohl. 1998. *Mylonites and other fault-related rocks of the Pine Mountain and Uchee belts of western Georgia and eastern Alabama.* Atlanta Geological Society field trip guidebook.

Hanley, T. B., et al., 1997. Constraints on the location of the Carolina/Avalon terrane boundary in the southernmost exposed Appalachians, western Georgia and eastern Alabama. In *Central and southern Appalachian sutures: Results of the EDGE project and related studies,* Geological Society of America special papers 314, p. 15–24.

Harris, R. S., and J. M. German, eds. 2012. *Aggregate resources and the Permo-Jurassic geology of the southeastern Georgia Piedmont.* Georgia Geological Society guidebooks 32 (1).

Harris, R. S., et al. 2011. The Woodbury Impact Structure. In *The geology of the inner Piedmont at the northeast end of the Pine Mountain window*, Georgia Geological Society guidebooks 9 (1), eds. M. T. Huebner and R. D. Hatcher, Jr., p. 91–102.

Hatcher, R. D., Jr. 2002. *An inner Piedmont primer*. Carolina Geological Society annual field trip.

Hatcher, R. D., Jr. 2011. New Pine Mountain window geologic map. In *The geology of the inner Piedmont at the northeast end of the Pine Mountain window*, Georgia Geological Society guidebooks 9 (1), eds. M. T. Huebner and R. D. Hatcher, Jr., p. 49–52.

Hatcher, R. D., Jr., Bream, B. R., and A. J. Merschat. 2007. Tectonic map of the southern and central Appalachians: A tale of three orogens and a complete Wilson cycle. In *The 4D framework of continental crust*, Geological Society of America memoir 200, eds. R. D. Hatcher, Jr., et al., p. 595–632.

Haynes, J. T. 1994. *The Ordovician Deicke and Millbrig K-bentonite beds of the Cincinnati Arch and southern Valley and Ridge Province*. Geological Society of America special paper 290.

Heatherington, A. L., et al. 1996. The Corbin Gneiss: Evidence for Grenvillian magmatism and older continental basement in the southernmost Blue Ridge. *Southeastern Geology* 36 (1):15–25.

Henderson, S. W. 1999. *The geology of Civil War battlefields in the Chattanooga and Atlanta campaigns in the Valley and Ridge of Georgia*. Georgia Geological Society guidebooks 19 (1), p. 53–78.

Hibbard, J., et al. 2006. *Lithotectonic map of the Appalachian orogen (south), Canada–United States of America*. Geological Survey of Canada, map 02096A, scale 1:1,500,000.

Hicks, D. W., and S. P. Opsahl. 2004. *The natural history of the Flint River*. The natural Georgia series: The Flint River, Sherpa Guides. Lenz Marketing. http://www.sherpaguides.com/georgia/flint_river/natural_history/index.html.

Higgins, M. W., and R. L. Atkins, 1981. The stratigraphy of the Piedmont southeast of the Brevard zone in the Atlanta, Georgia, area. In *Latest thinking on the stratigraphy of selected areas in Georgia*, vol. 1, Georgia Geologic Survey information circular 54-A, ed. P. B. Wigley, p. 3–40.

Higgins, M. W., and R. F. Crawford. 2006. *Geologic map of the Atlanta 30' x 60' quadrangle, Georgia*. The Geologic Mapping Institute and the Atlanta Geological Society.

Higgins, M. W., et al. 1988. *The structure, stratigraphy, tectonostratigraphy, and evolution of the southernmost part of the Appalachian orogen, Georgia and Alabama*. U.S. Geological Survey professional paper 1475.

Higgins, M. W., et al. 1996. Geology of the Cartersville District and the Cartersville fault problem: A progress report. In *The Cartersville fault problem revisited*, Georgia Geological Society guidebooks 16 (1), ed. R. L. Kath, p. 9–61.

Hooper, R. J., and R. D. Hatcher, Jr. 1989. *The geology of the east end of the Pine Mountain window and adjacent Piedmont, central Georgia.* Georgia Geological Society guidebooks 9 (2).

Hooper, R. J., et al. 1997. The character of the Avalon terrane and its boundary with the Piedmont terrane in central Georgia. In *Central and southern Appalachian sutures: Results of the EDGE project and related studies.* Geological Society of America special papers 314.

Hoyt, J. H. 1968. Geology of the Golden Isles and lower Georgia Coastal Plain. In *The future of the marshlands and Sea Islands of Georgia,* Georgia Natural Areas Council and Coastal Area Planning and Development Commission, ed. D. S. Maney, p. 18–32.

Huddlestun, P. F. 1988. *A revision of the lithostratigraphic units of the Coastal Plain of Georgia: The Miocene through Holocene.* Georgia Geologic Survey bulletin 104.

Huddlestun, P. F. 1993. *A revision of the lithostratigraphic units of the Coastal Plain of Georgia: The Oligocene.* Georgia Geologic Survey bulletin 105.

Huddlestun, P. F., and J. H. Hetrick. 1986. *Upper Eocene stratigraphy of central and eastern Georgia.* Georgia Geologic Survey bulletin 95.

Huddlestun, P. F., and J. H. Hetrick. 1991. *The stratigraphic framework of the Fort Valley Plateau and the central Georgia kaolin district.* Georgia Geological Society guidebooks 11 (1).

Huddlestun, P. F., and J. H. Summerour. 1996. *The lithostratigraphic framework of the uppermost Cretaceous and lower Tertiary of eastern Burke County, Georgia.* Georgia Geologic Survey bulletin 127.

Huebner, M. T., and R. D. Hatcher, Jr. 2011. Evidence for sinistral Mesozoic inversion of the dextral Alleghanian Towaliga fault, central Georgia. In *The geology of the inner Piedmont at the northeast end of the Pine Mountain window,* Georgia Geological Society guidebooks 9 (1), eds. M. T. Huebner and R. D. Hatcher, Jr., p. 55–72.

Huebner, M. T., Hatcher, R. D., Jr., and C. W. Howard. 2011. Geologic overview of the inner Piedmont at the northeast end of the Pine Mountain window. In *The geology of the inner Piedmont at the northeast end of the Pine Mountain window,* Georgia Geological Society guidebooks 9 (1), eds. M. T. Huebner and R. D. Hatcher, Jr., p. 1–28.

Hulbert, R. C., Jr., and A. E. Pratt. 1998. New Pleistocene (Rancholabrean) vertebrate faunas from coastal Georgia. *Journal of Vertebrate Paleontology* 18 (2):412–29.

Hulbert, R. C., Jr., et al. 1998. A new middle Eocene protocetid whale (mammalia: cetacea: archaeoceti) and associated biota from Georgia. *Journal of Paleontology* 72 (5):907–27.

Hurst, V. J. 1957. Prehistoric vertebrates of the Georgia Coastal Plain. *Georgia Mineral Newsletter* 10:77–93.

Isolated Wetlands and Carolina Bays Task Force. 2012. *Conservation of Carolina Bays in Georgia*. http://www.scstatehouse.gov/committeeinfo/IsolatedWetlandsandCarolinaBaysTaskForce/October172012Meeting/GA_Carolina%20Bays.pdf.

Jackson, O. 2001. *North Georgia traveler: A guidebook to scenic destinations, historic lodging, historic restaurants, and unique shopping opportunities in north Georgia*. Legacy Communications, Inc.

Johnson, A. S., et al. 1974. *An ecological survey of the coastal region of Georgia*. National Park Service scientific monograph 3. http://www.nps.gov/history/history/online_books/science/3/index.htm.

Johnson, T. 2002. *Georgia nature weekends: Fifty-two adventures in nature*. Globe Pequot Press.

Joyce, L. G. 1985. *Geologic guide to Providence Canyon State Park*. Georgia Geologic Survey geologic guide 9.

Kath, R. L., ed. 1996. *The Cartersville fault problem revisited*. Georgia Geological Society guidebooks 16 (1).

Kath, R. L., ed. 2008. *The Emerson-Talladega fault, the Great Smoky fault, and adjacent folding and faulting: Geology and historical interpretations based on detailed geologic mapping in Polk and Bartow counties, Georgia*. Georgia Geological Society guidebooks 28 (1).

Kath, R. L., et al. In press. *Geologic map of the Cartersville, Allatoona Dam, Burnt Hickory Ridge, and Acworth (northern) 7.5′ quadrangles, Georgia*. Georgia Geologic Survey open-file report.

King, D. T., and L. Petruny. 2008. *Impact stratigraphy of the U.S. gulf coastal states*. Gulf Coast Association of Geological Societies transactions 58, p. 503–16.

Kish, S. A., et al. 1985. *Geology of the southwestern Piedmont of Georgia*. Geological Society of America field trip guidebooks 5.

Kogel, J. E., et al. 2000. *Geology of the commercial kaolin mining district of central and eastern Georgia*. Georgia Geological Society guidebooks 20 (1).

Kogel, J. E., et al. 2002. *The Georgia kaolins: Geology and utilization*. Society for Mining, Metallurgy, and Exploration.

Lacefield, J. 2000. *Lost worlds in Alabama rocks: A guide to the state's ancient life and landscapes*. Paleo-Alabama Project.

Lawton, D. E. 1976. *Geologic map of Georgia*. Georgia Geologic Survey map SM-3, scale 1:500,000.

Lawton, D. E. 1977. *Geologic map of Georgia*. Georgia Geologic Survey map SM-3, scale 1:2,000,000.

Lenz, R. J. 2001. *Longstreet highroad guide to the Georgia coast and Okefenokee*. Taylor Trade Publishing.

Lininger, J. 2003. The mineral history of Georgia. *Matrix: A Journal of the History of Minerals* 11 (4).

Long, L. T. 1993. *Earthquake and seismic hazard in Georgia.* Georgia Institute of Technology.

Markewich, H. W., and W. Markewich. 1994. *An overview of Pleistocene and Holocene inland dunes in Georgia and the Carolinas: Morphology, distribution, age, and paleoclimate.* U.S. Geological Survey bulletin 2069.

Martin, A. J. 2012. *Life traces of the Georgia coast: Revealing the unseen lives of plants and animals.* Indiana University Press.

McBride, J. H., et al. 2005. Integrating seismic reflection and geological data and interpretations across an internal basement massif: The southern Appalachian Pine Mountain window, USA. *GSA Bulletin* 117 (5–6):669–86.

McConnell, K. I., and C. E. Abrams. 1984. *Geology of the greater Atlanta region.* Georgia Geologic Survey bulletin 96.

Mittenthal, M. D., and D. L. Harry. 2004. Seismic interpretation and structural validation of the southern Appalachian fold and thrust belt, northwest Georgia. In *Paleozoics, northwest Georgia: Structure, seismicity, geomorphology, hydrology, and economic geology,* Georgia Geological Society guidebooks 24 (1), eds. T. M. Chowns and R. L. Kath, p. 1–12.

Neathery, T. L., ed. 1986. *Southeastern section of the Geological Society of America.* Centennial field guide 6.

Pickering, S. M., Jr. 1970. *Stratigraphy, paleontology, and economic geology of portions of Perry and Cochran quadrangles, Georgia.* Georgia Geologic Survey bulletin 81.

Pilkey, O. H., and M. E. Fraser. 2003. *A celebration of the world's barrier islands.* Columbia University Press.

Reinhardt, J., and T. G. Gibson. 1980. Upper Cretaceous and lower Tertiary geology of the Chattahoochee River valley, western Georgia and eastern Alabama. In *Geological Society of America field guide* 2, p. 385–463.

Reinhardt, J., Prowell, D. C., and R. A. Christopher. 1984. Evidence for Cenozoic tectonism in the southwest Georgia Piedmont. *GSA Bulletin* 95 (10):1176–87.

Rich, F. J., and G. A. Bishop, eds. 1998. *Geology and natural history of the Okefenokee Swamp and Trail Ridge, southeastern Georgia–northern Florida.* Georgia Geological Society guidebooks 18 (1).

Richards, H. G. 1969. *Illustrated fossils of the Georgia Coastal Plain.* Georgia Department of Mines, Mining, and Geology (Georgia Geological Society).

Roden, M. F., et al., eds. 2005. *Investigations of Elberton Granite and surrounding rocks.* Georgia Geological Society guidebooks 25 (1).

Ross, T. E. 1987. A comprehensive bibliography of the Carolina bays literature. *Journal of the Elisha Mitchell Scientific Society* 103 (1):28–42.

Sanders, S. 2003. Providence Canyon. *New Georgia Encyclopedia.* www.georgiaencyclopedia.org.

Schoettle, T. 1987. *A field guide to Jekyll Island.* Georgia Sea Grant College Program, University of Georgia.

Schoettle, T. 1993. *A naturalist's guide to St. Simons Island.* Watermarks Publishing and Printing.

Schoettle, T. 1996. *A guide to a Georgia barrier island: Featuring Jekyll Island with St. Simons and Sapelo islands and a field guide to Jekyll Island.* Watermarks Publishing and Printing.

Schoettle, T. 2002. *A naturalist's guide to the Okefenokee Swamp.* Darien Printing and Graphics.

Schoettle, T. 2011. *A beachcomber's guide to Georgia's barrier islands.* Creative Printing, Inc.

Schwimmer, D. R. 1985. First pterosaur records from Georgia: Open marine facies, Eutaw Formation (Santonian). *Journal of Paleontology* 59 (3):674–76.

Schwimmer, D. R. 1986. Late Cretaceous fossils from the Blufftown Formation (Campanian) in western Georgia. *The Mosasaur* 3:109–23.

Schwimmer, D. R. 1993. Late Cretaceous dinosaurs from the Blufftown Formation in western Georgia and eastern Alabama. *Journal of Paleontology* 67 (2):288–96.

Schwimmer, D. R. 2006. Paleontology of the Coastal Plain Province. *New Georgia Encyclopedia.* www.georgiaencyclopedia.org.

Schwimmer, D. R., and R. H. Best. 1989. First dinosaur fossils from Georgia, with notes of additional Cretaceous vertebrates from the state. *Georgia Journal of Science* 47:147–57.

Schwimmer, D. R., et al. 1994. Giant fossil coelacanths of the Late Cretaceous in the eastern United States. *Geology* 22 (6):503–6.

Schwimmer, D. R., et al. 1997. *Xiphactinus vetus* and the distribution of *Xiphactinus* species in the eastern United States. *Journal of Vertebrate Paleontology* 17 (3):610–15.

Scotese, C. R. 1997. *Paleogeographic atlas, PALEOMAP progress report 90-0497.* Department of Geology, University of Texas.

Simkin, T., et al. 2006. *This dynamic planet: World map of volcanoes, earthquakes, impact craters, and plate tectonics.* U.S. Geological Survey geologic investigations series map I-2800, scale 1:30,000,000.

Size, W. B., and N. Khairallah. 1989. Geology of the Stone Mountain Granite and Mount Arabia Migmatite. In *Excursions in Georgia Geology,* Georgia Geological Society guidebooks 9 (1), ed. W. J Fritz.

Smith, P. 1962. *Aboriginal stone constructions in the southern Piedmont.* University of Georgia, Laboratory of Archaeology series report 4.

Smith, W. H. 1983. *Kaolin deposits of central Georgia: An introduction to their origin and use.* White Hall Press/Budget Publications.

Smith, W. H. 1985. *Guide to the geology of Bartow County Georgia.* White Hall Press/Budget Publications.

Sneed, J. M., and L. Blair. 2005. *The late Pleistocene record of Kingston Saltpeter Cave, Bartow County, Georgia.*

Stann, K. 1995. *Georgia handbook: Includes Atlanta, Savannah, and the Blue Ridge Mountains.* 2nd ed. Moon Publications, Inc.

Stewart, K. G., and M. R. Roberson. 2007. *Exploring the geology of the Carolinas: A field guide to favorite places from Chimney Rock to Charleston.* University of North Carolina Press.

Stormer, J. C., Jr., and J. A. Whitney, eds. 1980. *Geological, geochemical, and geophysical studies of the Elberton Batholith, eastern Georgia.* Georgia Geological Society 15th annual field trip. Georgia Geologic Survey guidebook 19.

Thigpen, J. R., and R. D. Hatcher, Jr. 2009. *Geologic map of the western Blue Ridge and portions of the eastern Blue Ridge and Valley and Ridge provinces in southeast Tennessee, southwest North Carolina, and northern Georgia.* Geological Society of America map and chart MCH097.

Thomas, W. A. 2006. Tectonic inheritance at a continental margin. *GSA Today* 16 (2):4–11.

Tull, J. F., ed. 2007. *Tectonics of the Georgia Blue Ridge: Basement/cover rift architecture, important aspects of overlying drift and clastic wedge facies, and the westernmost accretionary terrane.* Georgia Geological Society guidebooks 27 (1).

Tull, J. F., and C. S. Holm. 2005. Structural evolution of a major Appalachian salient-recess junction: Consequences of oblique collisional convergence across a continental margin transform fault. *GSA Bulletin* 117 (3–4):482–99.

University of Georgia, Department of Geology. 2003. *Georgia Geology.* http://geology.uga.edu/index.php/about/941.

Vincent, H. R., McConnell, K. I., and P. C. Perley. 1990. *Geology of selected mafic and ultramafic rocks of Georgia: A review.* Georgia Geologic Survey information circular 82.

Wenner, D. B., ed. 1997. *Geology of the Georgia Piedmont in the vicinity of Athens and eastern metropolitan Atlanta area.* Georgia Geological Society guidebooks 17 (1).

Wharton, C. H. 1977. *The natural environments of Georgia.* Georgia Geologic Survey bulletin 114.

Whitney, J. A., and G. O. Allard. 1990. *Structure, tectonics, and ore potential along a transect across the inner Piedmont, Charlotte Belt, and Slate Belt of eastern Georgia.* Georgia Geologic Society guidebooks 10 (1).

Whitney, J. A., Dennison, J. M., and P. J. W. Gore. 1999. *Geology and geomorphology of Stone Mountain, Georgia.* Geological Society of America, southeastern section.

Winslett, L. 2007. *Stone Mountain: A walk in the park.* Dahlonega, GA: Bright Hawk Press.

Winslett, L., and J. Winslett. 2004. *Wildflowers of Stone Mountain: A field guide.* Bright Hawk Press.

Woodward, N. B. 1985. *Valley and Ridge thrust belt: Balanced structural sections, Pennsylvania to Alabama.* Studies in geology 12. University of Tennessee Department of Geological Sciences.

INDEX

Page numbers in bold indicate an illustration or information in a caption.

acid precipitation, 201
Adairsville, 145
Adel, 82
Aequipecten spillmani, **65**
agate, 10, 84
agatized coral, 64, **65**, 83, 84, **84**
age dating, 12
agriculture, 89, 103, 113
Alapaha River, 83, 106
Albany, **56**, 113, 116
Albany Civic Center, 113
Albany State University, 113, **115**
algae, 142
Allatoona Dam, 197–98
Alleghany event, 185
allochthon, 246
Alma, 111
Altamaha Formation, 49, 54, **55**
Altamaha Grit, 97, **97**, 106, **106**, 107
Altamaha River, **30**, 32, 106, 111;
    delta of, 33, 34; sand dune fields
    and, 60
aluminum, 1, 86
Amelia Island (Florida), 46
Amicalola Creek, **9**, 214
Amicalola Falls, 1, 8, **9**, 213, 214,
    214, 215, **215**, **216**
Amicalola Falls State Park, 214–15
Amphianthus pusillus, 272
amphibolite, 175, 260, **260**, **261**,
    293, 295–96, **296**
amygdules, 200
Anderson Creek, 9, 214, **216**
Anderson Memorial Gardens, 139
Andersonville, 86, 87
Andersonville Bauxite Mining
    District, 86

Andersonville fault, 86
Andersonville National Historic
    Site, **87**
Anna Ruby Falls Scenic Area, 223
Antebellum Trail, 276
anticlines, 122, **122**
anticline valleys, 164
aplite dikes, 286
Appalachian Mountains, 5, 47
Appalachian Plateau: amount of
    tectonic deformation of, 153;
    Blue Ridge–Piedmont rocks
    and, 126; cross section of,
    124–25; description of the, 121;
    distinguishing characteristics
    of, 6; faults of the, 124; fossils
    of the, 128; relationship of
    the topography to underlying
    bedrock, **120**, 122, **122**; rock
    formations of, **127**
Appalachian Trail, 203, 219, 223
Appalachiosaurus, 68
Appling Granite, 272
aquifers, 56
Arabia Mountain, 281, 286, 288,
    288, 295
Arabia Mountain Migmatite, **178**,
    286–87, **287**
Arabia Mountain National Heritage
    Area, 286
aragonite, 10
archaeological sites, 96, 108, 148,
    213, 276–77, **277**
Archelon, **41**, 42
Archimedes, 130
Argentinosaurus, 1
Argyle terrace, 107

armoring of beaches, 33
Armuchee Ridges, 121, 123, **124–25,**
    157
ash, volcanic, 135
Ashburn, 83
asteroids, 63
asthenosphere, 3
Athens, 236–38, 241
Athens Gneiss, 237, **237**
Atlanta: amphibolite of, **293,**
    295–96, **296;** elevation of, 278;
    geologic map of, **280;** gilded
    dome of state capitol, 221,
    279; gneiss and schist of, 292,
    293; granite intrusions of, 289;
    landscape of, 279; mountains
    of, 281; mylonite of, 297–98,
    298; quarries and road cuts
    of, 291–98; quartzite of, 295;
    reasons for early growth of, 279;
    rock types beneath, 267, 279;
    siting of, 3, 279; soapstone of, ix,
    296; as a tourist destination, 225;
    weathering related to climate
    of, 282
Atlanta Sand and Supply, 103
Atlantic Ocean, 4, 181, 190
Atlantic seafloor, 6
attapulgite, 79
Attapulgus, 79
augen gneiss, 211, **211**
Augusta, 3, 272, 274–76
Augusta Canal, 274, 275
Auraria, 221
Austell Gneiss, 194, 195

back-barrier islands, 20
Bacon Terraces, 56, 59, **59,** 107
Bainbridge, 78, 79
Bakers Rock, 236
Banks Lake, 83
barite, 146, **147**
barium sulfate, 146
Barnesville, 264
barrier islands, 15, 21, 23, 27.
    See also Sea Islands

Barrier Island Sequence District, 57,
    58, 73, 117
barrier island shoreline complexes,
    57, **58, 59**
Bartletts Ferry Dam, 247
Bartram, William, 238
basement rock, 181, 185, 230, 231,
    257. *See also* Laurentia
basins, buried, 189
basins, hidden, 92, **93**
Battle of Resaca, 143, 145, 158
Battle of Ringgold, 139
bauxite, 86
Baxley, 107, 111
beaches: armoring of, 33; black
    sand on, 73; description of,
    17; erosion of, 35; intertidal
    zone of, 17; migration of, 35;
    organisms that can tolerate,
    19; renourishment of, 25, 35;
    ridge and runnel systems of, 17;
    ripples on, 17, **18,** 19, **19;** sand of
    Georgia's, 35; squeaky sand of,
    17; types of sand, 17; widening
    of, 41
bed, 10
bedding planes, 10, **72, 255**
bed, graded, 176
Bee Mountain, 146
Ben Hill Granite, 289, **289,** 297
bentonite, 135
Big Pond, 107
biological succession, 288
Birmingham, AL, 133
bivalves, 130
Blackbeard Island, **29**
Black Rock Mountain, 229
Black Rock Mountain State Park, 229
Black Rock Overlook, 229, **229**
black-spored quillwort, 236
blackwater lakes, 83
blackwater swamp, 112, **112.** *See
    also* Okefenokee Swamp
Blairsville, 217
Bloody Marsh, 33
Bloody Pond, 167

blueberries, 111
blue hole springs, 115
Blue Ridge, 6, 7, 8, 10, 17, 207
Blue Ridge Mountains, 203
Blue Ridge–Piedmont: ages of terranes of, 181; amount of marble within, 206; distinct rocks of the, 175; diversity of minerals and rocks of, 176; evidence for deepwater origins of, 175; folding of its rocks, 177; granite intrusions of, 179; igneous intrusions of, 179; metasedimentary rocks of, 175; metavolcanic rocks of, 175; most common stone of, 177; oldest basement rocks of, 195; premetamorphic character of its rocks, 175–76; significant boundaries within, 181; terranes of, 181–83, **184**, **186**; thrust faulting and, 185; Valley and Ridge and, 185; youngest rocks of, 190, 285
Blue Ridge thrust front, 126, 185
Blue Ridge (town), 203
Blufftown Formation, 67, 74
Boat Rock Preserve, 289, **289**
boudins, **293**, 294
boulder fields, 212, 223
brachiopods, 128, **129**
Bradley Peak, 288
Brasstown Bald, 217, **217**
Brasstown Bald window, 219
breccia, impact, 64
Brevard fault zone, 187, **187**, 226, 233, **234**
Broxton Rocks Preserve, 106, **106**
Brunswick, 34, 37, 69
Brunswick River, 33, 37
Brunswick terrane, 185
bryozoans, 129, 130
Buford Dam, 233, **234**
Bull River, 22
"burning lake of fire," 235
button schist, 298

Cabretta Island, **29**
Calamites, 131
calcining, 91
calcite, 10, **94**, 158, 159, 177
calcium carbonate, 10
Calhoun, 145
Callichirus major, 19
Cambrian period, 14, 126
Caney Bay, 117
Canoochee River, 60, 98
Canton, 203, 208, **208**
Canton Formation, 208
Cape Charles, VA, 63
carbonate, 177
carbonate minerals, 10
carbonate rocks, 206
carbonates, 10
Carcharodon megalodon, 66
Carnegie, Thomas, 42
Carolina bays, 60, 61, 80, 82, 85
Carolina superterrane, 185, 267
Carolinia, 185
Carters Dam, 159
Cartersville Mining District, 145, 146
Cartoogechaye terrane, 183
cataclasite, 251, **251**
"cat's paws," 282, **283**, 294
Cat Square terrane, 183, 185, 267
caves, 135, **136**, 155, 171, **172**
Cave Spring, 171–72
Cave Spring Cave, 171, **172**
Cecil, 82
cedar glades, 166
Cedar Rock, 272, **273**
Cedar Shoals, 236
Cedartown, 171, 172
Celeoth Creek, 257
cement, 103
cement, natural, 262
CEMEX quarry, 103, **104**
Cenozoic era, 14
Central Georgia Seismic Zone, 277
cephalopods, 130
chalcedony, 84
Charleston earthquake (August 31, 1886), 25, 275

charnockite, 257, **257**

Chatsworth, 209

Chattahoochee National Forest, 225

Chattahoochee Palisades, 295

Chattahoochee River, **232**, 233, 247, 249, 260–61, **260**, 262

Chattahoochee River National Recreation Area, 295

Chattahoochee River valley, 223, 225

Chattanooga Shale, 168–70, **169**

Chattooga River, **232**, 233

Chennault House, 238

Cherokee, 225

Cherokee Falls, 153

Cherokee marble, 204

Cherokee Nation, 190, 204

Cherokee villages, 270

chert, 10, 79, **87**, 154

"chert and dirt," 10

Chesapeake Bay, 63

Chickamauga, 168

Chickamauga Battlefield, 166–68

Chickamauga Group, 128, 135

Chickamauga Valley, 121, **124–25**

Chilhowee Group, 146

chloritoid, 203

Civilian Conservation Corps, 219, 256, 269

Civil War: Atlanta and, 290; battle at Missionary Ridge, 164; Battle of Resaca, 145; Battle of Ringgold, 139; cannons manufactured for, 262; Chickamauga Battlefield, 166; Columbus textile mills and, 262; ironclad warships produced for, 262; last major Confederate victory of the, 166; May 1864 campaign, 144, **144**; naval machinery of, 262; prisoner of war camp, 86, 96; sandstone boulders as ammunition during the, 143. See also March to the Sea

Clam Creek, 39

clams, **65**, 130

clams, boring, **84**

Clarkesville, 225

Clarks Hill Dam, 240

Clarks Hill Lake, 272

Claxton terrace, 107

clay, 176

Clayton Formation, 77, **77, 78**

cleavage, slaty, 173, 177

Cleburne, General Patrick, 144

Climax Caverns, 79

Clinchfield, 49, 103, 104

Cloudland Canyon State Park, 150–53

coal, 131, 162

coal swamps, 130

Coastal Plain: age dating rocks of the, 103; asteroid impacts and, 63, 64; cross section, **49**; depth of Piedmont rocks beneath the, 89, 92; description of, 47; distinguishing characteristics of, 6; fossils, 64, 67; geologic map of, **48**; hidden basins beneath the, 92, **93**, 189; kaolin resources in, 70, **70**, 73; landforms of the, 60; physiographic districts of the, 52, **52**; Piedmont metamorphic rocks exposed in the, 100, **102**; sedimentary formations of the, 47; structural features and the, 86; thickness of sedimentary formations of the, 47; topography of the, 8

Cobb Energy Performing Arts Center, 295

Coca-Cola, 148

Cockspur Island, 23

coelacanths, 66

Cohutta Mountain, 209

Columbus, 3, 260, 262

Columbus Iron Works, 262

Conasauga Group, 128, **128**, 149

concretion, 194, **194**

Confederacy, 238

Confederate Army of Tennessee, 144

Confederate Hall, **2**, 285

Confederate Naval Iron Works, 262

Confederate Powder Works, 275

conglomerate, 10, 198, **198**
Consolidated Gold Mine, 220, **221**
continental crust, 4
continents, 3, 6
Cooper's Furnace Day Use Area,
    197–98
Coosa River, 10
Copper Basin, 201, **201**
copper ore, 201
coral, 10
coral, agatized, 64, **65**, 83, 84, **84**
corals, horn, 130
corals, rugose, 130
corals, tabulate, 130
Corbin Metagranite, 197, **197**, 199,
    199
Cordele, 82
cordgrass, smooth, 20
core, Earth's, 3
Cove, the, 249–53
Cove Dome, 252, 258
Coweta Town, 261
Cowrock terrane, 183
crab, ghost, 19
Crawfish Spring, 168
Crawfordville, 272
creep (erosion), 215
Creole marble, 204
crinoids, 129, **129**
Crisson Gold Mine, 220
Crockford-Pigeon Mountain
    Wildlife Management Area, 155
crossbedding, 11, **11**
crossbeds, 11, **11**, 151–52, **151**, **152**
crust, Earth's, 3
Crystal Springs, 170
CSX, 137, 270
Cumberland Island, 21, 42, **44**, **45**, 46
Cumberland Island National
    Seashore, 42
Cumberland Plateau, 121, 122, 123.
    See also Applachian Plateau
Cumberland Sound, 46
Cumming, 222
Curtis Creek, 225
Cusseta, 74, 76

Cusseta Town, 261
Cuthbert, 78, 113

Dacula, 236
Dahlonega, 1, 190, 216, 220–21, **279**
Dahlonega gold belt, 183, 190–91,
    200
Dahlonega Gold Museum, 221, **222**
Dahlonega terrane. See Dahlonega
    gold belt
daisy, yellow Confederate, 288
Daniel Creek, 150, **150**, 151, 153
Darien, 30, 89
Davidson-Arabia Nature Preserve,
    286–89
Davis, Confederate President
    Jefferson, 238, 282
Dawsonville, 221
debris flow, 173
Deep Creek, West Fork, 82
Deer-Lick Springs, 195
Deicke bentonite, 135
Deinosuchus, 68
DeKalb County Courthouse, 297
deltas, 15
deposition, 26
Devils Racecourse Shoals, 295
Devonian period, 126, 169, 170
diabase, **188**, 284, **284**
diabase dikes, **188**, 189, 190, 284–
    85, **284**
Diamorpha smallii, 288, **288**
Dick Ridge, 156
dikes: aplite, 286; diabase, 189, 190,
    284, 285; pegmatite, 179, 230,
    230, **285**, 286
dinosaurs, 14, 64, 67, 68, 90
dip, 12
"dish gardens," 272
Dog River window, 194. See also
    tectonic windows
dolomite, 10, 177
dolostone, 10, **147**
Dougherty Plain, 55–56, **56**, 79, **79**,
    113
Dowdell Knob, 256, **256**

dredging, 15, 25, 36
drilling mud, 146
drusy quartz, 168
Dry Branch (town), 97
duck-billed dinosaurs, 68
Ducktown Basin Museum, **201**
Dug Gap, 142, 144, 145
Dug Gap Battlefield Park, 142–43, **143**
Dukes Creek, 223
Dukes Creek Gold and Ruby Mines, 225
Duluth, 297, **298**
Dunbarton Basin, 94
dune fields. *See* sand dune fields
Dungeness, 42
Durham, 162
Dutchy, 244, **244**

Eagle and Phenix Dam, 261, **261**, 262
Eagle and Phenix Mill, 262
Eagle Mill, 262
Earth, 3, 4, **4**, 12, 133, 253
earthquakes, 3, 25, 124, 185, 188, 236, 240, 277
East Beach (St. Simons Island), 33
East Chickamauga Creek, 137, 139
East Ellijay, 203, 213
Eastern Continental Divide, 229, 233, 278
East Palisades overlook, **177**, 295
East Quarry (Stone Mountain), 282, **283**
Echeconnee Creek valley, 80
Edwin I. Hatch Nuclear Electric Generating Plant, 111
Elachee Nature Science Center, 233
Elbert Boat Ramp, 246
Elberton, 1, 243–45
Elberton batholith, 243
Elberton Blue. *See* Elberton Granite
Elberton Granite, **2**, 179, 183, 238, 243–45, **243**, **245**
Elberton Granite Museum, 244, **244**
elephant (*Elephas columbi*), 31
Ellijay, 203, 213

Ellisons Cave, 155
Emuckfaw Formation, 194, **194**
English settlers, 36
environmental degradation, 201, 223, 271
eons, 14
epidote, 286
epochs, 14
eras, 14
Eremotherium laurillardi, 68
Eremotherium mirabile, 31
erosion: erosional remnants, 8, 281; erosion surfaces, **72**, 76; headward, 8; meandering streams and, 39; related to farming practices, 76; resistance to, 8; soil, 76, 271; stream, 8, 9, 10; topography and, 8; wild horses and, 42
escarpments, 8, 21, 62
Estelle, 154
Etowah, 198
Etowah Indian Mounds Historic Site, 148
Etowah marble, 204, **206**
Eutaw Formation, 68, 74
eutrophication, 169
exfoliation, 235, **283**
exfoliation dome, 282
exfoliation joints, 282
extinction events, 14, 133, 170

fairy crosses, 207
Fall Line: on the Chattahoochee River, 261, **261**; definition of, 3; growth of cities and, 261; Native Americans and, 275; reasons for development along, 3; rock formations at, 47; on the Savannah River, 275, **275**
Fall Line Hills, 52
Fall Line Red Hills District, 97
Fall Line Sand Hills District, 80, 89
Fall Line unconformity, 100, **102**
faults: Coastal Plain and, 86, 89; normal, **182**; strike-slip, 181,

182; thrust, 124, **124–25**, 126, 135, **136**, **182**; types of, **182**
fault zones, 187
F. D. Roosevelt State Park, 256, **256**
feldspar, 179, 197, **197**, 199, **199**
Fernbank Museum of Natural History, 1
Fernbank Science Center, 297
ferns, 130
ferns, seed, 131
Fifteenmile Creek, 98
Fitzsimmons, Henry, 204
Flat Creek, 216
Flat Rock Park, 260
Flat Tub Wildlife Management Area, 107
flint, 10, 79
Flint River, 60, 78, 86, 113, 249, 258
flooding, 86, 113, 115, 227
Floridan aquifer, 46, 56, 78, 115
folds, isoclinal, 177, **178**, 203, **208**
foliation, 177, **177**, **180**
Folkston, 113
foraminifera, 64, **65**
forams. *See* foraminifera
formations, explanation of, 2
Forsyth, 267
Fort Benning, 74
Fort Frederica National Monument, 36, **36**
Fort Mountain, 209, 211, 212, **212**, 213
Fort Mountain State Park, 212–13
Fort Payne Chert, 170, **174**
Fort Pulaski National Monument, 23
Fort Screven, 25
Fort Valley, 84, 85, **85**
Fort Valley Group, 85, **85**
Fort Valley Plateau, 53–54, **54**, 80
fossils: amphibian tracks, 132, **132**; bony fish, 66; burrows, 130, 139, **141**; Cambrian-age, 128; clam, **65**; Coastal Plain, 64, **65**, 66, 67; collecting of, 14; dinosaur, 67, 68; Eocene-age, 103; first, 12; foraminifera, **65**; geologic time scale and, 12; limestone and, 130; "living," 66; marine reptile, 66, 67; Mississippian-age, 130; molds, **84**; official state fossil, 66; Ordovician-age, 128; oyster, **65**; Pennsylvanian-age, 130, 131, **131**, 132, **132**; plant, 130, 131, **131**; Pleistocene-age, 31, 68, 149; reptile, 68; reptile tracks, 132; sand dollar, **65**; scallop, **65**; sea turtle, **41**, 42; sediments likely to contain, 64; shark teeth, 66, **66**, 68; snail, **65**; trace, 130; types of fossilization, 64; vertebrate, 68; whale, 80, 95, **95**
Fox Mountain, 135
fracking, 168
fuller's earth, 71, 79

Gainesville, 233
garnet, 176, **176**
Gascoigne Bluff, 31
gas, natural, 168
gas seeps, 105
gastropods, 130
geodes, 84, **84**
geologic time, 12, **13**, 14
geologic time scale, 12, **13**, 14
geophagia, 73
Georgetown, 113
Georgia: development of bedrock of, 5; famous red clay of, 181; most seismically active part of, 277; satellite image of the coast of, **16**; Seven Natural Wonders of, 76, 115, 150, 214; State Capitol of, 279, **279**; state mineral of, 176; state seal of, **2**; terranes of, **184**
Georgia Aquarium, 1
Georgiacetus vogtlensis, 95, **95**
Georgia College Natural History Museum and Planetarium, 278
Georgia Guidestones, 245–46, **246**
Georgia Institute of Technology, 236
georgiaites, 63

Georgia Legislature, 261
Georgia Marble Company, 205
Georgia Marble Festival, 204
Georgia Railroad, 270, 279
Georgia Sea Turtle Center, 40, **41**
giant club moss, 131
gibbsite, 86
Gilmer County prison quarry, 203
Girvanella, 142
glaciation, 31, 57
glaciers, 21
Glade Farm, 216
glauconite, 104, 170
global cooling, 49, 56
gnat line, 62
gneiss, 177, **178**, 179, **180**, 290
gneiss, augen, 211, **211**
gneiss, biotite, 229, **229**
gneiss, granitic, 179, **188**, **266**
Goat Rock Dam, 247
Goat Rock mylonite, 247
Goat Rock mylonite zone, 247
goethite, 77
gold: coins, 221; concentration of in
    streams, 191; formation of, 190;
    largest nuggets found in Georgia,
    203; in man-made structures,
    221, 222, **279**; mining of, 191;
    nuggets, 221; panning, 14; placer
    deposits of, 191, **191**. *See also*
    mining, gold
gold rush, 1, 190, 191, 193, 221
Golucks-Rhodes Mine, 272
Gondwana, 5, 181, 182
Gore, 168, 169, 170
graded bed, 176
Graham, 107
Grand Bay, 61, 82
Grand Bay Wetland Education
    Center, 82
Grand Bay Wildlife Management
    Area, 82
granite: cutting of, 244, 245;
    hardness of, 244; kaolin and, 71;
    largest body of in Georgia, 243;
    mineral composition of, 245;

of the Piedmont, 81, 179; plate
    tectonics and, 4; production in
    Elberton, 244; quarrying of, **273**;
    rare and endangered plants and,
    274; uses of, 243
granite, black, 245
Granite Capital of the World, 1, 243
granite industry, 244
granite, red, 245
granitic gneiss, 8, 179, **266**
Graves Mountain, 238–39, **239**
Gray's Reef National Marine
    Sanctuary, 31
graywacke, 176
Great Buffalo Lick, 238
Great Depression, 269, 271
Great Valley, 121, **124–25**, 159
green marble, 208
Griffin, 264
Griffin Ridge Wildlife Management
    Area, 108
groins, 25, **26**
ground sloth, giant, 68–69, **69**
groundwater, 53, 253
groups, explanation of, 2
Guidestones, Georgia, 245, 246
Gulf of Mexico, 47, 64, 169, 181
gunpowder, 275
gymnosperms, 131

hadrosaurs, 68
hammocks, 20, 39
Happy Valley, 164, **166**, **167**
Hard Labor Creek State Park, 269
Hartsfield-Jackson International
    Airport, 279
Hartwell Lake, 235
Hawkinsville, 104
Hawthorne Group, 115
Hazlehurst, 21, 104, 106, 107
Hazlehurst terrace, 107
headward erosion, 8
Heggies Rock, 272, 274, **274**
Helen, 223, 225
Helianthus porteri, 288
hematite, 133

Hemlock Falls, 153
Hiawassee River valley, 223
High Falls, 265, **266**
High Falls Granite, 265
High Falls Lake, 264
High Falls State Park, 264–65, **265**
High Shoals of the Apalachee, 270
Hightower Indian Trail, 270, 274
hinge line, 122
Hollis Quartzite, 249, 251–52, **252**,
    253, 258–59, **258**, **259**
Holly Springs Quarry, 208
hoodoos, 162
horn corals, 130
Horn Mountain, 157
horses, wild, 42, 44
horsetails, 131
Huber, **53**
Huber Formation, 73
Hurricane Cave, **136**
Hurricane Dora, 33
Hurricane Falls, 232
Hurricane Shoals Park, 235

ice age, 2
ice sheets, 21, 56
igneous rock, 12, 175
ilmenite, 73
Imerys Marble, Inc., 205
impact structures, 258
impressions, fossil, 64
Indian mounds, 96, 225
"Indian paint pots," 78
Indians: and Cumberland
    Island, 42; middens of, 27, 30;
    Mississippian culture, 148;
    settlements of, 96; Skidaway
    Island and, 27; soapstone
    industry and, 296; use of chert, 78
Indian Seats, 222
Indian Springs Granite, 264, 265
Indian Springs State Park, 264
Indian Territory, 89
Indian trails, 85, 89, 271
Indian villages, 261
interglacial periods, 21

intertidal zone, 17
Intracoastal Waterway, 69
intrusions, igneous, 179
iron, 1
ironclad warships, 262
iron geodes, 77
iron industry, 146
iron nodules, 77, **77**
iron ore, 77, 78, **78**, 146
iron oxide, 133, 161
island complexes, 15
isoclinal fold, 177, **178**, 203, 208
Isoetes melanospora, 289
Isoetes tegetiformans, 272, **274**
Itacolumi, Brazil, 251
itacolumite, 251, **252**

Jackson, Thomas J. "Stonewall," 282
jaguar, **149**
Jane Hurt Yarn Interpretive Center,
    231
Jasper, 204
Jekyll Island: erosion of, 33, 37, 39,
    41; geologic map of, 32; growth
    of, 41; migration of, 37; ridge
    and runnel system of, **18**; rising
    sea level and, 41; satellite image
    of, **38**; shape of, 37; spit, 37;
    structure of, 39; type of barrier
    island, 21
Jesup, 60, 107, 108
Johns Mountain, 156
Johnson, President Lyndon B., 33
Johnson Rocks, 33
J. Strom Thurmond Lake, 240, 246,
    272, 276

kaolin: amount shipped from
    Port of Savannah, 98; colors
    within kaolin pits, 71; differing
    quality of, 71; estimated amount
    in Georgia, 69; excavation
    techniques for removing, 70;
    formation of, 71; Georgia and,
    1; history of, 70; impact on
    Georgia's economy, 70; location

of, 73; mines, **72**, **90**; natural exposure of, **72**; number of mines in Georgia, 70; plant tours, 91; processing of, 91; reasons for different colors of, 89; richest concentration of, 71; stored, **91**; uses for, 1, 70

Kaolin Belt, 70, **70**, 97

Kaolin Capital of the World, 1, 89

Kaolin Festival, 91

kaolinite, 71, 179

karst, 113

Kennesaw, 200, 295, 297

Kennesaw Mountain, 281, 290

Kennesaw Mountain National Battlefield Park, 290

Keown Falls, 157

King and Prince resort, 33

Kingston Saltpeter Cave, 275

Kingston thrust fault, **124–25**, 156

Kiokee belt, 272

kyanite, 239

Ladds Quarry, 148, **148**, 149, **149**

LaFarge Aggregates, 222

LaFayette, 154, 168

Lake Allatoona, 199–200

Lake Chatuge, 217, **217**

Lakeland, 82

Lake Lanier, 216, 233, **234**

Lake Louise, 83

Lake Oconee, 269, 270, **270**, 271, 277

Lake Sinclair, 277

Lamar Mounds, 96

landforms, 60

landscape provinces, 6–10

Lanier Mountain, 236

Laurentia, 181, 182, 231

Laurentia terrane, 182

Lawrenceville, 204, 295

Lazaretto Creek, 23

LBJ Wall, 33

Lee, Robert E., 23, 282

Len Foote Hike Inn, 215

Lepidocyclina, 64, **65**

Lepidodendron, 131, **131**

Licklog Creek, **9**

Liesegang banding, 161, **161**

lighthouses, 25, 33, **35**

lime sink lakes, 83

lime sinks, 55–56, **56**, 83, **83**, 91, 113

limestone, 10, 153, 204

limonite, 77, 146

Lincolnton Metadacite, 239–40, **240**

Lithia Springs, 195

lithium, 195

Lithonia, 267, 295

Lithonia Gneiss, **269**, 286, 287, 294

Lithophaga, **84**

lithosphere, 3

Little Cumberland Island, **45**

Little Grand Canyon, 1, 76

Little Kennesaw Mountain, 290

Little Ocmulgee River, 60, 105

Little Ocmulgee State Park, 105, **105**

Little Ogeechee River, 98

Little Ohoopee River, 110

Little River, 272

Little St. Simons Island, 33, **34**

Little Towaliga River, **266**

Little White House Historic Site, 254

live oaks, 31

longshore drift, 15, 17

Lookout Mountain, **124–25**, 132, 135, 153–54, 159, 161–64

Lookout Valley, 122–23, **122**, 133, 150

Lookout Valley anticline, 122, **122**, **124–25**, 126, 133, 135, 137

Lophorhothon, 68

Louisville, 94

Lumber City, **55**, 105

Lumpkin, 76, 78

Lyell, Charles, 96

Macon, 3, 80, 267

Madison, President James, 23

magma, 4

magnetite, 73

Magnolia Springs State Park, 96

Mammut americanum, 31
Manchester, 253, 258
manganese, 146
mantle, 3
mapping, geological, 2
marble, 1, 204–7, **205**, **206**, **208**
marble mining district, 204–7
Marble Museum, 205
March to the Sea, 90, 94, 198, 271
Mark Anthony Cooper's Iron
    Works, 198
Marshallville, 85
marshlands, 20
MARTA, 292
Martin, John, 225
Martin Marietta Aggregates Quarry,
    273
mastodons, 31, 69, **69**
Mattox, 112
Maury Shale Member, 170
McCaysville, 201
McLemore Cove, 123, **124**, 154, 162,
    164
McLemore Cove anticline, 162
McPherson, General, 145
McQueens Island Historic Trail, 22
McRae, **55**
mélange, 292
Meridian, 30, **30**
Mesozoic era, 14, 188–90
metaconglomerate, 198
metadacite, 239, 240
metagraywacke, 8, 176
metamorphic rocks, 177
metamorphism, 176–77
metasedimentary rocks, 175
metavolcanic rocks, 175, **200**
Metter, 60
mica, 175
Miccosukee Formation, 82
Michael C. Carlos Museum, **204**
microbes, 10
Mid-Atlantic Ridge, 3, 4, **4**, 6
middens, 27, 30, 42, 44
migmatite, 179, 287
Mill Creek, 142

Mill Creek Gap, 145
Milledgeville, 3, 89, 277–78
minerals, 14, 17, 73, 176
minerals, heavy, 17, 73
mines, Georgia's richest, 1
mining: advantages of weathered
    ores, 155; bauxite, 86; coal, 162;
    copper, 201, **201**, 203; Georgia
    and, 1; gold, 191, 203, 216, 220,
    221, 223; of heavy minerals, 73,
    74; of heavy mineral sand, 117;
    of hematite, 133; hydraulic, 191,
    223; of iron ore, 78; kaolin, 86,
    90, **90**; kyanite, 239; marble, 204,
    205; metagranite, 197; mica, 264;
    ochre, 146, **146**; open-pit, 191;
    placer, 191; of sand, 103; of sand
    dunes, 108, 116; tunnel, 193
Missionary Ridge, 145, 164, **166**
Mississippian Indian culture, 96,
    148
Mississippian period, 126, 129
Modoc fault zone, 240, 272
molds, fossil, 64, **65**, **84**
mollusks, 10
monadnocks, 8, 281
monazite, 74
Monteagle Limestone, 154
Montezuma, **54**, 85, 86
Montezuma Bluffs Natural Area, 85
montmorillonite, 71
Moody Air Force Base, 82
mosasaurs, 66
moss animals, 129
moss, giant club, 131
mountain building events, 183, 185
Mt. Oglethorpe, 203
Mt. Wilkinson, 295
mudstone, 10
"mud volcano," 235
Mulcoa Plant #5, 86
Murphy belt, 203, 206, 207, **207**
Murphy belt syncline, 203
muscovite mica, 179, 264
museums, 1, 299–304
mushroom rocks, 162, **163**

mylonite, 247, 253, 297, 298
mylonite zones, 179, **182**, 187, **187**, 247

Nacoochee Valley, 225
Nanafalia Formation, **65**, 86
natural gas, 168
Needle's Eye, 160, **160**
Neels Gap, 219
Nelson, 205
Neo-Acadian event, 183, 185
neodymium, 74
Neptune Park, 33
Neuropteris, 131
New Deal, 271
New England, 133
New Manchester, 290
New Riverside Ochre Mine, 146, **146**
Nodoroc, 235
nodules, 10
No Mans Friend Pond, 82
Nora Mill Granary, 225
normal faults, **182**
North End Beach (Jekyll Island), 39, **40**
North Georgia College, 221
North Highlands Dam, **178**, 260, 261
North Newport River, **117**
nuclear power plants, 95, 111

Oak Mountain, 256, 257
oaks, live, 31
oaks, turkey, **110**
Oaky Woods Wildlife Management Area, 103
Ocala Limestone, 78
ochre, 146, **146**
Ocmulgee National Monument, 96
Ocmulgee River, 83, 96, 105, 249, 267
Oconee National Forest, 271
Oconee River, **72**, 89, 271
Ocypode quadrata, 19
Ogeechee River, 60, 98, **99**
Oglethorpe, 85, 86
Oglethorpe, General James, 98, 274

Ohoopee Dunes Natural Area, 98, 110, **110**
Ohoopee River, 60, 98, 110
oil, 168
oil seeps, 105
oil shale, 168
Okefenokee Basin, 57
Okefenokee Swamp, 1, 21, 57, 74, 111, 112, **112**
Okefenokee Swamp Park, 111, 113
Okefenokee terrace, 98
Old Ironsides, 31
Old Stone Church Museum, 139
oncolites, 142
ooids, 154
Ophiomorpha, 74
Orangeburg Escarpment, 56, 62, 98
Ordovician period, 126
ornithischians, 67, 68
Ornithomimus, 68
Ossabaw Island, 27, **29**
ostracods, 130
Ostrea crenulimarginata, 85
Ostrea gigantissima, **65**
oyster middens, 27, 30, **44**
oysters, **39**, **65**
oysters, giant, 85

Paleozoic era, 14, 181
Palmetto Granite, 297
Pangaea, 3, 5, 92, 181, 188
Panola Granite, 286
Panola Mountain, 281, 286
Panola Mountain State Park, 286
patch reefs, 84
peaches, 80, 84, 103
Peachtree Center Station, 292
Pearson terrace, 107
peat, 162, 236
pecans, 82, 84, 103
pedestal rocks, 155, **155**, 162
pegmatite dikes, 179, 230, 285, 286
Pelham Escarpment, 55, 62, 79
Penholoway shoreline complex, 117
Pennsylvanian period, 126, 162
periods, geologic, 14

permineralization, 64
Perry Escarpment, 104
phosphate, 170
phyllite, 198, **198**
physiographic districts, 52, **52**
physiographic provinces, 6–10, **7**, 121, **121**
Piedmont: beach sand and, 17; definition of, 6; distinguishing characteristics of, 6, 7; easternmost exposure of in Georgia, 274; faulting within the, 255; lakes and rivers of, 270; rocks beneath the Coastal Plain, 89, 92; topography of, 8
Pierre Shale, 42
Pigeon Hill, 290, **290**
Pigeon Mountain, 123, **124**, 154–56, **155**, **156**, **164**
Pine Mountain, 253, **254**, 255, **255**, 256, **256**
Pine Mountain Gold Museum, 193
Pine Mountain terrane, 181, 183, 259, 265, **266**
pines, slash, 107
pine trees, 288
Pio Nono Formation, **53**, **102**
placer deposits, 191, **191**
Plant Hatch, 111
plants, rare and endangered, 60, 111–12, 236, 272, **274**, 289
Plant Vogtle, 95
plate tectonics, 3–6
Pleistocene epoch, 21, 57, 60, 68
plesiosaurs, 67, **67**
plunging folds, 123, **123**, 124, **124**
Pocket, the, 157
pocosins, 60
Point, the, 151, **151**
pool sprite, 236
pool sprite, tiny, 272
Port Columbus National Civil War Naval Museum, 262
Port of Savannah, 25, 98
potholes, 291
Precambrian, 14

Price Memorial Hall, 221
Providence Canyon, 76–77, **77**
Providence Canyon State Outdoor Recreation Area, 76
Providence Sand, 77, **77**, **78**
provinces, landscape, 6
provinces, physiographic, 6, 7
Pruitt Creek Wall, 219
pterosaurs, 68

Quarry Mountain, 148, **148**, 213
quartz, 8, 10, 73, 177, 179. See also chert; flint
quartz, botryoidal, 84, **84**
quartz, cryptocrystalline, 79
quartz, drusy, 168
quartzite, 8, 175, **177**, 253
quartz sand, 17
quartz, shocked, 64
quillwort, black-spored, 236, 289
quillwort, mat-forming, 272, **274**
Quincy, FL, 79

Rabun Gneiss, 227
radioactive decay, 12
radiometric dating, 12
radium, 116
Radium Springs, 115, 116, **116**
Raising a Ledge, 282
Reagin Quarry, 267
Red Mountain Formation, 133
Red Mountain sandstone, 139
Red Top Mountain, 199
Red Top Mountain Formation, 199
Red Top Mountain State Park, 199–200
reef, live-bottom, 31
reefs, patch, 84
Reilly, Major J. W., 276
renourishment, beach, 25, 35, 36
replacement (fossils), 64
Resaca, 159
reservoir-induced seismicity, 185, 240, 246, 276
residuum, 10
Rhodes Hall, **293**, 294

Richard B. Russell Lake, 246
Riddleville Basin, 92
ridge and runnel systems, 17, **18**
rifting, 188–90, **189**
Ringgold Gap, 137, 139, **140**, 144
Ringgold Railroad Depot, 139, **141**
Ring of Fire, 4
ripples, 11, 17, **18**, 19, **19**
Rising Fawn, 133, 135, **136**
rivers, 60
rivers, meandering, 74
Riverwalk, 261, **261**
Rock and Shoals State Natural
    Area, 237
Rock City Gardens, 1, **11**, 159–61
Rockdale County, **269**
Rock Eagle, 213, 276–77, **277**
Rock Hawk, 213, 276
Rockmart Slate, 173, **173**, **174**
rock pedestals, 155
rocks, formal naming of, 2
Rocktown, 155, **155**, **156**
Rocky Face Mountain, 142, 144, 213
Rocky Shoals, 261, **261**
Rome, GA, 1, 170
Rome fault, **124–25**
Rome Formation, 170, **171**
Rome thrust fault, 158
Roosevelt, President Franklin D.,
    253, 254, 256
Roosevelt Warm Springs Institute, 254
Rosendale, New York, 262
Rosendale Natural Cement, 262
Rossville, 164
Rowland Springs Formation, 197
Ruby Falls, 160
rugose corals, 130
runnels, 17, 18
Russell Dam, 246
Russell Lake, 246
Russell Lake allochthon, 246
rust, 133
rutile, 73, 238, **239**

salt marsh, 17, 20, **23**, 117
San Andreas Fault, 4

sand, black, 73
sand, squeaking of, 17
sandbars, 11, 17
sand dollars, **65**
sand dunes, 11, 25, **26**, **44**, 98, **99**, 108
sand dunes, parabolic, 42, 108
sand dune fields, 60, 110
sand dunes, riverine, 105, **105**, 110
Sandersville, 49, 89
Sandersville Limestone, **65**, 91
Sand Hills, 52
Sand Mountain, **124–25**, 149, **150**,
    151
sandstone, 10, 82, **94**
sandstone, dolomitic, 31
Sandy Springs, 295
Sangamon interglacial, 21
Sapelo Island, **29**, 30, 31
saprolite, 71, **102**, 179, **180**, 270,
    270, 284
Satilla River, 111
Savannah: City Hall, **99**; Fort
    Pulaski and, 23; geology beneath,
    21; port of, 25, 98; siting of, 2,
    98; as a tourist destination, 225;
    vertebrate fossils found near, 68
Savannah River: Augusta Canal and,
    274; Cockspur Island and, 23;
    deepening of, 100; depth of, 100;
    dredging of, 25, 100; Fall Line at,
    275, **275**; large sediment load of,
    25; length of, 10; satellite image
    of, **24**
Savannah-Tybee Railroad, 22
Sawnee Mountain Preserve, 222
scale-bark tree, 131
scallops, **65**
Scapanorhynchus texanus, **66**
scarps, 21
schist, 177
schist, button, 298
Schwimmer, Dr. David, 68
Scotland, 105
seafloor, 4
Sea Islands, 15, **16**, 17, 21, 23, 27, 33
sea level, 21, 56, 57

sea level, high, 47, 49, 103, 111
sea level, low, 49, 60
seawalls, 33, 35
sediment, 10
sedimentary rocks, 10, 12, **94**, 104
sedimentation, 10–12
Sedum pusillum, 272
seed ferns, 130
seismicity, reservoir-induced, 185, 240, 246, 276
Seminole Landfill, 297
Sequatchie Formation, 128, **129**
Sequatchie Valley, 122, 123
SETI, 253
Seven Natural Wonders of Georgia, 76, 115, 150, 214
Shady Dolomite, 145, **147**
Shaking Rock Park, 238
shale, 10, 168, 169
shale, oil, 168
sharks, 68
shark teeth, 66, **66**, 68
Sharp Mountain, 203
Shell Bluff, 49, 96
Shell Bluff Landing, 96
Sherman, General William T., 90, 198
shoreline complexes, 21
shorelines, 21
shrimp, ghost, 19, **20**
silica, 10, 177
Siloam, 271
Siloam Granite, 271, **271**
siltstone, 10, **94**
Silurian period, 126
sinkholes, 55
60s Cliff, 219, **219**
Skidaway Island, **24**, 27, 68, 69
Skyride (Stone Mountain), 282
slate, 173
slaty cleavage, 173, **173**, 177
sloths, giant ground, 31, 68–69, **69**
sloughs, 41
Smithgall, Charles, 223
Smithgall Woods State Park, 223
Smith House Restaurant, 221
snails, **65**

Snake Creek Gap, 158
Snellville, 236, 295
Snellville Mountain, 236
snow fences, 25, 26
soapstone, 296, 297
Soapstone Ridge, 1, 296, 297
solution pits, 272, 274, 282, 288, **288**
Soperton, 97
sorting, differential, 73
Soto, Hernando de, 78, 148, 225
Southeastern Cave Conservancy, 135
Southeastern Federation of Mineralogical Societies, 14
Sparks, 82
Sparta Granite, 272, **273**
Spartina alterniflora, 20
specific gravity, 73
Sphenophyllum, 131
spheroidal weathering, 257
sponges, 10
Sprewell Bluff State Outdoor Recreation Area, 258–59, **258**, **259**
Springer Mountain, 203, 215
Springs, 264
springs, blue hole, 115
springs, natural, 115
stalactites, 172, **172**
State Botanical Garden, 237
state parks. *See individual state parks*
Statesboro, 21, 94
staurolite, 176, 207, **207**
St. Catherines Island, 21, 27, **29**
steel, 1
Stephenson, Dr. M. F., 190
Steven C. Minkin Paleozoic Footprint Site, 132
St. Helens, Mount, 239
Stockmar Gold Mine, 193, **193**
stonecrop, 288
stonecrop, dwarf granite, 272
Stone Mountain, 8, 213, 281–86, **281**, 288, 295
Stone Mountain Granite, 8, **99**, 281, 282, **283**, 294
Stone Mountain Park, **2**, 281–82, 284–86

Stone Pile Gap, 219
stream capture: Alapaha River and, 83; Amicalola Falls and, 214, **216**; Blue Ridge landscape and, 10; process of, 8–9, **9**; Tallulah Gorge and, 232–33, **232**
strike-slip faults, 181, **182**
St. Simons Island, 21, 31, 32, 33, **34**, 36, **38**, 40
St. Simons Island Lighthouse, 35, **35**
St. Simons Sound, 37, **40**
subduction zones, 4
sulfuric acid, 201
Summerville, 168
supercontinents, 5
superterrane, 185
surf zone, 15
Surrency, 107
Suwannee Canal Recreation Area, 113
Suwannee terrane, 185
Sweetwater Creek, 87, 291
Sweetwater Creek State Park, 290–91
Sylvester, 116
synclines, 122

tabby, 36, **36**, 37
tabulate corals, 130
Taconic event, 183
talc, 209
Tallulah Falls Dome, 229, 231
Tallulah Falls Lake, 231
Tallulah Falls Quartzite, 230, **230**
Tallulah Falls (town), 230
Tallulah Gorge, 1, 231–33, **231**, **232**
Tallulah Gorge State Park, 231–33, **232**
Tallulah River, 229, **232**, 233
Talona Mountain, 203
Tate, 204, **205**
Tate, Colonel Sam, 205
Tate House, 205, **206**
Taylor Ridge, 139, 156
tectonic plate boundaries, 3, 4–5, **4**, **5**, **6**

tectonic plates, 4, **5**
tectonic windows, 159, 183, **184**, 217
tektites, 63, **63**
Tellus Science Museum, 1
Tennessee River, 10
Tennessee Valley Divide, 223
Tennille Lime Sinks, 91
terraces, 21, 56
terranes, 181, **184**, **186**. *See also individual terranes*
theropods, 67, 68
Thomaston, 264
thrust faults, 124, **124–25**, 126, 135, **136**, **182**
thrust sheets, 124
tidal channels, 20
tidal range, 15
tides, 15, 17
Tifton, 82
Tifton Upland, 55, 79
time scale. *See* geologic time scale
titanium, 73
Tivola Limestone, **65**, 103, **104**
Tobesofkee Creek valley, 80
Toccoa, 227
Toccoa Creek, 227
Toccoa Falls, **226**, 227
Toccoa Falls College, 227
Toccoa Reservoir, 227
topography, 8
Tortoise Shell Rock, **11**
tourmaline, **99**
Towaliga River, **266**
trace fossils, 130
Trahlyta, 219
Trail of Tears, 190
Trail Ridge, 21, 57, 74, 107, 113
travertine, 172
Tray Mountain, 223
trenches, deep-sea, 4
Triassic period, 92
trilobites, 14, 128, **128**
Tropical Storm Alberto, 85, 113, 267
tsunami, 63
Tugaloo River, **232**, 233
Tugaloo terrane, 183, 185

Tunnel Hill, **142**, 145
Tunnel Hill Heritage Center, 141
Tunnel Hill Ridge, 141
turbidites, 176
turbidity currents, 176
Turritella, **65**
turtlebacks, **156**, 162
turtles, sea, 33, 41, 42
Tuscaloosa Formation, 74
Twiggs Clay, 71, 103
Tybee Island, 23, **24**, 25, **26**
Tybee Island Marine Science
    Center, 27
Tybee Lighthouse, 25
Tybee pavilion and pier, 25, **26**
Tylosaurus proriger, 66
Tyrannosaurus rex, 68

unconformity, 100
Unicoi Gap, 223
Unicoi State Park, 223
University of Georgia, 236, **237**
Upatoi Creek, 74
uranium, 169
U.S. Army Corps of Engineers,
    20, 25
U.S. Mint, 1, 221
USS *Constitution*, 31

Valdosta, 83, **83**
Valdosta Limesink District, 55, 83
Valley and Ridge: Blue Ridge–
    Piedmont rocks and the,
    126; cross section of the,
    124–25; description of the, 121;
    distinguishing characteristics
    of the, 6; faults of the, 124,
    124–25; fossils of the, 128;
    landscape districts of the, 121,
    121; metamorphosed rocks
    of the, 146; relationship of
    underlying bedrock to the
    topography of the, **120**, 122,
    122; rock formations of the, **127**;
    tectonic deformation of the, 153;
    topography of the, 8

veins, 176
verde antique, 208
vernal pools, 236, 288, 289
Vidalia, 110
Vidalia onions, 110
Vidalia Upland, 54, **55**, 111
Villa Rica, 193
Vinings, 295
volcanic ash, 135
volcanic chains, 4, 181
volcanoes, 3
Vulcan Materials Company, 197
Vulcan Materials Kennesaw Quarry,
    292
Vulcan Materials Norcross Quarry,
    188, **291**, 292

Walasi-Yi Interpretive Center, 219
Walker Branch, 226
Walker Mountain, 170
Walk of Flags and Stones, 254
Walk-Up Trail (Stone Mountain), 285
Wallace Dam, 269
Wallenda, Karl, 232
Warm Springs (spring), 253, 254, **254**
Warm Springs (town), 253, 254
War of 1812, 23
Warrenton, 272
Washington, 238
Wassaw Island, **24**, 27, **29**
waterfall retreat, 153
water gap, 137
Watermelon Capital of the World, 82
watermelons, 82
watersheds, 8
Watson Mill Bridge State Park, 241,
    241
Waycross, 21, 111
Waycross terrace, 111
Waynesboro, 94
weathering, 179, 181, 257, 282. *See
    also* solution pits
weathering rim, 284
West Armuchee Valley, 156
Western and Atlantic Railroad,
    137, 279

whales, fossil, 80, 95, **95**
whales, modern, 95
White Oak Mountain, 139
White Path, 203
White Pond, 89
Whitestone, 203
White thrust fault, 145
Wicomico shoreline complex, 117
Wilder Brigade Monument, 167
Willacoochee, 116
Wills Valley, 123, 135
Wills Valley anticline, 135, 137
Wilmington Island, **24**
Winder, 235
Wine Highway, 220, 225
Withlacoochee River, 84
Woodbury Impact Structure, 252
Woodland Indians, 296

Woody Gap, 220
Works Progress Administration, 27, 269
wrack line, 17

xenoliths, **237**, **243**, **266**, **271**, 285
Xiphactinus, 66

Yamacraw Bluff, 98, **100**
Yonah Mountain, 225
York Creek, 225

Zahnd Natural Area, 162, **163**
"Ziggy" (fossil whale), 80
zigzag map patterns, 123, **123**
zigzag ridges, 157
zircon, 73
Zygorhiza, 80

—COURTESY PAUL SCHROEDER   —COURTESY RINA ROSENBERG

PAMELA J. W. GORE, professor of geology at Perimeter College of Georgia State University, has taught geology in Georgia for nearly thirty years. She served as president of the Georgia Geological Society and secretary-treasurer of the Southeastern Section of the National Association of Geoscience Teachers for more than ten years. She received her PhD in geology from George Washington University in Washington DC, where her dissertation was on the sedimentology, stratigraphy, and paleontology of a Triassic basin. She has also done graduate research about the Piedmont in Maryland.

WILLIAM (BILL) WITHERSPOON retired in 2014 after teaching K–12 students and their teachers for seventeen years at Fernbank Science Center, part of the DeKalb County (GA) School District. He now leads geology walks and talks that you can find listed at www.georgiarocks.us/events as well as at the *Roadside Geology of Georgia* Facebook page. He was named the Georgia Outstanding Earth Science Teacher in 2007 by the National Association of Geoscience Teachers. He worked in the petroleum industry and taught at the University of Tennessee at Chattanooga. His PhD topic at the University of Tennessee was the structural geology of the Blue Ridge thrust front in Tennessee.

347